Student's Solutions Manual
Part I

Calculus and
Analytic Geometry
8th Edition

Student's Solutions Manual
Part I

Calculus and
Analytic Geometry
8th Edition

Thomas/Finney

Thomas L. Cochran
Michael Schneider

Belleville Area College

ADDISON-WESLEY PUBLISHING COMPANY

Reading, Massachusetts   Menlo Park, California   New York
Don Mills, Ontario   Wokingham, England   Amsterdam   Bonn
Sydney   Singapore   Tokyo   Madrid   San Juan   Milan   Paris

ISBN   0-201-53305-7

3 4 5 6 7 8 9 10 AL 959493

The authors would like to dedicate this book to their children, Jane, J.T., and Connie Cochran and Michael and John Schneider

Tom Cochran
Michael Schneider

The authors would like to thank Bruce Sisko, Belleville Area College, for the fine job he did in proofreading and checking our work.

The authors have attempted to make this manual as error free as possible but nobody's "perfeck". If you find errors, we would appreciate knowing them. You are more than welcome to write us at:
Belleville Area College
2500 Carlyle Road
Belleville, Illinois 62221

TABLE OF CONTENTS

Student's Solutions Manual
Part I

Calculus and
Analytic Geometry
8th Edition

CHAPTER 1

THE RATE OF CHANGE OF A FUNCTION

SECTION 1.1 CARTESIAN COORDINATES AND EQUATIONS FOR LINES

1. $\Delta x = -1 - (-3) = 2$

    $\Delta y = -2 - 2 = -4$

3. $\Delta x = -8.1 - (-3.2) = -4.9$

    $\Delta y = -2 - (-2) = 0$

5. $m = \dfrac{\Delta y}{\Delta x} = \dfrac{-1-2}{-2-(-1)} = 3$

    $m_\perp = -\dfrac{1}{3}$

7. $m = \dfrac{\Delta y}{\Delta x} = \dfrac{3-3}{-1-2} = 0$

    $m_\perp$ does not exist.

9. a) $x = -1$

    b) $y = \dfrac{4}{3}$

11. a) $x = 0$

    b) $y = -\sqrt{2}$

13. $P(-1,1), m = 1 \Rightarrow y - 1 = 1(x - (-1)) \Rightarrow y = x + 2$

15. $P(0,b), m = 2 \Rightarrow y - b = 2(x - 0) \Rightarrow y = 2x + b$

17. $(1,1), (2,1) \Rightarrow m = 0 \Rightarrow$ Horizontal line $\Rightarrow y = 1$

19. $(0,0), (2,3) \Rightarrow m = \dfrac{3}{2} \Rightarrow y - 0 = \dfrac{3}{2}(x - 0) \Rightarrow y = \dfrac{3}{2}x$

21. $y = x + \sqrt{2}$

23. $y = -5x + 2.5$

25. $y = 0 \Rightarrow x = 4$

    $x = 0 \Rightarrow y = 3$

27. $y = 0 \Rightarrow x = \sqrt{3}$

    $x = 0 \Rightarrow y = -\sqrt{2}$

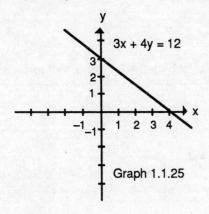

Graph 1.1.25

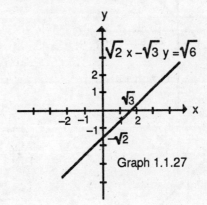

Graph 1.1.27

29. a) $P(2,1), L: y = x + 2 \Rightarrow m_L = 1 \Rightarrow$ // line is $y - 1 = 1(x - 2) \Rightarrow y = x - 1$

    b) $m_\perp = -1 \Rightarrow y - 1 = -1(x - 2) \Rightarrow y = -x + 3$ is $\perp$ line.

31. a) P(1,2), L: $x + 2y = 3 \Rightarrow y = -\frac{1}{2}x + \frac{3}{2} \Rightarrow m_L = -\frac{1}{2} \Rightarrow$ // line is $y - 2 = -\frac{1}{2}(x - 1) \Rightarrow y = -\frac{1}{2}x + \frac{5}{2}$

    b) $m_\perp = 2 \Rightarrow y - 2 = 2(x - 1) \Rightarrow y = 2x$ is $\perp$ line.

33. a) P(−2,4), L: $x = 5 \Rightarrow$ // line is $x = -2$ b) $\perp$ line is $y = 4$

35. A(−2,3), $\Delta x = 5$, $\Delta y = -6 \Rightarrow x_2 = 5 + (-2) = 3$ and $y_2 = -6 + 3 = -3 \Rightarrow (x_2, y_2) = (3,-3)$

37. $\Delta x = 5$, $\Delta y = 6$, B(3,−3) Let A = (x,y) Then $\Delta x = x_2 - x_1 \Rightarrow 5 = 3 - x \Rightarrow x = -2$. $\Delta y = y_2 - y_1 \Rightarrow$

    $6 = -3 - y \Rightarrow y = -9$ ∴ A = (−2,−9)

39. a) A ≈ (0 in,69°), B ≈ (0.4 in, 68°) $\Rightarrow m = \dfrac{68° - 69°}{.4 - 0} = -2.5°/\text{in}$

    b) A ≈ (0.4 in,69°), B ≈ (4 in,10°) $\Rightarrow m = \dfrac{10° - 68°}{4 - .4} = -16.1°/\text{in}$

    c) A ≈ (4 in,10°), B ≈ (4.6 in,5°) $\Rightarrow m = \dfrac{5° - 10°}{4.6 - 4} = -8.3°/\text{in}$

41. $p = kd + 1$ $(d_1,p_1) = (0,1)$, $(d_2,p_2) = (100, 10.94) \Rightarrow \dfrac{\Delta p}{\Delta d} = \dfrac{10.94 - 1}{100 - 0} = .0994$ atm/m. If $d = 50$, then $\dfrac{\Delta p}{\Delta d}$

    $= \dfrac{p - 1}{50 - 0} = .0994 \Rightarrow p - 1 = 50(.0994) \Rightarrow p = 5.97$ atm

43. a)

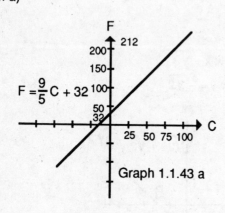

$F = \frac{9}{5}C + 32$

Graph 1.1.43 a

b) (0,32) and (100,212) $\Rightarrow m = \dfrac{212 - 100}{32 - 0} = \dfrac{9}{5}$. The F intercept is

    $32 \Rightarrow F = \frac{9}{5}C + 32$

c) $F = \frac{9}{5}C + 32$. Let F = C. Then $C = \frac{9}{5}C + 32 \Rightarrow C = -40° = F$

45. a) (−4,−2) and (3,−7)

    b) Width = |−4 − 3| = 7, height = |−2 − (−7)| = 5 $\Rightarrow$ Area = 35

47.

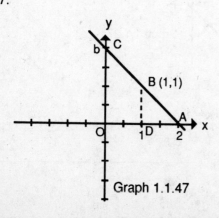

Graph 1.1.47

$\dfrac{|AD|}{|BD|} = \dfrac{|AO|}{|CO|} \Rightarrow \dfrac{1}{1} = \dfrac{2}{b} \Rightarrow b = 2$

49. Let $A = (-1,1)$, $B = (2,3)$, $C = (2,0)$. Since BC is vertical and $m(BC) = 3$, let AD_1 be vertical (upward)

$\Rightarrow D_1 = (-1,4)$. Let AD_2 be vertical (downward) $\Rightarrow D_2 = (-1,-2)$. Let $D_3 = (x,y)$. m of AB = m of CD_3 $\Rightarrow$

$\dfrac{y-3}{x-2} = -\dfrac{1}{3}$ and m of AC = m of BD_3 $\Rightarrow \dfrac{y-0}{x-2} = \dfrac{2}{3}$. Solve the system of equations to get $x = 5$, $y = 2$ $\Rightarrow$

$D_3 = (5,2)$

51. $2x + ky = 3 \Rightarrow ky = -2x + 3 \Rightarrow y = -\dfrac{2}{k}x + \dfrac{3}{k} \Rightarrow m = -\dfrac{2}{k}$. $x + y = 1 \Rightarrow y = -x + 1 \Rightarrow m = -1$.

$\therefore m_{//} = -1 \Rightarrow -\dfrac{2}{k} = -1 \Rightarrow k = 2$. $m_\perp = 1 \Rightarrow -\dfrac{2}{k} = 1 \Rightarrow k = -2$.

SECTION 1.2   FUNCTIONS AND THEIR GRAPHS

1. D: All Reals
   R: $y \geq 1$

3. D: $-x \geq 0 \Rightarrow x \leq 0$
   R: $y \geq 0$

5. D: All Reals except $x = \dfrac{(2n+1)\pi}{2}$, n an integer
   R: $y \geq 0$

7. D: All Reals
   R: $y = -1, 1, 0$

9. Odd

11. Odd

13. Neither

15. Even

17. Even

19. Odd

21. D: $-\infty < x < \infty$
    R: $y \geq 0$
    Symmetric to y axis

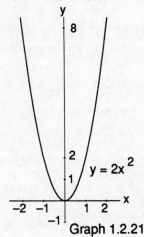

Graph 1.2.21

23. D: $-\infty < x < \infty$
    R: $y \geq -9$
    Symmetric to y axis

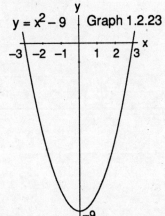

Graph 1.2.23

25. D: All Reals
    R: All Reals
    Symmetric to origin

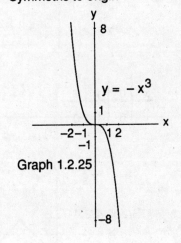

Graph 1.2.25

27.   D: $x \neq 0$

   R: $y \neq 0$

   Symmetric to origin

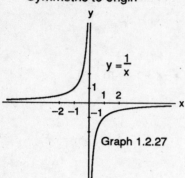

Graph 1.2.27

29. Symmetric to origin

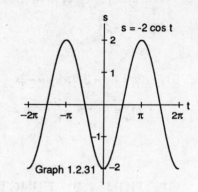

Graph 1.2.29

31. Symmetric to s-axis

$s = -2 \cos t$

Graph 1.2.31

33.   a) No. $\sqrt{s}$ not a Real Number if $s < 0$ b) No. Division by 0 is undefined   c) D: $s > 0$

35.   a) i, $y = x^2 - 1$ symmetric to y axis and R: $y \geq -1$ b) iv, R: $y \geq 0$ and $x = 1 \Rightarrow y = 0$

37.   a) $0 \leq x < 1$ b) $-1 < x \leq 0$

39.   a)

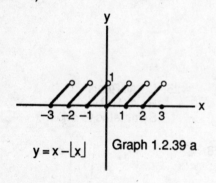

$y = x - \lfloor x \rfloor$ Graph 1.2.39 a

39.   b)

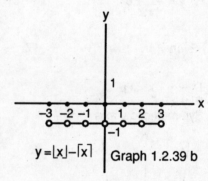

$y = \lfloor x \rfloor - \lceil x \rceil$ Graph 1.2.39 b

41.

x	0	1	2
y	0	1	0

Graph 1.2.41

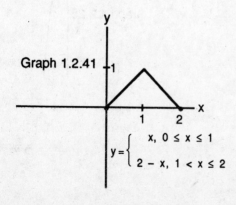

$y = \begin{cases} x, \ 0 \leq x \leq 1 \\ 2 - x, \ 1 < x \leq 2 \end{cases}$

43.

$y = \begin{cases} 3 - x, \ x \leq 1 \\ 2x, \ 1 < x \end{cases}$

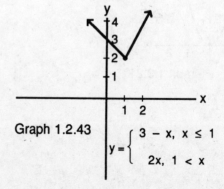

Graph 1.2.43 $y = \begin{cases} 3 - x, \ x \leq 1 \\ 2x, \ 1 < x \end{cases}$

45. $y = \begin{cases} 1, & x < 5 \\ 0, & 5 \le x \end{cases}$

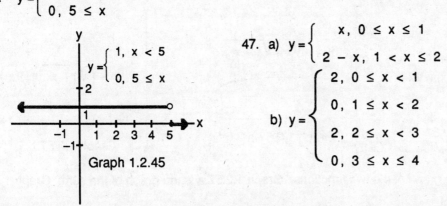

Graph 1.2.45

47. a) $y = \begin{cases} x, & 0 \le x \le 1 \\ 2 - x, & 1 < x \le 2 \end{cases}$

b) $y = \begin{cases} 2, & 0 \le x < 1 \\ 0, & 1 \le x < 2 \\ 2, & 2 \le x < 3 \\ 0, & 3 \le x \le 4 \end{cases}$

49. $D_f: -\infty < x < \infty,\ D_g: x \ge 1 \Rightarrow D_{f+g} = D_{f-g} = D_{fg} = D_{g/f}: x \ge 1;\ D_{f/g}: x > 1$

51. a) $f(g(0)) = 2$

b) $g(f(0)) = 22$

c) $f(g(x)) = (x^2 - 3) = x^2 + 2$

d) $g(f(x)) = (x + 5)^2 - 3 = x^2 + 10x + 22$

e) $f(f(-5)) = 5$

f) $g(g(2)) = -2$

g) $f(f(x)) = (x + 5) + 5 = x + 10$

h) $g(g(x)) = (x^2 - 3)^2 - 3 = x^4 - 6x^2 + 6$

53.

	$g(x)$	$f(x)$	$(f \circ g)(x)$
a)	$x - 7$	$\sqrt{x}$	$\sqrt{x - 7}$
b)	$x + 2$	$3x$	$3(x + 2) = 3x + 6$
c)	x^2	$\sqrt{x - 5}$	$\sqrt{x^2 - 5}$
d)	$\dfrac{x}{x - 1}$	$\dfrac{x}{x - 1}$	$\dfrac{\frac{x}{x-1}}{\frac{x}{x-1} - 1} = \dfrac{x}{x - (x - 1)} = x$
e)	$\dfrac{1}{x - 1}$	$1 + \dfrac{1}{x}$	$1 + \dfrac{1}{\frac{1}{x-1}} = 1 + (x - 1) = x$
f)	$\dfrac{1}{x}$	$\dfrac{1}{x}$	$\dfrac{1}{\frac{1}{x}} = x$

55. a) $y = \sin x \cos x$: $y(-x) = \sin(-x) \cos(-x) = (-\sin x)(\cos x) = -\sin x \cos x \Rightarrow y = \sin x \cos x$ is odd

b) Let f be even $\Rightarrow f(-x) = f(x)$. Let g be odd $\Rightarrow g(-x) = -g(x)$. Then, if $h(x) = f(x)g(x)$, $h(-x) =$
$f(-x)g(-x) = (f(x))(-g(x)) = -f(x)g(x) = -h(x) \Rightarrow h$ is odd.

57.

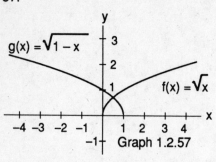

Graph 1.2.57

a)

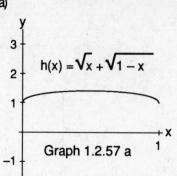

Graph 1.2.57 a

b)

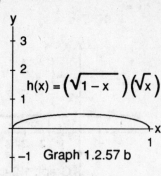

Graph 1.2.57 b

Note: Graph 1.2.57 is the graph of the two functions. Graph 1.2.57 a is the graph of the sum. Graph 1.2.57 b is the graph of the product.

c)

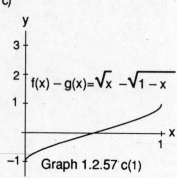

Graph 1.2.57 c(1)

Graph 1.2.57 c(2)

d)

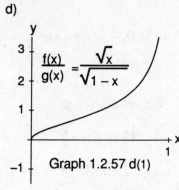

Graph 1.2.57 d(1)

57. d) (Continued)

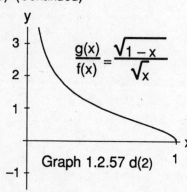

Graph 1.2.57 d(2)

Note: Graph 1.2.57 c(1) is the graph of f − g; Graph 1.2.57c(2) is the graph of g − f. Graph 1.2.57 d(1) is the graph of f/g; Graph 1.2.57 d(2) is the graph of g/f.

59.

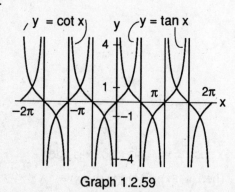

Graph 1.2.59

Graphs intersect at $x = \dfrac{(2n + 1)\pi}{4}$, n an integer; tan x = 0 when cot x is undefined; cot x = 0 when tan x is undefined.

SECTION 1.3 CALCULUS AND COMPUTATION

1. a) Answers vary

    b) $e^{-1} = 0.367879441$

    $e^{-10} = 0.000045399$

    $e^{-100} = 3.7200759 \times 10^{-44}$

    $e^{-1000} = 0$ (Answers vary)

3. Answers vary

    $x^* = \pi$

5. $y = x$

7. $\dfrac{x}{x+1} - \left(\dfrac{-1}{x+1}\right) = c \Rightarrow \dfrac{x+1}{x+1} = c \Rightarrow c = 1$ if $x \neq -1$

9. $\tan x \sin 2x - (-2\cos^2 x) = \tan x \sin 2x + 2\cos^2 x = (\tan x)(2\sin x \cos x) + 2\cos^2 x = 2\sin^2 x + 2\cos^2 x = 2$
    for every $x \neq \dfrac{n\pi}{2}$, n an odd integer.

11. $\ln(2x) - \ln x = \ln\left(\dfrac{2x}{x}\right) = \ln 2 \therefore$ Constant for $x > 0$

13. a) $\sqrt{3} = 1.732050808$

    $\sqrt{1.732050808} = 1.316074013$

    $\sqrt{1.316074013} = 1.14720269$, etc

    b) $\sqrt{5} = 2.236067978$

    $\sqrt{2.236067978} = 1.495348781$

    $\sqrt{1.495348781} = 1.222844545$, etc

15. $\sqrt[10]{2} = 1.071773463$; $\sqrt[10]{1.071773463} = 1.00695555$, $\sqrt[10]{1.00695555} = 1.000693387$; $\sqrt[10]{1.000693387} = 1.000069317$, etc

17. $x_0 = 1$

    $x_1 = 1.540302306$

    $x_2 = 1.570791601$

    $x_3 = 1.570796327$

    $x_4 = 1.570796327$, etc

19.

x	0.1	0.01	0.001	0.0001	0.00001	0.000001
x^x	0.7943282	0.9549925	0.993116	0.9990793	0.9998848	0.9999861

21. $m_{sec} = \dfrac{\ln(1 + \Delta x)}{\Delta x}$ Slope = 1 (See table below)

Δx	0.1	0.01	0.001	0.0001	0.00001	0.000001	0.0000001
m_{sec}	0.9531018	0.995033	0.9995003	0.99995	0.999995	0.9999995	0.9999999

SECTION 1.4   ABSOLUTE VALUE (MAGNITUDE) AND TARGET VALUES

1. $d = \sqrt{(0-1)^2 + (1-0)^2} = \sqrt{2}$ 3. $d = \sqrt{(2\sqrt{3} - (-\sqrt{3}))^2 + (4-1)^2} = \sqrt{27+9} = 6$

5. $d = \sqrt{(0-a)^2 + (0-b)^2} = \sqrt{a^2 + b^2}$ 7. 3   9. 5

11. a) False

    b) True

    c) True

    d) True

    e) True $2 < x < 6 \Rightarrow \frac{1}{3} < \frac{x}{6} < 1 \Rightarrow 3 > \frac{6}{x} > 1$

    f) True $|x-4| < 2 \Rightarrow -2 < x-4 < 2 \Rightarrow 2 < x < 6$

    g) True $2 < x < 6 \Rightarrow -2 > -x > -6 \Rightarrow -6 < -x < -2 \Rightarrow -6 < -x < 2$

    h) True

13. $x = \pm 2$

15. $|2t + 5| = 4 \Rightarrow$

    $2t + 5 = 4$ or $2t + 5 = -4$

    $\Rightarrow t = -\frac{1}{2}$ or $t = -\frac{9}{2}$

17. $|8 - 3s| = 9 \Rightarrow$

    $8 - 3s = 9$ or $8 - 3s = -9 \Rightarrow$

    $s = -\frac{1}{3}$ or $s = \frac{17}{3}$

19. $|x + 3| < 1 \Rightarrow$

    $-1 < x + 3 < 1$

    $\Rightarrow -4 < x < -2 \Rightarrow e$

21. $|1 - x| < 2 \Rightarrow$

    $-2 < 1 - x < 2 \Rightarrow$

    $3 > x > -1 \Rightarrow b$

23. $\left|\frac{x-1}{2}\right| < 1 \Rightarrow$

    $-1 < \frac{x-1}{2} < 1 \Rightarrow$

    $-1 < x < 3 \Rightarrow b$

25. $-2 < y < 2$

27. $|r - 1| \leq 2 \Rightarrow$

    $-2 \leq r - 1 \leq 2 \Rightarrow$

    $-1 \leq r \leq 3$

29. $|3s - 7| < 2 \Rightarrow$

    $-2 < 3s - 7 < 2 \Rightarrow$

    $\frac{5}{3} < s < 3$

31. $\left|\frac{t}{2} - 1\right| \leq 1 \Rightarrow$

    $-1 \leq \frac{t}{2} - 1 \leq 1 \Rightarrow$

    $0 \leq t \leq 4$

33. $3 < x < 9 \Rightarrow$

    $-3 < x - 6 < 3 \Rightarrow$

    $|x - 6| < 3$

35. $-5 < x < 3 \Rightarrow$

    $-4 < x + 1 < 4 \Rightarrow$

    $|x + 1| < 4$

37. $|x^2 - 100| < 0.1 \Rightarrow$

    $-0.1 < x^2 - 100 < 0.1 \Rightarrow$

    $99.9 < x^2 < 100.1 \Rightarrow$

    $\sqrt{99.9} < x < \sqrt{100.1}$

39. $\left|\sqrt{x - 7} - 4\right| < 0.1 \Rightarrow$

    $-0.1 < \sqrt{x - 7} - 4 < 0.1 \Rightarrow$

    $3.9 < \sqrt{x - 7} < 4.1 \Rightarrow$

    $15.21 < x - 7 < 16.81 \Rightarrow$

    $22.21 < x < 23.81$

41. $\left|\frac{120}{x} - 5\right| < 1 \Rightarrow$

    $-1 < \frac{120}{x} - 5 < 1 \Rightarrow$

    $4 < \frac{120}{x} < 6 \Rightarrow$

    $\frac{1}{4} > \frac{x}{120} > \frac{1}{6} \Rightarrow$

    $30 > x > 20$

43. $|(x + 1) - 4| < 0.5 \Rightarrow |x - 3| < 0.5$

45. $\left|\left(-\dfrac{x}{2} + 1\right) - (-2)\right| < \dfrac{1}{2} \Rightarrow \left|-\dfrac{x}{2} + 3\right| < \dfrac{1}{2} \Rightarrow -\dfrac{1}{2} < -\dfrac{x}{2} + 3 < \dfrac{1}{2} \Rightarrow -\dfrac{7}{2} < -\dfrac{x}{2} < -\dfrac{5}{2} \Rightarrow -7 < -x < -5 \Rightarrow$

$7 > x > 5 \Rightarrow 1 > x - 6 > -1 \Rightarrow |x - 6| < 1$

47. $|A - 9| \le 0.01 \Rightarrow \left|\pi\left(\dfrac{x}{2}\right)^2 - 9\right| \le 0.01 \Rightarrow -0.01 \le \pi\left(\dfrac{x}{2}\right)^2 - 9 \le 0.01 \Rightarrow 8.99 \le \pi\left(\dfrac{x}{2}\right)^2 \le 9.01 \Rightarrow$

$\dfrac{8.99}{\pi} \le \left(\dfrac{x}{2}\right)^2 \le \dfrac{9.01}{\pi} \Rightarrow \sqrt{\dfrac{8.99}{\pi}} \le \dfrac{x}{2} \le \sqrt{\dfrac{9.01}{\pi}} \Rightarrow 2\sqrt{\dfrac{8.99}{\pi}} \le x \le 2\sqrt{\dfrac{9.01}{\pi}} \Rightarrow 3.383 \le x \le 3.387 \Rightarrow$

$-0.002 \le x - 3.385 \le 0.002 \Rightarrow |x - 3.385| \le 0.002$

49. a) a any negative Real

   b) $a \ge 0$

51. $y = \sqrt{x^2} \Rightarrow D: -\infty < x < \infty$ and $R: y \ge 0$

   $y = \left(\sqrt{x}\right)^2 \Rightarrow D: x \ge 0$ and $R: y \ge 0$

53. $f(x) = x^2 + 2x = 1$ and $(g \circ f)(x) = |x + 1| \Rightarrow (g \circ f)(x) = \sqrt{(x + 1)^2} = \sqrt{x^2 + 2x + 1} = \sqrt{f(x)} \Rightarrow g(x) = \sqrt{x}$

SECTION 1.5   SHIFTS, CIRCLES, AND PARABOLAS

1. a) $y = (x + 4)^2$

   b) $y = (x - 7)^2$

3. a) Position 4   c) Position 2

   a) Position 1   d) Position 3

5. $C(0,2), a = 2 \Rightarrow x^2 + (y - 2)^2 = 4$

7. $C(3,-4), a = 5 \Rightarrow (x - 3)^2 + (y + 4)^2 = 25$

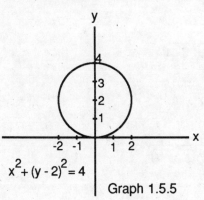

$x^2 + (y - 2)^2 = 4$

Graph 1.5.5

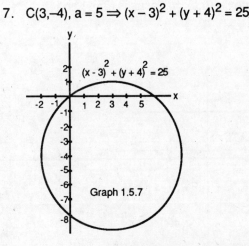

$(x - 3)^2 + (y + 4)^2 = 25$

Graph 1.5.7

9. $x^2 + y^2 = 4$

11. $(x - 3)^2 + (y - 3)^2 = 9$

13. a) Exterior of circle with center (0,0) and radius = 1

   b) Interior of circle with center (0,0) and radius = 2

   c) Interior of concentric ring centered at (0,0) with interior radius = 1 and exterior radius = 2

15. $(x + 2)^2 + (y + 1)^2 < 6$

17. F(0,4), Directrix: $y = -4 \Rightarrow$

$$y = \frac{x^2}{4(4)} \Rightarrow y = \frac{1}{16}x^2$$

19. F(0,–3), Directrix: $y = 3 \Rightarrow$

$$y = -\frac{x^2}{4(3)} \Rightarrow y = -\frac{1}{12}x^2$$

21. $y = \frac{x^2}{2} \Rightarrow 4p = 2 \Rightarrow p = \frac{1}{2}$. Parabola opens

up and vertex = (0,0) $\Rightarrow$ Focus is $\left(0, \frac{1}{2}\right)$,

and vertex = (0,0) $\Rightarrow$ Focus is $\left(0, -\frac{1}{4}\right)$ and

Directrix is $y = -\frac{1}{2}$.

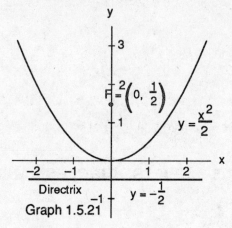

Graph 1.5.21

23. $y = -\frac{x^2}{4} \Rightarrow 4p = 4 \Rightarrow p = 1$. Parabola opens

up and vertex = (0,0) $\Rightarrow$ Focus is (0,–1) and

Directrix is $y = 1$.

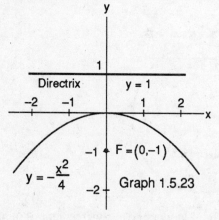

Graph 1.5.23

25. $y + 3 = (x + 2)^2$

27. $y = x^2$

29. $y = \sqrt{x + 4}$

31. $y = \frac{1}{2}x$

33. $y = \sqrt{-(x - 9)}$

35.

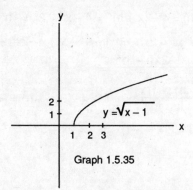

Graph 1.5.35

37.

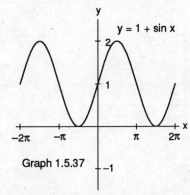

Graph 1.5.37

39.

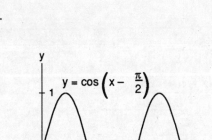

Graph 1.5.39

41. a) iii   b) ii   c) i   d) iv

43. $y - y_0 = m(x - x_0)$

45. $y = p \Rightarrow p = \dfrac{x^2}{4p} \Rightarrow 4p^2 = x^2 \Rightarrow x = \pm 2p$

$(p > 0) \Rightarrow$ Points on the graph are $(\pm 2p, p)$

$\Rightarrow$ distance between the points is

$|2p - (-2p)| = |4p| = 4p$ for $p > 0$.

SECTION 1.6   SLOPES, TANGENT LINES, AND DERIVATIVES

1.

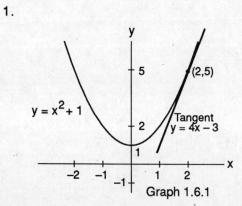

Graph 1.6.1

STEP 1: $f(x) = x^2 + 1$, $f(x + h) = (x + h)^2 + 1 = x^2 + 2xh + h^2 + 1$

STEP 2: $f(x + h) - f(x) = (x^2 + 2xh + h^2 + 1) - (x^2 + 1) = 2xh + h^2$

STEP 3: $\dfrac{f(x + h) - f(x)}{h} = \dfrac{2xh + h^2}{h} = 2x + h$

STEP 4: $f'(x) = \lim_{h \to 0} \dfrac{f(x + h) - f(x)}{h} = \lim_{h \to 0} (2x + h) = 2x$

$\therefore m_{tan} = f'(2) = 4 \Rightarrow y - 5 = 4(x - 2) \Rightarrow y = 4x - 3$

3.

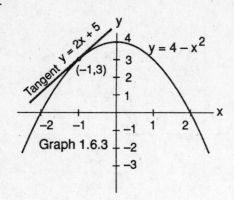

Graph 1.6.3

STEP 1: $f(x) = 4 - x^2$, $f(x+h) = 4 - (x+h)^2 = 4 - x^2 - 2xh - h^2$

STEP 2: $f(x+h) - f(x) = (4 - x^2 - 2xh - h^2) - (4 - x^2) = -2xh - h^2$

STEP 3: $\dfrac{f(x+h) - f(x)}{h} = \dfrac{-2xh - h^2}{h} = -2x - h$

STEP 4: $f'(x) = \lim\limits_{h \to 0} \dfrac{f(x+h) - f(x)}{h} = \lim\limits_{h \to 0} (-2x - h) = -2x$

$\therefore m_{tan} = f'(-1) = 2 \Rightarrow y - 3 = 2(x - (-1)) \Rightarrow y = 2x + 5$

5.

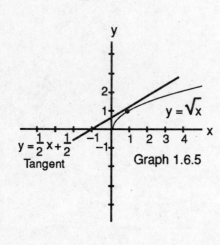

Graph 1.6.5

STEP 1: $f(x) = \sqrt{x}$, $f(x+h) = \sqrt{x+h}$

STEP 2: $f(x+h) - f(x) = \sqrt{x+h} - \sqrt{x}$

STEP 3: $\dfrac{f(x+h) - f(x)}{h} = \dfrac{\sqrt{x+h} - \sqrt{x}}{h} =$

$\dfrac{\sqrt{x+h} - \sqrt{x}}{h} \cdot \dfrac{\sqrt{x+h} + \sqrt{x}}{\sqrt{x+h} + \sqrt{x}} = \dfrac{(x+h) - x}{h\left(\sqrt{(x+h)} + \sqrt{x}\right)}$

$= \dfrac{h}{h\left(\sqrt{x+h} + \sqrt{x}\right)} = \dfrac{1}{\sqrt{x+h} + \sqrt{x}}$

STEP 4: $f'(x) = \lim\limits_{h \to 0} \dfrac{f(x+h) - f(x)}{h} = \lim\limits_{h \to 0} \dfrac{1}{\sqrt{x+h} + \sqrt{x}} =$

$\dfrac{1}{2\sqrt{x}}$ $\therefore m_{tan} = f'(1) = \dfrac{1}{2\sqrt{1}} = \dfrac{1}{2} \Rightarrow y - 1 = \dfrac{1}{2}(x - 1) \Rightarrow$

$y = \dfrac{1}{2}x + \dfrac{1}{2}$

7. STEP 1: $f(x) = x^2$, $f(x+h) = (x+h)^2$

STEP 2: $f(x+h) - f(x) = (x+h)^2 - x^2 = (x^2 + 2xh + h^2) - x^2 = 2xh + h^2$

STEP 3: $\dfrac{f(x+h) - f(x)}{h} = \dfrac{2xh + h^2}{h} = 2x + h$

STEP 4: $f'(x) = \lim\limits_{h \to 0} \dfrac{f(x+h) - f(x)}{h} = \lim\limits_{h \to 0} (2x + h) = 2x.$ $\therefore m_{tan} = f'(1) = 2.$

9. STEP 1: $f(x) = \dfrac{2}{x}$, $f(x+h) = \dfrac{2}{x+h}$

STEP 2: $f(x+h) - f(x) = \dfrac{2}{x+h} - \dfrac{2}{x} = \dfrac{2x}{x(x+h)} - \dfrac{2(x+h)}{x(x+h)} = \dfrac{-2h}{x(x+h)}$

STEP 3: $\dfrac{f(x+h) - f(x)}{h} = \dfrac{\frac{-2h}{x(x+h)}}{h} = \dfrac{-2h}{hx(x+h)} = \dfrac{-2}{x(x+h)}$

STEP 4: $f'(x) = \lim\limits_{h \to 0} \dfrac{f(x+h) - f(x)}{h} = \lim\limits_{h \to 0} \dfrac{-2}{x(x+h)} = \dfrac{-2}{x^2}.$ $\therefore m_{tan} = f'(-2) = \dfrac{-2}{(-2)^2} = -\dfrac{1}{2}$

11. STEP 1: $f(x) = x + \dfrac{9}{x}$, $f(x+h) = (x+h) + \dfrac{9}{x+h}$

STEP 2: $f(x+h) - f(x) = \left((x+h) + \dfrac{9}{x+h}\right) - \left(x + \dfrac{9}{x}\right) = h + \dfrac{9}{x+h} - \dfrac{9}{x} = h + \dfrac{9x - 9(x+h)}{x(x+h)} =$

$h + \dfrac{-9h}{x(x+h)}$

STEP 3: $\dfrac{f(x+h) - f(x)}{h} = \dfrac{h + \dfrac{-9h}{x(x+h)}}{h} = 1 - \dfrac{9}{x(x+h)}$

STEP 4: $f'(x) = \lim\limits_{h\to 0} \dfrac{f(x+h) - f(x)}{h} = \lim\limits_{h\to 0} \left(1 - \dfrac{9}{x(x+h)}\right) = 1 - \dfrac{9}{x^2}$. $\therefore m_{tan} = f'(-3) = 1 - \dfrac{9}{(-3)^2} =$

$1 - 1 = 0$

13. STEP 1: $f(x) = \sqrt{2x}$, $f(x+h) = \sqrt{2(x+h)}$

STEP 2: $f(x+h) - f(x) = \sqrt{2(x+h)} - \sqrt{2x}$

STEP 3: $\dfrac{f(x+h) - f(x)}{h} = \dfrac{\sqrt{2(x+h)} - \sqrt{2x}}{h} = \dfrac{\sqrt{2x+2h} - \sqrt{2x}}{h} \cdot \dfrac{\sqrt{2x+2h} + \sqrt{2x}}{\sqrt{2x+2h} + \sqrt{2x}} =$

$\dfrac{(2x+2h) - 2x}{h\left(\sqrt{2x+2h} + \sqrt{2x}\right)} = \dfrac{2}{\sqrt{2x+2h} + \sqrt{2x}}$

STEP 4: $f'(x) = \lim\limits_{h\to 0} \dfrac{f(x+h) - f(x)}{h} = \lim\limits_{h\to 0} \dfrac{2}{\sqrt{2x+2h} + \sqrt{2x}} = \dfrac{1}{\sqrt{2x}}$

$\therefore m_{tan} = f'\left(\dfrac{1}{2}\right) = \dfrac{1}{\sqrt{2\left(\dfrac{1}{2}\right)}} = 1$

15. STEP 1: $f(x) = \dfrac{1}{\sqrt{2x+3}}$, $f(x+h) = \dfrac{1}{\sqrt{2(x+h)+3}} = \dfrac{1}{\sqrt{2x+2h+3}}$

STEP 2: $f(x+h) - f(x) = \dfrac{1}{\sqrt{2x+2h+3}} - \dfrac{1}{\sqrt{2x+3}} = \dfrac{\sqrt{2x+3} - \sqrt{2x+2h+3}}{\sqrt{2x+3}\sqrt{2x+2h+3}} =$

$\dfrac{\sqrt{2x+3} - \sqrt{2x+2h+3}}{\sqrt{2x+3}\sqrt{2x+2h+3}} \cdot \dfrac{\sqrt{2x+3} + \sqrt{2x+2h+3}}{\sqrt{2x+3} + \sqrt{2x+2h+3}} =$

$\dfrac{(2x+3) - (2x+2h+3)}{\sqrt{2x+3}\sqrt{2x+2h+3}\left(\sqrt{2x+3} + \sqrt{2x+2h+3}\right)} =$

$\dfrac{-2h}{\sqrt{2x+3}\sqrt{2x+2h+3}\left(\sqrt{2x+3} + \sqrt{2x+2h+3}\right)}$

STEP 3: $\dfrac{f(x+h) - f(x)}{h} = \dfrac{\dfrac{-2h}{\sqrt{2x+3}\sqrt{2x+2h+3}\left(\sqrt{2x+3} + \sqrt{2x+2h+3}\right)}}{h} =$

$\dfrac{-2}{\sqrt{2x+3}\sqrt{2x+2h+3}\left(\sqrt{2x+3} + \sqrt{2x+2h+3}\right)}$

STEP 4: $f'(x) = \lim\limits_{h\to 0} \dfrac{f(x+h) - f(x)}{h} = \lim\limits_{h\to 0} \dfrac{-2}{\sqrt{2x+3}\sqrt{2x+2h+3}\left(\sqrt{2x+3} + \sqrt{2x+2h+3}\right)}$

$= \dfrac{-2}{(2x+3)\left(2\sqrt{2x+3}\right)} = \dfrac{-1}{(2x+3)\sqrt{2x+3}}$

$\therefore m_{tan} = f'(-1) = \dfrac{-1}{(1)\sqrt{1}} = -1$

17. STEP 1: $f(s) = \dfrac{s^2}{2}$, $f(s+h) = \dfrac{(s+h)^2}{2} = \dfrac{s^2 + 2sh + h^2}{2}$

    STEP 2: $f(s+h) - f(s) = \dfrac{(s^2 + 2sh + h^2)}{2} - \dfrac{s^2}{2} = \dfrac{2sh + h^2}{2}$

    STEP 3: $\dfrac{f(s+h) - f(s)}{h} = \dfrac{\dfrac{2sh + h^2}{2}}{h} = \dfrac{2sh + h^2}{2h} = \dfrac{2s + h}{2}$

    STEP 4: $f'(s) = \lim\limits_{h \to 0} \dfrac{f(s+h) - f(s)}{h} = \lim\limits_{h \to 0} \dfrac{2s + h}{2} = s$ $\therefore \left. \dfrac{dr}{ds} \right|_{s = \sqrt{2}} = \sqrt{2}$

19. STEP 1: $f(t) = \dfrac{1}{t + 1}$, $f(t+h) = \dfrac{1}{(t+h) + 1} = \dfrac{1}{t + h + 1}$

    STEP 2: $f(t+h) - f(t) = \dfrac{1}{t + h + 1} - \dfrac{1}{t + 1} = \dfrac{t + 1}{(t+1)(t+h+1)} - \dfrac{t + h + 1}{(t+1)(t+h+1)} =$

    $\dfrac{-h}{(t+1)(t+h+1)}$

    STEP 3: $\dfrac{f(t+h) - f(t)}{h} = \dfrac{\dfrac{-h}{(t+1)(t+h+1)}}{h} = \dfrac{-h}{h(t+1)(t+h+1)} = \dfrac{-1}{(t+1)(t+h+1)}$

    STEP 4: $f'(t) = \lim\limits_{h \to 0} \dfrac{f(t+h) - f(t)}{h} = \lim\limits_{h \to 0} \dfrac{-1}{(t+1)(t+h+1)} = \dfrac{-1}{(t+1)^2}$ $\therefore \left. \dfrac{ds}{dt} \right|_{t=2} = \dfrac{-1}{(2+1)^2} = -\dfrac{1}{9}$

21. STEP 1: $f(q) = \sqrt{q + 1}$, $f(q+h) = \sqrt{(q+h) + 1} = \sqrt{q + h + 1}$

    STEP 2: $f(q+h) - f(q) = \sqrt{q + h + 1} - \sqrt{q + 1}$

    STEP 3: $\dfrac{f(q+h) - f(q)}{h} = \dfrac{\sqrt{q + h + 1} - \sqrt{q + 1}}{h} =$

    $\dfrac{\sqrt{q + h + 1} - \sqrt{q + 1}}{h} \cdot \dfrac{\sqrt{q + h + 1} + \sqrt{q + 1}}{\sqrt{q + h + 1} + \sqrt{q + 1}} = \dfrac{(q + h + 1) - (q + 1)}{h\left(\sqrt{q + h + 1} + \sqrt{q + 1}\right)} =$

    $\dfrac{h}{h\left(\sqrt{q + h + 1} + \sqrt{q + 1}\right)} = \dfrac{1}{\sqrt{q + h + 1} + \sqrt{q + 1}}$

    STEP 4: $f'(q) = \lim\limits_{h \to 0} \dfrac{f(q+h) - f(q)}{h} = \lim\limits_{h \to 0} \dfrac{1}{\sqrt{q + h + 1} + \sqrt{q + 1}} = \dfrac{1}{2\sqrt{q + 1}}$

$\therefore \left. \dfrac{dp}{dq} \right|_{q=3} = \dfrac{1}{2\sqrt{3 + 1}} = \dfrac{1}{4}$

23. a) The derivative of f is not defined at x = 0, 1, and 4.

    b) On (−4,0), $m_{tan} = m_{segment} = \dfrac{2 - 0}{0 - (-4)} = \dfrac{1}{2} \Rightarrow f'x) = \dfrac{1}{2}$ on (−4,0). On (0,1), $m_{tan} = m_{segment} = \dfrac{-2 - 2}{1 - 0} =$

    $-4 \Rightarrow f'(x) = -4$ on (0,1). On (1,4), $m_{tan} = m_{segment} = \dfrac{-2 - (-2)}{4 - 1}$

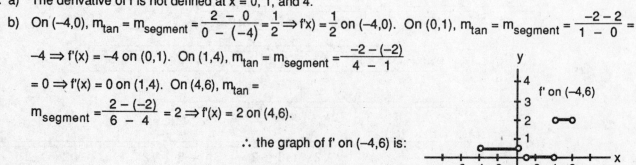

    $= 0 \Rightarrow f'(x) = 0$ on (1,4). On (4,6), $m_{tan} =$

    $m_{segment} = \dfrac{2 - (-2)}{6 - 4} = 2 \Rightarrow f'(x) = 2$ on (4,6).

                                   $\therefore$ the graph of f' on (−4,6) is:

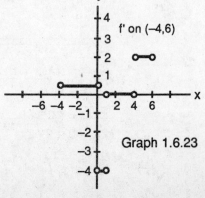

Graph 1.6.23

25. Conclusion: The derivative of $f(x) = |x|$ is $f'(x) = \frac{|x|}{x}, x \neq 0$

Graph of f'
y
x
1
−1

Graph 1.6.25 a

y
x
y = |x|/x, x ≠ 0
1
−1

Graph 1.6.25 b

27.

y
h > 0
1.5
0.5
−0.5
y = 1/(2√x)
h = 0.5
h = 1
h = 2
h = 4
1
2
x

Graph 1.6.27a

As h approaches 0, the difference quotient approaches $y = \frac{1}{2\sqrt{x}}$

y
3.5
3
2.5
2
1.5
1
0.5
h < 0
y = 1/(2√x)
h = −0.5
h = −1
h = −2
h = −4
1
2
3
4
5
x

Graph 1.6.27b

SECTION 1.7 LIMITS OF FUNCTION VALUES

1. 2 3. 25 5. −5 7. 45 9. −15

11. $\frac{5}{8}$ 13. −2 15. 5 17. 0 19. $\frac{11}{4}$

21. $\lim_{x\to -5} \frac{x^2 + 3x - 10}{x + 5} = \lim_{x\to -5} \frac{(x+5)(x-2)}{x+5} = \lim_{x\to -5} (x-2) = -7$

23. $\lim_{x\to 1} \frac{x - 1}{x^2 - 1} = \lim_{x\to 1} \frac{x-1}{(x+1)(x-1)} = \lim_{x\to 1} \frac{1}{x+1} = \frac{1}{2}$

25. $\lim_{x\to -3} \frac{x + 3}{x^2 + 4x + 3} = \lim_{x\to -3} \frac{x+3}{(x+3)(x+1)} = \lim_{x\to -3} \frac{1}{x+1} = -\frac{1}{2}$

27. $\lim_{x\to 5} \frac{x - 5}{x^2 - 25} = \lim_{x\to 5} \frac{x-5}{(x-5)(x+5)} = \lim_{x\to 5} \frac{1}{x+5} = \frac{1}{10}$

29. $\lim_{x\to 2} \frac{2x - 4}{x^3 - 2x^2} = \lim_{x\to 2} \frac{2(x-2)}{x^2(x-2)} = \lim_{x\to 2} \frac{2}{x^2} = \frac{1}{2}$

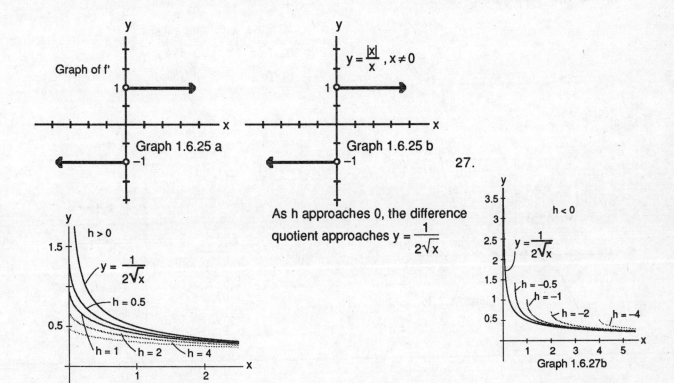

31. a) $\lim_{x \to 2^+} f(x) = 2$, $\lim_{x \to 2^-} f(x) = 1$

    b) No, $\lim_{x \to 2^+} f(x) \neq \lim_{x \to 2^-} f(x)$

33. a) True   f) True

    b) True   g) False

    c) False   h) False

    d) True   i) False

    e) True   j) False

35. a)

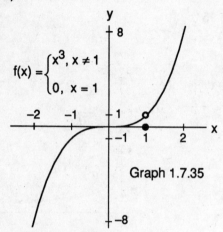

$$f(x) = \begin{cases} x^3, & x \neq 1 \\ 0, & x = 1 \end{cases}$$

Graph 1.7.35

   b) $\lim_{x \to 1^+} f(x) = 1$
      $\lim_{x \to 1^-} f(x) = 1$

   c) Yes, $\lim_{x \to 1^-} f(x) = \lim_{x \to 1^+} f(x) = 1$

37.

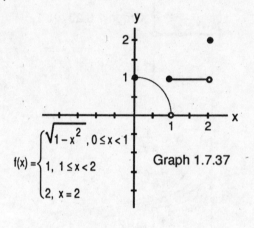

$$f(x) = \begin{cases} \sqrt{1-x^2}, & 0 \leq x < 1 \\ 1, & 1 \leq x < 2 \\ 2, & x = 2 \end{cases}$$

Graph 1.7.37

39.

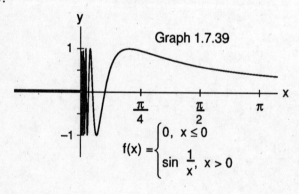

Graph 1.7.39

$$f(x) = \begin{cases} 0, & x \leq 0 \\ \sin \dfrac{1}{x}, & x > 0 \end{cases}$$

a) $\lim_{x \to 0^+} f(x)$ does not exist

b) $\lim_{x \to 0^-} f(x) = 0$

c) $\lim_{x \to 0} f(x)$ does not exist because $\lim_{x \to 0^+} f(x)$ does not exist.

41. a) $\lim_{x \to 0^+} \lfloor x \rfloor = 0$ b) $\lim_{x \to .5} \lfloor x \rfloor = 0$ c) $\lim_{x \to 2^-} \lfloor x \rfloor = 1$

43. a) $\lim_{x \to 4} [g(x) + 3] = 0$ c) $\lim_{x \to 4} [g(x)]^2 = (-3)^2 = 9$

    b) $\lim_{x \to 4} [xf(x)] = 4(0) = 0$ d) $\lim_{x \to 4} \dfrac{g(x)}{f(x) - 1} = \dfrac{-3}{0 - 1} = 3$

45. Right Hand Derivative: $\lim\limits_{h \to 0^+} \dfrac{f(1+h) - f(1)}{h} = \lim\limits_{h \to 0^+} \dfrac{1 + 2h - 1}{h} = \lim\limits_{h \to 0^+} 2 = 2$ since $f(1+h) =$

$2(1+h) - 1 = 1 + 2h$ and $f(1) = 1$.

Left Hand Derivative: $\lim\limits_{h \to 0^-} \dfrac{f(1+h) - f(1)}{h} = \lim\limits_{h \to 0^-} \dfrac{\sqrt{1+h} - 1}{h} =$

$\lim\limits_{h \to 0^-} \dfrac{\sqrt{1+h} - 1}{h} \cdot \dfrac{\sqrt{1+h} + 1}{\sqrt{1+h} + 1} = \lim\limits_{h \to 0^-} \dfrac{(1+h) - 1}{h\left(\sqrt{1+h} + 1\right)} = \lim\limits_{h \to 0^-} \dfrac{1}{\sqrt{1+h} + 1} = \dfrac{1}{2}$ since

$f(1+h) = \sqrt{1+h}$ and $f(1) = 1$. ∴ the derivative doesn't exist since the left hand and right hand derivatives

are not equal.

Note: On exercises 47 through 53, let $\delta = \min(|x_0 - a|, |x_0 - b|)$ (the smaller of the distances from x_0 to a and b)

47.

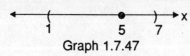

Graph 1.7.47

Let $\delta = 2$. Then $|x - 5| < 2 \Rightarrow$

$-2 < x - 5 < 2 \Rightarrow 3 < x < 7 \Rightarrow$

$1 < x < 7$

49.

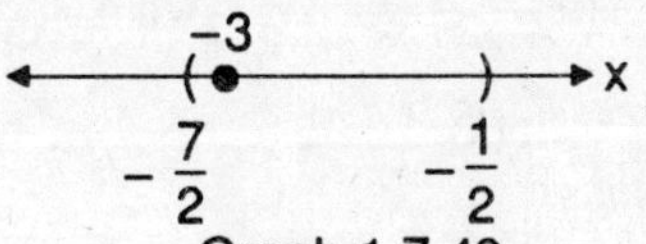

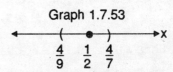

Graph 1.7.49

Let $\delta = \dfrac{1}{2}$. Then $|x - (-3)|$

$< \dfrac{1}{2} \Rightarrow -\dfrac{1}{2} < x + 3 < \dfrac{1}{2} \Rightarrow$

$-\dfrac{7}{2} < x < -\dfrac{5}{2} \Rightarrow -\dfrac{7}{2} < x < -\dfrac{1}{2}$

51.

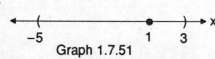

Graph 1.7.51

Let $\delta = 2$. Then $|x - 1| < 2 \Rightarrow$

$-2 < x - 1 < 2 \Rightarrow -1 < x < 3 \Rightarrow$

$-5 < x < 3$

53.

Graph 1.7.53

Let $\delta = \dfrac{1}{18}$. Then $|x - \dfrac{1}{2}| < \dfrac{1}{18} \Rightarrow$

$-\dfrac{1}{18} < x - \dfrac{1}{2} < \dfrac{1}{18} \Rightarrow \dfrac{4}{9} < x < \dfrac{5}{9} \Rightarrow$

$\dfrac{4}{9} < x < \dfrac{4}{7}$

55. $|(2x - 4) - 6| < 0.2 \Rightarrow |2x - 10| < 0.2 \Rightarrow 2|x - 5| < 0.2 \Rightarrow |x - 5| < 0.1$ ∴ let $\delta = 0.1$

57. $|x^2 - 4| < 1 \Rightarrow -1 < x^2 - 4 < 1 \Rightarrow 3 < x^2 < 5 \Rightarrow \sqrt{3} < x < \sqrt{5}$ Now $\sqrt{5} - 2 \approx 0.236$ and $2 - \sqrt{3} \approx 0.268$.

∴ let $\delta = \sqrt{5} - 2$

59. $|\sqrt{x} - 1| < \dfrac{1}{4} \Rightarrow -\dfrac{1}{4} < \sqrt{x} - 1 < \dfrac{1}{4} \Rightarrow \dfrac{3}{4} < \sqrt{x} < \dfrac{5}{4} \Rightarrow \dfrac{9}{16} < x < \dfrac{25}{16}$ Now $\dfrac{25}{16} - 1 = \dfrac{9}{16}$ and $1 - \dfrac{9}{16} = \dfrac{7}{16}$ ∴ let $\delta = \dfrac{7}{16}$

61. $|(x + 1) - 4| < 0.01 \Rightarrow |x - 3| < 0.01 \Rightarrow -0.01 < x - 3 < 0.01 \Rightarrow 2.99 < x < 3.01$

63. $|(x^2 - 5) - 4| < 0.05 \Rightarrow |x^2 - 9| < 0.05 \Rightarrow -0.05 < x^2 - 9 < 0.05 \Rightarrow 8.95 < x^2 < 9.05 \Rightarrow \sqrt{8.95} < |x| < \sqrt{9.05}$

65. $|\sqrt{19-x}-4| < 0.03 \Rightarrow -0.03 < \sqrt{19-x}-4 < 0.03 \Rightarrow 3.97 < \sqrt{19-x} < 4.03 \Rightarrow$

   $15.7609 < 19-x < 16.2409 \Rightarrow -3.2391 < -x < -2.7591 \Rightarrow 3.2391 > x > 2.7591$

67. $\left|\dfrac{1}{x}-4\right| < 0.1 \Rightarrow -0.1 < \dfrac{1}{x}-4 < 0.1 \Rightarrow 3.9 < \dfrac{1}{x} < 4.1 \Rightarrow \dfrac{1}{3.9} > x > \dfrac{1}{4.1} \Rightarrow \dfrac{10}{39} > x > \dfrac{10}{41}$

69. $\lim_{x \to 1}(2x+3) = 5.$ $|(2x+3)-5| < 0.01 \Rightarrow |2x-2| < 0.01 \Rightarrow 2|x-1| < 0.01 \Rightarrow |x-1| < 0.005$ $\therefore$ let $\delta \le 0.005$

71. $\lim_{x \to 1/2}(4x-2) = 0.$ $|(4x-2)-0| < 0.02 \Rightarrow |4x-2| < 0.02 \Rightarrow 4\left|x-\dfrac{1}{2}\right| < 0.02 \Rightarrow \left|x-\dfrac{1}{2}\right| < 0.005$

   $\therefore$ let $\delta \le 0.005$

73. $\lim_{x \to 2}\left(\dfrac{x^2-4}{x-2}\right) = \lim_{x \to 2}(x+2) = 4.$ $\left|\dfrac{x^2-4}{x-2}-4\right| < 0.05 \Rightarrow |(x+2)-4| < 0.05 \Rightarrow |x-2| < 0.05$

   $\therefore$ let $\delta \le 0.05$

75. $\lim_{x \to 11}\sqrt{x-7} = 2.$ $\left|\sqrt{x-7}-2\right| < 0.01 \Rightarrow -0.01 < \sqrt{x-7}-2 < 0.01 \Rightarrow 1.99 < \sqrt{x-7} < 2.01 \Rightarrow$

   $3.9601 < x-7 < 4.0401 \Rightarrow 10.9601 < x < 11.0401$ Now $11-10.9601 = 0.0399$ and $11.0401-11 = 0.0401$

   $\therefore$ let $\delta \le .0399$

77. $\lim_{x \to 2}\dfrac{4}{x} = 2.$ $\left|\dfrac{4}{x}-2\right| < 0.4 \Rightarrow -0.4 < \dfrac{4}{x}-2 < 0.4 \Rightarrow 1.6 < \dfrac{4}{x} < 2.4 \Rightarrow \dfrac{1}{1.6} > \dfrac{x}{4} > \dfrac{1}{2.4} \Rightarrow \dfrac{4}{1.6} > x > \dfrac{4}{2.4} \Rightarrow$

   $\dfrac{5}{2} > x > \dfrac{5}{3}$ Now $\dfrac{5}{2}-2 = \dfrac{1}{2}$ and $2-\dfrac{5}{3} = \dfrac{1}{3}$ $\therefore$ let $\delta \le \dfrac{1}{3}$

79. For $\varepsilon = 0.01$: $|(9-x)-5| < 0.01 \Rightarrow |4-x| < 0.01$ For $\varepsilon = 0.0001$: $|(9-x)-5| < 0.0001 \Rightarrow$

   $\Rightarrow |x-4| < 0.01$ $|4-x| < 0.0001 \Rightarrow |x-4| < 0.0001$

   $\therefore$ let $\delta = 0.01$ $\therefore$ let $\delta = 0.0001$

   For $\varepsilon = 0.001$: $|(9-x)-5| < 0.001 \Rightarrow |4-x| <$ For arbitrary ε: $|(9-x)-5| < \varepsilon \Rightarrow |4-x| < \varepsilon$

   $0.001 \Rightarrow |x-4| < 0.001$ $\Rightarrow |x-4| < \varepsilon$

   $\therefore$ let $\delta = 0.001$ $\therefore$ let $\delta = \varepsilon$

81. Let $\sqrt{x-5} < \varepsilon$ for $x > 5$. Then $x-5 < \varepsilon^2 \Rightarrow x < 5+\varepsilon^2 \Rightarrow 5 < x < 5+\varepsilon^2$ $\therefore$ let $\delta = \varepsilon^2 \Rightarrow I = (5, 5+\varepsilon^2)$

   $\lim_{x \to 5^+}\sqrt{x-5} = 0$

83. a)

x	1.1	1.01	1.001
$\dfrac{x^2-1}{x-1}$	2.1	2.01	2.001

   b) $\lim_{x \to 1}\dfrac{x^2-1}{x-1} = \lim_{x \to 1}\dfrac{(x-1)(x+1)}{x-1} =$
   etc. $\lim_{x \to 1}(x+1) = 2$

85.

x	1.1	1.01	1.001
$\dfrac{\ln(x^2)}{\ln x}$	2	2	2

87. The difference quotient is $\dfrac{\sqrt{9 - (0 + h)^2} - \sqrt{9 - 0^2}}{h} = \dfrac{\sqrt{9 - h^2} - 3}{h}$

h	0.1	0.01	0.001	−0.1	−0.01	−0.001
$\dfrac{\sqrt{9 - h^2} - 3}{h}$	−0.01667	−0.00167	−0.00017	0.01667	0.00167	0.00017

The value of the derivative at x = 0 appears to be 0.

89. The limit appears to be $\dfrac{1}{2}$

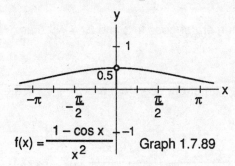

$f(x) = \dfrac{1 - \cos x}{x^2}$ Graph 1.7.89

SECTION 1.8   LIMITS INVOLVING INFINITY

1. a) $\dfrac{2}{5}$
   b) $\dfrac{2}{5}$

3. a) 0
   b) 0

5. a) −3
   b) −3

7. a) $+\infty$
   b) $-\infty$

9. a) 0
   b) 0

11. a) 7
    b) 7

13. a) $-\dfrac{2}{3}$
    b) $-\dfrac{2}{3}$

15. a) $(-1)(1) = -1$
    b) $(-1)(1) = -1$

17. a) $\left(-\dfrac{1}{2}\right)(2) = -1$
    b) $\left(-\dfrac{1}{2}\right)(2) = -1$

19. $+\infty$

21. $+\infty$

23. $+\infty$

25. $+\infty$

27. $-\infty$

29. $\dfrac{1}{2}$

31. 1

33. $+\infty$ **35.** $\dfrac{1}{3} - \dfrac{5}{6} = -\dfrac{1}{2}$

37. a) $\displaystyle\lim_{x \to 2^+} \frac{1}{x^2 - 4} = \lim_{x \to 2^+} \frac{1}{(x-2)(x+2)} = +\infty$ since the denominator goes to 0 and $x - 2 > 0$, $x + 2 > 0$ as

$x \to 2^+$.

b) $\displaystyle\lim_{x \to 2^-} \frac{1}{x^2 - 4} = \lim_{x \to 2^-} \frac{1}{(x-2)(x+2)} = -\infty$ since the denominator goes to 0 and $x - 2 < 0$, $x + 2 > 0$ as

$x \to 2^-$.

c) $\displaystyle\lim_{x \to -2^+} \frac{1}{x^2 - 4} = \lim_{x \to -2^+} \frac{1}{(x-2)(x+2)} = -\infty$ since the denominator goes to 0 and $x - 2 < 0$, $x + 2 > 0$

as $x \to -2^+$.

d) $\displaystyle\lim_{x \to -2^-} \frac{1}{x^2 - 4} = \lim_{x \to -2^-} \frac{1}{(x-2)(x+2)} = +\infty$ since the denominator goes to 0 and $x - 2 < 0$, $x + 2 < 0$

as $x \to -2^-$.

39. a) $\displaystyle\lim_{x \to -2^+} \frac{x^2 - 1}{2x + 4} = \infty$ since the denominator goes to 0, the numerator goes to 3 and $2x + 4 > 0$ as

$x \to -2^+$.

b) $\displaystyle\lim_{x \to -2^-} \frac{x^2 - 1}{2x + 4} = -\infty$ since the denominator goes to 0, the numerator goes to 3 and $2x + 4 < 0$ as

$x \to -2^-$.

c) $\displaystyle\lim_{x \to \infty} \frac{x^2 - 1}{2x + 4} = \lim_{x \to \infty} \frac{x - \frac{1}{x}}{2 + \frac{4}{x}} = \infty$ **d)** $\displaystyle\lim_{x \to -\infty} \frac{x^2 - 1}{2x + 4} = \lim_{x \to -\infty} \frac{x - \frac{1}{x}}{2 + \frac{4}{x}} = -\infty$

41. $\displaystyle\lim_{x \to 0^+} \frac{\lfloor x \rfloor}{x} = \lim_{x \to 0^+} \frac{0}{x}$ **43.** $\displaystyle\lim_{x \to \infty} \frac{|x|}{|x| + 1} = \lim_{x \to \infty} \frac{x}{x + 1} = 1$

$= 0$

45. $\displaystyle\lim_{x \to -\infty} f(x) = \lim_{x \to -\infty} \frac{1}{x} = 0$ $\displaystyle\lim_{x \to 0^+} f(x) = \lim_{x \to 0^+} -1 = -1$

$\displaystyle\lim_{x \to 0^-} f(x) = \lim_{x \to 0^-} \frac{1}{x} = -\infty$ $\displaystyle\lim_{x \to \infty} f(x) = \lim_{x \to \infty} -1 = -1$

47. $\displaystyle\lim_{x \to \infty} \frac{f(x)}{g(x)} = \lim_{x \to \infty} \frac{a_n x^n + a_{n-1} x^{n-1} + \cdots + a_1 x + a_0}{b_m x^m + b_{m-1} x^{m-1} + \cdots + b_1 x + b_0} =$

$\displaystyle\lim_{x \to \infty} \frac{a_n \frac{x^n}{x^m} + a_{n-1} \frac{x^{n-1}}{x^m} + \cdots + a_1 \frac{x}{x^m} + \frac{a_0}{x^m}}{b_m + b_{m-1} \frac{x^{m-1}}{x^m} + \cdots + b_1 \frac{x}{x^m} + \frac{b_0}{x^m}}$. We can see that the term that matters is $\dfrac{x^n}{x^m}$.

If $n = m$, then $\displaystyle\lim_{x \to \infty} \frac{x^n}{x^m} = 1$, and the limit of all other terms in the numerator and denominator (except the b_m)

47. (Continued)

is 0. $\therefore \lim\limits_{x \to \infty} \dfrac{f(x)}{g(x)} = \dfrac{a_n(1) + 0 + 0 + \cdots + 0 + 0}{b_m + 0 + 0 + \cdots + 0 + 0} = \dfrac{a_n}{b_m}$.

If $n < m$, then $\lim\limits_{x \to \infty} \dfrac{x^n}{x^m} = 0$ and $\lim\limits_{x \to \infty} \dfrac{f(x)}{g(x)} = \dfrac{0 + 0 + \cdots + 0 + 0}{b_m + 0 + \cdots + 0 + 0} = 0$.

If $n > m$, then $\lim\limits_{x \to \infty} \dfrac{x^n}{x^m} = \infty \Rightarrow \lim\limits_{x \to \infty} \dfrac{f(x)}{g(x)} = \infty$.

49. The limit is 0.4.

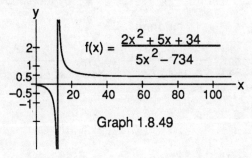

$$f(x) = \dfrac{2x^2 + 5x + 34}{5x^2 - 734}$$

Graph 1.8.49

51. The limit appears to be 0.

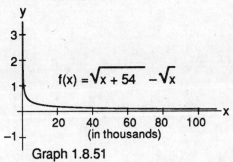

$$f(x) = \sqrt{x + 54} - \sqrt{x}$$

(in thousands)

Graph 1.8.51

SECTION 1.9 THE SANDWICH THEOREM AND $\dfrac{\sin \Theta}{\Theta}$

1. $\dfrac{1}{2}$

3. $\lim\limits_{x \to 0} \dfrac{\sin x}{2x^2 - x} = \lim\limits_{x \to 0} \left(\dfrac{\sin x}{x}\right)\left(\dfrac{1}{2x - 1}\right)$

$= (1)(-1) = -1$

5. $\lim\limits_{\theta \to 0^+} \dfrac{\theta}{\sin \theta} = \lim\limits_{\theta \to 0^+} \dfrac{1}{\dfrac{\sin \theta}{\theta}} = 1$

7. $\lim\limits_{x \to 0} \dfrac{\sin 2x}{x} = \lim\limits_{x \to 0} 2\left(\dfrac{\sin 2x}{2x}\right) = 2$

9. $\lim\limits_{x \to 0} \dfrac{\tan 2x}{2x} = \lim\limits_{x \to 0} \left(\dfrac{\sin 2x}{2x}\right)\left(\dfrac{1}{\cos 2x}\right)$

$= (1)(1) = 1$

11. $\lim\limits_{x \to \infty} \dfrac{\sin 2x}{x} = \lim\limits_{x \to \infty} \dfrac{2 \sin 2x}{2x} =$

$\lim\limits_{\theta \to \infty} \dfrac{2 \sin \theta}{\theta} = 0$. (Let $\theta = 2x$)

13. $\lim\limits_{t \to \infty} \left(2 + \dfrac{\sin t}{t}\right) = 2 + 0 = 2$

15. $\lim\limits_{x \to \infty} x \sin \dfrac{1}{x} = \lim\limits_{\theta \to 0^+} \dfrac{1}{\theta} \sin \theta = \lim\limits_{\theta \to 0^+} \dfrac{\sin \theta}{\theta}$

$= 1$. (Let $\theta = \dfrac{1}{x} \Rightarrow \theta \to 0^+$ as $x \to \infty$)

17. $\lim\limits_{x \to -\infty} \dfrac{\cos \dfrac{1}{x}}{1 + \dfrac{1}{x}} = \lim\limits_{\theta \to 0^-} \dfrac{\cos \theta}{1 + \theta} = \dfrac{\cos 0}{1 + 0} = 1$. (Let $\theta = \dfrac{1}{x} \Rightarrow \theta \to 0^-$ as $x \to -\infty$)

19. $\lim\limits_{x \to 0^+} \dfrac{\sin x}{|x|} = \lim\limits_{x \to 0^+} \dfrac{\sin x}{x} = 1$ since $x \to 0^+ \Rightarrow x > 0 \Rightarrow |x| = x.$ $\lim\limits_{x \to 0^-} \dfrac{\sin x}{|x|} = \lim\limits_{x \to 0^-} \dfrac{\sin x}{-x} = -1$ since

$x \to 0^- \Rightarrow x < 0 \Rightarrow |x| = -x.$ $\therefore \lim\limits_{x \to 0} \dfrac{\sin x}{|x|}$ does not exist.

21. a)

x	.1	.01	.001
$\dfrac{1-\cos x}{x^2}$	.4995583474	.4999958	.5

b) If $\dfrac{1}{2} - \dfrac{x^2}{24} < \dfrac{1 - \cos x}{x^2} < \dfrac{1}{2}$, then since

$\lim_{x \to 0}\left(\dfrac{1}{2} - \dfrac{x^2}{24}\right) = \dfrac{1}{2}$ and $\lim_{x \to 0} \dfrac{1}{2} = \dfrac{1}{2}$,

$\lim_{x \to 0} \dfrac{1 - \cos x}{x^2} = \dfrac{1}{2}$ by the sandwich theorem

c) $\lim\limits_{x \to 0} \dfrac{1 - \cos x}{x^2} = \lim\limits_{x \to 0} \dfrac{1 - \cos x}{x^2} \cdot \dfrac{1 + \cos x}{1 + \cos x} = \lim\limits_{x \to 0} \dfrac{1 - \cos^2 x}{x^2(1 + \cos x)} = \lim\limits_{x \to 0} \dfrac{\sin^2 x}{x^2(1 + \cos x)} =$

$\lim\limits_{x \to 0} \dfrac{\sin x}{x} \cdot \dfrac{\sin x}{x} \cdot \dfrac{1}{1 + \cos x} = (1)(1)\left(\dfrac{1}{1 + 1}\right) = \dfrac{1}{2}$

23. $\lim\limits_{x \to \pm\infty} \dfrac{2x^2}{x^2 + 1} = 2,\ \lim\limits_{x \to \pm\infty} \dfrac{2x^2 + 5}{x^2} = 2 \Rightarrow \lim\limits_{x \to \pm\infty} f(x) = 2$ since $\dfrac{2x^2}{x^2 + 1} < f(x) < \dfrac{2x^2 + 5}{x^2}.$

SECTION 1.10   CONTINUOUS FUNCTIONS

1. a) Yes   c) Yes   3. a) No
   b) $\lim_{x \to -1^+} f(x) = 0$, yes   d) Yes   b) No

5. a) $\lim_{x \to 2} f(x) = 0$ 7. All except $x = 2$ 9. $[-1,0) \cup (0,1) \cup (1,2]$
   b) $f(2) = 0$

11. $x - 2 = 0 \Rightarrow x = 2$ 13. $x + 1 = 0 \Rightarrow x = -1$ 15. Continuous everywhere

17. $x^2 - 1 = 0 \Rightarrow x = \pm 1$ 19. $x = 0$

21. Yes; $\lim_{x \to 1} \dfrac{x^2 - 1}{x - 1} = \lim_{x \to 1} \dfrac{(x + 1)(x - 1)}{x - 1} = \lim_{x \to 1}(x + 1) = 2.$ $f(1) = 2$ $\therefore \lim_{x \to 1} f(x) = f(1)$

23. $h(x) = \dfrac{x^2 + 3x - 10}{x - 2} = \dfrac{(x + 5)(x - 2)}{x - 2} = x + 5$ if $x \neq 2.$ $\therefore$ let $h(2) = 7$

25. $g(x) = \dfrac{x^2 - 16}{x^2 - 3x - 4} = \dfrac{(x - 4)(x + 4)}{(x - 4)(x + 1)} = \dfrac{x + 4}{x + 1}$ if $x \neq 4.$ $\therefore$ let $g(4) = \dfrac{8}{5}$

27. $f(x) \to 8$ when $x \to 3^- \Rightarrow 2ax = 8$ when $x = 3 \Rightarrow 2a(3) = 8 \Rightarrow a = \dfrac{4}{3}$

29. a) $\lim\limits_{x \to 0} \dfrac{\sqrt{1 + \cos x}}{2} = \dfrac{\sqrt{1 + 1}}{2} = \dfrac{\sqrt{2}}{2}$ b) $\lim_{x \to 0}\left(\cos\left(1 - \dfrac{\sin x}{x}\right)\right) = \cos(1 - 1) = 1$

31. $x = 2$ and 3. No minimum because $y \to 0$ as $x \to 1^-$ but $y > 0$ for any x.

33. No maximum because $y \to 1$ as $x \to 1^-$ or $x \to -1^+$ but $y < 1$ for any $x \in (-1,1)$

35. Assume $f(a) > 0$ for some value of x, $x = a$. If $f(b) < 0$ for some value of x, $x = b$, then f must take on all values between $f(a)$ and $f(b)$ since f is continuous on $[a,b]$ (Intermediate Value Theorem). Thus $f(x) = 0$ for some $x \in (a,b)$. But f is never $0 \Rightarrow f(x) > 0$ for all x. The proof is the same if we assume $f(a) < 0$ for some value of x, $x = a$. $\therefore$ if f is continuous and never 0, then it is always positive or always negative.

37. Derivatives of functions have the intermediate value property. $f(0) = 0$ and $f(1) = 1$ but f does not take on the value $\dfrac{1}{2}$ anywhere in $[0,1] \Rightarrow f$ does not have the intermediate value property. $\therefore$ f cannot be the derivative of any function on $[0,1] \Rightarrow f$ cannot be the derivative of any function on $(-\infty, \infty)$.

39. Let $\varepsilon > 0$ be given. Let $\delta = \varepsilon^2$. Then when $0 < x - 0 < \delta$, $0 < x < \varepsilon^2 \Rightarrow \sqrt{x} < \varepsilon \Rightarrow \sqrt{x} - 0 < \varepsilon \Rightarrow \left|\sqrt{x} - 0\right| < \varepsilon$. $\therefore \lim\limits_{x \to 0^+} \sqrt{x} = 0$.

41. $x \approx 1.87939, -1.53209, -0.347296$ 43. $x \approx 0.73908513$ 45. $x = 3.515625$

SECTION 1.M   MISCELLANEOUS EXERCISES

1. If $A(-2,5)$ moves to B on the y–axis then $B = (0,y) \Rightarrow \Delta x = 0 - (-2) = 2 \Rightarrow \Delta y = 3\Delta x = 6 \Rightarrow \Delta y = y - 5 = 6 \Rightarrow y = 11$. $\therefore B$ is $(0,11)$.

3. a)

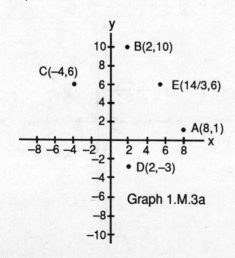

Graph 1.M.3a

b) $m_{AB} = \dfrac{10 - 1}{2 - 8} = \dfrac{9}{-6} = -\dfrac{3}{2}$; $m_{BC} = \dfrac{10 - 6}{2 - (-4)} = \dfrac{2}{3}$; $m_{CD} = \dfrac{-3 - 6}{2 - (-4)} = -\dfrac{3}{2}$; $m_{DA} = \dfrac{1 - (-3)}{8 - 2} = \dfrac{2}{3}$; $m_{CE} = \dfrac{6 - 6}{(14/3) - (-4)} = 0$; $m_{BD} = \dfrac{10 - (-3)}{2 - 2}$ is undefined.

c) Yes, A, B, C, and D since the opposite sides are parallel (opposite sides have equal slopes).

d) Yes, since $m_{AE} = \dfrac{1 - 6}{8 - (14/3)} = \dfrac{-5}{10/3} = -\dfrac{3}{2}$ and $m_{AB} = -\dfrac{3}{2} \Rightarrow$ A, B, and E are collinear.

e) CD, since $m_{CD} = -\dfrac{3}{2}$, $C = (-4,6) \Rightarrow y - 6 = -\dfrac{3}{2}(x - (-4)) \Rightarrow y = -\dfrac{3}{2}x \Rightarrow$ the y–intercept is 0.

5.

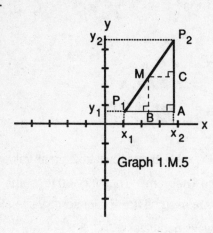

Graph 1.M.5

Let A be (x_2,y_1), M be the midpoint of P_1P_2. Then ΔP_1AP_2 is a right triangle. Drop the perpendicular from M to P_1A at B. Then B is the point (x,y_1) since MB is vertical. Also ΔMBP_1 is a right triangle $\Rightarrow$ $\Delta P_1BM \approx \Delta P_1AP_2 \Rightarrow l(P_1B) = \frac{1}{2}l(P_1A)$ since $l(P_1M) = \frac{1}{2}l(P_1P_2)$. $\therefore$ B is the midpoint of $P_1A \Rightarrow$ the x–coordinate of B is the average of x_1 and $x_2 \Rightarrow B = \left(\dfrac{x_1 + x_2}{2}, y_1\right) \Rightarrow$ x–coordinate of M is $\dfrac{x_1 + x_2}{2}$ since MB is vertical. Drop the perpendicular from M to P_2A at $C \Rightarrow \Delta MCP_2 \approx \Delta P_1AP_2 \Rightarrow l(P_2C) = \frac{1}{2}l(P_2A)$ for similar reasons. $\therefore$

C is the midpoint of $P_2A \Rightarrow$ y–coordinate of C is the average of y_1 and $y_2 \Rightarrow C = \left(x_2, \dfrac{y_1 + y_2}{2}\right) \Rightarrow$ y–coordinate of M is $\dfrac{y_1 + y_2}{2}$ since MC is horizontal. $\therefore M = \left(\dfrac{x_1 + x_2}{2}, \dfrac{y_1 + y_2}{2}\right)$.

7. $A = \pi r^2$ and $C = 2\pi r \Rightarrow r = \dfrac{C}{2\pi} \Rightarrow A = \pi\left(\dfrac{C}{2\pi}\right)^2 = \dfrac{C^2}{4\pi}$

9.   On $[0,1)$, $m = \dfrac{0 - 1}{1 - 0} = -1 \Rightarrow$ the line is $y - 1 = -1(x - 0) \Rightarrow y = -x + 1$. On $[1,2]$, $m = \dfrac{0 - 1}{2 - 1} = -1 \Rightarrow$ the line is $y - 1 = -1(x - 1) \Rightarrow y = -x + 2$. $\therefore f(x) = \begin{cases} -x + 1, & 0 \le x < 1 \\ -x + 2, & 1 \le x \le 2 \end{cases}$

11. $l(AB) = \sqrt{(6 - 4)^2 + (4 - (-3))^2} = \sqrt{2^2 + 7^2} = \sqrt{53}$. $l(AC) = \sqrt{(6 - (-2))^2 + (4 - 3)^2} = \sqrt{8^2 + 1^2} = \sqrt{65}$. $l(BC) = \sqrt{(4 - (-2))^2 + (-3 - 3)^2} = \sqrt{6^2 + (-6)^2} = \sqrt{72}$. $\therefore \Delta ABC$ is not isosceles since no two sides are equal in length and is not a right triangle since the sum of the squares of the two shorter sides is not equal to the square of the the longest side.

13. a)   b)   c)

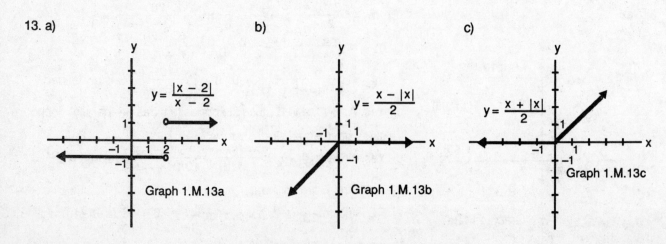

$y = \dfrac{|x - 2|}{x - 2}$ $y = \dfrac{x - |x|}{2}$ $y = \dfrac{x + |x|}{2}$

Graph 1.M.13a   Graph 1.M.13b   Graph 1.M.13c

15. a)

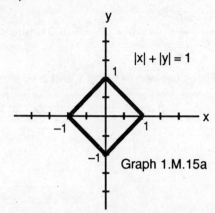

$|x| + |y| = 1$

Graph 1.M.15a

b)

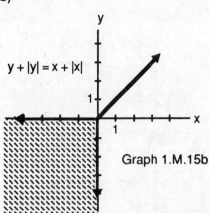

$y + |y| = x + |x|$

Graph 1.M.15b

17. $\left|\dfrac{\sqrt{x}}{2} - 1\right| < 0.2 \Rightarrow -0.2 < \dfrac{\sqrt{x}}{2} - 1 < 0.2 \Rightarrow 0.8 < \dfrac{\sqrt{x}}{2} < 1.2 \Rightarrow 1.6 < \sqrt{x} < 2.4 \Rightarrow 2.56 < x < 5.76.$

$\left|\dfrac{\sqrt{x}}{2} - 1\right| < 0.1 \Rightarrow -0.1 < \dfrac{\sqrt{x}}{2} - 1 < 0.1 \Rightarrow 0.9 < \dfrac{\sqrt{x}}{2} < 1.1 \Rightarrow 1.8 < \sqrt{x} < 2.2 \Rightarrow 3.24 < x < 4.84$

19. $\lim\limits_{h \to 0^+} \dfrac{|0 + h| - |0|}{h} = \lim\limits_{h\to 0^+} \dfrac{|h|}{h} = \lim\limits_{h \to 0^+} \dfrac{h}{h} = 1$ since $|h| = h$ if $h \to 0^+$. $\lim\limits_{h \to 0^-} \dfrac{|0 + h| - |0|}{h} = \lim\limits_{h\to 0^-} \dfrac{|h|}{h}$

$= \lim\limits_{h\to 0^-} \dfrac{-h}{h} = -1$ since $|h| = -h$ if $h \to 0^-$. $\therefore \lim\limits_{h \to 0} \dfrac{|0 + h| - |0|}{h}$ does not exist $\Rightarrow f'(0)$ does not exist for

$f(x) = |x|$.

21. a) $\lim\limits_{x \to c} 3f(x) = (3)(-7) = -21$

    b) $\lim\limits_{x \to c} (f(x))^2 = (-7)^2 = 49$

    c) $\lim\limits_{x \to c} (f(x))(g(x)) = (-7)(0) = 0$

    d) $\lim\limits_{x \to c} \dfrac{f(x)}{g(x) - 7} = \dfrac{-7}{0 - 7} = 1$

    e) $\lim\limits_{x \to c} \cos(g(x)) = \cos 0 = 1$

    f) $\lim\limits_{x \to c} |f(x)| = |-7| = 7$

23.

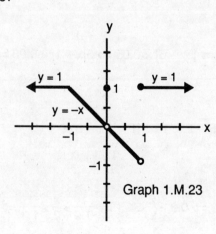

$y = 1$

$y = 1$

$y = -x$

Graph 1.M.23

a) $\lim\limits_{x \to -1^+} f(x) = 1$ $\lim\limits_{x \to 1^+} f(x) = 1$

$\lim\limits_{x \to -1^-} f(x) = 1$ $\lim\limits_{x \to 1^-} f(x) = -1$

$\lim\limits_{x \to 0^+} f(x) = 0$ $\lim\limits_{x \to 0^-} f(x) = 0$

b) $\lim\limits_{x \to -1} f(x) = 1$ $\lim\limits_{x \to 0} f(x) = 0$

$\lim\limits_{x \to 1} f(x)$ does not exist since $\lim\limits_{x \to 1^+} f(x) \neq \lim\limits_{x \to 1^-} f(x)$

c) f is continuous at $x = -1$.

25. If $0 \le \left|\sqrt{x} \sin \dfrac{1}{x}\right| \le \sqrt{x}$ and since $\lim\limits_{x \to 0^+} 0 = 0$ and $\lim\limits_{x \to 0^+} \sqrt{x} = 0$, then $\lim\limits_{x \to 0^+} \sqrt{x} \sin \dfrac{1}{x} = 0$
by the sandwich theorem.

27. a)

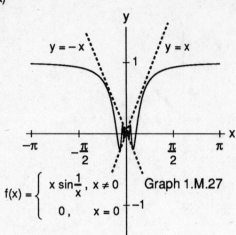

$$f(x) = \begin{cases} x\sin\dfrac{1}{x}, & x \neq 0 \\ 0, & x = 0 \end{cases}$$ Graph 1.M.27

b) $\displaystyle\lim_{x \to 0} x\sin\frac{1}{x} = \lim_{x \to 0} \frac{\sin\frac{1}{x}}{\frac{1}{x}} = \lim_{b \to \infty} \frac{\sin b}{b} = 0$ where

$b = \dfrac{1}{x}$ and sin b oscillates between -1 and 1 and $b \to \infty$.

$f(0) = 0.$ $\therefore \displaystyle\lim_{x \to 0} x\sin\frac{1}{x} = 0 = f(0) \Rightarrow f$ is continuous at $x = 0$.

29. a) $\lim_{x \to 0} x^2 = 0$. But as $x \to 0$, x^2 gets closer to -1 (yes it does--think about it). But $L \neq -1$. (There are

other answers to this exercise.)

b) Let $f(x) = \begin{cases} 0, & x \geq 0 \\ x + 1, & x < 0 \end{cases}$ and let $L = 0$. Then $|f(x) - 0| = |0 - 0| = 0 < \varepsilon$ (where $\varepsilon > 0$ is given) for every

$x \geq 0$. But $\lim_{x \to 0} f(x)$ does not exist because $\lim_{x \to 0^+} f(x) = 0 \neq 1 = \lim_{x \to 0^-} f(x)$.

31. $f(x) = \dfrac{|x - 5|}{x - 5} = 1$ for $x > 5$, $f(x) = \dfrac{|x - 5|}{x - 5} = -1$ for $x < 5$ For $\varepsilon = 4$, $1 - \varepsilon = -3$ and $1 + \varepsilon = 5 \Rightarrow$

$-3 < f(x) < 5$ which is true for all x.   For $\varepsilon = 2$, $1 - \varepsilon = -1$ and $1 + \varepsilon = 3 \Rightarrow -1 < f(x) < 3$ which is true

for all $x > 5$.   For $\varepsilon = 1$, $1 - \varepsilon = 0$ and $1 + \varepsilon = 2 \Rightarrow 0 < f(x) < 2$ which is true for all $x > 5$.   For $\varepsilon = \dfrac{1}{2}$,

$1 - \varepsilon = \dfrac{1}{2}$ and $1 + \varepsilon = \dfrac{3}{2} \Rightarrow \dfrac{1}{2} < f(x) < \dfrac{3}{2}$ which is true for all $x > 5$

33. $\lim_{x \to 3}(5x - 10) = 5.$ $|f(x) - 5| < 0.05 \Rightarrow |(5x - 10) - 5| < 0.05 \Rightarrow |5x - 15| < 0.05 \Rightarrow 5|x - 3| < 0.05 \Rightarrow$

$|x - 3| < 0.01$ $\therefore$ let $\delta \leq 0.01$

35. $\lim_{x \to 1}(5x - 10) = -5.$ $|f(x) - (-5)| < 0.05 \Rightarrow |(5x - 10) - (-5)| < 0.05 \Rightarrow |5x - 5| < 0.05 \Rightarrow 5|x - 1| < 0.05 \Rightarrow$

$|x - 1| < 0.01$ $\therefore$ let $\delta \leq 0.01$

37. $\lim_{x \to 9} \sqrt{x - 5} = 2.$ $|f(x) - 2| < 1 \Rightarrow |\sqrt{x - 5} - 2| < 1 \Rightarrow -1 < \sqrt{x - 5} - 2 < 1 \Rightarrow 1 < \sqrt{x - 5} < 3 \Rightarrow$

$1 < x - 5 < 9 \Rightarrow -3 < x - 9 < 5. \Rightarrow -3 < x - 9 < 3$ $\therefore$ let $\delta \leq 3$

39. $\lim_{x \to 2} \dfrac{2}{x} = 1.$ $|f(x) - 1| < 0.1 \Rightarrow \left|\dfrac{2}{x} - 1\right| < 0.1 \Rightarrow -0.1 < \dfrac{2}{x} - 1 < 0.1 \Rightarrow 0.9 < \dfrac{2}{x} < 1.1 \Rightarrow \dfrac{1}{0.9} > \dfrac{x}{2} > \dfrac{1}{1.1} \Rightarrow$

$\dfrac{10}{9} > \dfrac{x}{2} > \dfrac{10}{11} \Rightarrow \dfrac{20}{9} > x > \dfrac{20}{11} \Rightarrow \dfrac{2}{9} > x - 2 > -\dfrac{2}{11} \Rightarrow \dfrac{2}{11} > x - 2 > -\dfrac{2}{11}$ $\therefore$ let $\delta \leq \dfrac{2}{11}$

41. $\displaystyle\lim_{x \to \infty} \frac{x + \sin x}{2x + 5} = \lim_{x \to \infty} \frac{x}{2x + 5} + \lim_{x \to \infty} \frac{\sin x}{2x + 5} = \frac{1}{2} + 0 = \frac{1}{2}$

43. $\lim\limits_{x \to \infty} \dfrac{x \sin x}{x + \sin x} = \lim\limits_{x \to \infty} \dfrac{(x \sin x)/x}{(x + \sin x)/x} = \lim\limits_{x \to \infty} \dfrac{\sin x}{1 + \dfrac{\sin x}{x}}$ does not exist since $\dfrac{\sin x}{x} \to 0$ as $x \to \infty$ and

sin x oscillates between -1 and 1 as $x \to \infty$.

45. $\lim\limits_{x \to a} \dfrac{x^2 - a^2}{x - a} = \lim\limits_{x \to a} \dfrac{(x - a)(x + a)}{x - a} = \lim\limits_{x \to a} (x + a) = 2a$

47. $\lim\limits_{h \to 0} \dfrac{(x + h)^2 - x^2}{h} = \lim\limits_{h \to 0} \dfrac{x^2 + 2xh + h^2 - x^2}{h} = \lim\limits_{h \to 0} \dfrac{2xh + h^2}{h} = \lim\limits_{h \to 0} (2x + h) = 2x$

49. Let $f(x) = a_n x^n + a_{n-1} x^{n-1} + \cdots + a_1 x + a_0$ be a polynomial function of odd degree (n is odd) $\Rightarrow$ f is continuous everywhere. Assume $a_n > 0$. The proof is similar for $a_n < 0$. Then $\lim\limits_{x \to \infty} f(x) = \infty$ and $\lim\limits_{x \to -\infty} f(x) = -\infty$ since

$f(x) = x^{n-1}\left(a_n x + a_{n-1} + \dfrac{a_{n-2}}{x} + \cdots + \dfrac{a_0}{x^{n-1}}\right)$ and $n - 1$ is even. $\therefore$ there must exist an x_1 so that

$f(x_1) < 0$ and an x_2 so that $f(x_2) > 0 \Rightarrow$ there is a value b between x_1 and x_2 so that $f(b) = 0$ by the intermediate

value theorem.

51. No, because the interval is not closed.

53. True. The Intermediate Value Theorem says since f is continuous on the interval [1,2] and since $f(1) = 0$, $f(2) = 3$, then f takes on all values between 0 and 3, including 2.5 for some $x \in (1,2)$.

55. f continuous at c $\Rightarrow \lim\limits_{x \to c} f(x) = f(c) > 0$. Let $\varepsilon = \dfrac{f(c)}{2}$. Then there exists a $\delta > 0$ so that $|f(x) - f(c)| < \varepsilon$ when

$|x - c| < \delta$ or $-\delta < x - c < \delta \Rightarrow c - \delta < x < c + \delta$. Then for $c - \delta < x < c + \delta$, $|f(x) - f(c)| < \dfrac{f(c)}{2} \Rightarrow$

$-\dfrac{f(c)}{2} < f(x) - f(c) < \dfrac{f(c)}{2} \Rightarrow f(c) - \dfrac{f(c)}{2} < f(x) < f(c) + \dfrac{f(c)}{2} \Rightarrow \dfrac{1}{2} f(c) < f(x) < \dfrac{3}{2} f(c) \Rightarrow f(x) > 0$ since $\dfrac{1}{2} f(c) > 0$.

57. a) $a < b \Rightarrow b - a > 0$. $b - a = (b + c) - (a + c) \Rightarrow (b + c) - (a + c) > 0 \Rightarrow b + c > a + c$ or $a + c < b + c$. Also $b - a = (b - c) - (a - c) \Rightarrow (b - c) - (a - c) > 0 \Rightarrow b - c > a - c$ or $a - c < b - c$.

b) If $a < b$, then $a + c < b + c$. If $c < d$, then $b + c < b + d$. $\therefore a + c < b + d$. For the second part, let $a = 5$, $b = 6$, $c = -5$, $d = 10$. then $a - c = 5 - (-5) = 10 \not< b - d = 6 - 10 = -4$.

c) $\dfrac{1}{b} - \dfrac{1}{a} = \dfrac{a - b}{ab}$. $a < b \Rightarrow a - b < 0$ and a and b both of the same sign $\Rightarrow ab > 0$. $\therefore \dfrac{a - b}{ab} < 0 \Rightarrow \dfrac{1}{b} - \dfrac{1}{a} < 0 \Rightarrow \dfrac{1}{a} > \dfrac{1}{b}$.

d) $a < 0 \Rightarrow \dfrac{1}{a} < 0$. $b > 0 \Rightarrow \dfrac{1}{b} > 0$. $\therefore \dfrac{1}{a} < \dfrac{1}{b}$.

e) $a < b \Rightarrow a - b < 0$. $c > 0 \Rightarrow c(a - b) < 0 \Rightarrow ac - bc < 0 \Rightarrow ac < bc$.

f) $a < b \Rightarrow a - b < 0$. $c < 0 \Rightarrow c(a - b) > 0 \Rightarrow ac - bc > 0 \Rightarrow ac > bc$.

CHAPTER 2

DERIVATIVES

2.1 THE RULES OF DIFFERENTIATION

1. $\dfrac{dy}{dx} = -20x, \dfrac{d^2y}{dx^2} = -20$

3. $\dfrac{dy}{dx} = x^3 - x^2 + x - 1, \dfrac{d^2y}{dx^2} = 3x^2 - 2x + 1$

5. $y' = 2x, y'' = 2$

7. $y' = 2x^3 - 3x - 1, y'' = 6x^2 - 3, y''' = 12x, y^{(4)} = 12$

9. a) $y = (x + 1)(3 - x^2), y' = (1)(3 - x^2) + (-2x)(x + 1) = -3x^2 - 2x + 3$

   b) $y = -x^3 - x^2 + 3x + 3, y' = -3x^2 - 2x + 3$

11. a) $y = (x - 1)(x^2 + x + 1), y' = 1(x^2 + x + 1) + (2x + 1)(x - 1) = 3x^2$ b) $y = x^3 - 1, y' = 3x^2$

13. $y = \dfrac{2x + 5}{3x - 2}, \dfrac{dy}{dx} = \dfrac{2(3x - 2) - (2x + 5)(3)}{(3x - 2)^2} = \dfrac{-19}{(3x - 2)^2}$

15. $y = \dfrac{x^2 - 4}{x + 0.5}, \dfrac{dy}{dx} = \dfrac{2x(x + 0.5) - (x^2 - 4)(1)}{(x + 0.5)^2} = \dfrac{x^2 + x + 4}{(x + 0.5)^2}$

17. $y = (1 - x)(1 + x^2)^{-1} = \dfrac{1 - x}{1 + x^2}, \dfrac{dy}{dx} = \dfrac{-1(1 + x^2) - (2x)(1 - x)}{(1 + x^2)^2} = \dfrac{x^2 - 2x - 1}{(1 + x^2)^2}$

19. $y = \dfrac{1}{(x^2 - 1)(x^2 + x + 1)}, \dfrac{dy}{dx} = \dfrac{(0)(x^2 - 1)(x^2 + x + 1) - (1)\left[2x(x^2 + x + 1) + (x^2 - 1)(2x + 1)\right]}{(x^2 - 1)^2(x^2 + x + 1)^2} =$

    $\dfrac{1 - 3x^2 - 4x^3}{(x^2 - 1)^2(x^2 + x + 1)^2}$

21. $y = \dfrac{3}{x^2} = 3x^{-2}, \dfrac{dy}{dx} = -6x^{-3} = \dfrac{-6}{x^3}, \dfrac{d^2y}{dx^2} = 18x^{-4} = \dfrac{18}{x^4}$

23. $y = \dfrac{5}{3x} = \dfrac{5}{3}x^{-1}, \dfrac{dy}{dx} = \dfrac{-5}{3}x^{-2} = \dfrac{-5}{3x^2}, \dfrac{d^2y}{dx^2} = \dfrac{10}{3}x^{-3} = \dfrac{10}{3x^3}$

25. $y = x + 1 + \dfrac{1}{x} = x + 1 + x^{-1}, \dfrac{dy}{dx} = 1 - x^{-2} = 1 - \dfrac{1}{x^2}, \dfrac{d^2y}{dx^2} = 2x^{-3} = \dfrac{2}{x^3}$

27. $y = \dfrac{x^3 + 7}{x} = x^2 + 7x^{-1}, \dfrac{dy}{dx} = 2x - 7x^{-2} = 2x - \dfrac{7}{x^2}, \dfrac{d^2y}{dx^2} = 2 + 14x^{-3} = 2 + \dfrac{14}{x^3}$

29. $y = \dfrac{(x-1)(x^2+x+1)}{x^3} = \dfrac{x^3-1}{x^3} = 1 - x^{-3}, \dfrac{dy}{dx} = 3x^{-4} = \dfrac{3}{x^4}), \dfrac{dy}{dx} = 3x^{-4} = \dfrac{3}{x^4}, \dfrac{d^2y}{dx^2} = -12x^{-5} = -\dfrac{12}{x^5}$

31. $y = x(x-1)(x+1), \dfrac{dy}{dx} = x(x-1)\dfrac{d}{dx}(x+1) + x(x+1)\dfrac{d}{dx}(x-1) + (x+1)(x-1)\dfrac{d}{dx}(x) =$

$x(x-1) + x(x+1) + (x+1)(x-1) = 3x^2 - 1$; alternatively, $y = x(x-1)(x+1) = x^3 - x, \dfrac{dy}{dx} = 3x^2 - 1$

33. $y = (1-x)(x+1)(3-x^2), \dfrac{dy}{dx} = (1-x)(x+1)\dfrac{d}{dx}(3-x^2) + (1-x)(3-x^2)\dfrac{d}{dx}(x+1) +$

$(x+1)(3-x^2)\dfrac{d}{dx}(1-x) = (1-x)(x+1)(-2x) + (1-x)(3-x^2) + (x+1)(3-x^2)(-1) =$

$4x^3 - 8x$; alternatively, $y = (1-x)(x+1)(3-x^2) = 3 - 4x^2 + x^4, \dfrac{dy}{dx} = 4x^3 - 8x$

35. $y = (x-1)(x+1)(x-2)(x+2), \dfrac{dy}{dx} = (x-1)(x+1)(x-2)\dfrac{d}{dx}(x+2) + (x-1)(x+1)(x+2)\dfrac{d}{dx}(x-2) +$

$(x-1)(x-2)(x+2)\dfrac{d}{dx}(x+1) + (x+1)(x-2)(x+2)\dfrac{d}{dx}(x-1) = (x-1)(x+1)(x-2) +$

$(x-1)(x+1)(x+2) + (x-1)(x-2)(x+2) + (x+1)(x-2)(x+2) = 4x^3 - 10x$; alternatively, $y =$
$(x-1)(x+1)(x-2)(x+2) = x^4 - 5x^2 + 4, \dfrac{dy}{dx} = 4x^3 - 10x$

37. a) $s = \dfrac{t}{1+t^2}, \dfrac{ds}{dt} = \dfrac{(1)(1+t^2) - t(2t)}{(1+t^2)^2} = \dfrac{1-t^2}{(1+t^2)^2}$

  b) $s = t^2(t+1)(1-t^2) = t^2 + t^3 - t^4 - t^5, \dfrac{ds}{dt} = 2t + 3t^2 - 4t^3 - 5t^4$

39. a) $w = \dfrac{z^2+1}{z^2+2z+1}, \dfrac{dw}{dz} = \dfrac{2z(z^2+2z+1) - (2z+2)(z^2+1)}{(z^2+2z+1)^2} = \dfrac{2(z^2-1)}{(z+1)^4}$

  b) $w = (z + z^2)(z^{-1} + z^{-2}) = 1 + z^{-1} + z + 1, \dfrac{dw}{dz} = -z^{-2} + 1$

41. a) $\dfrac{d(u\,v)}{dx} = u\,v' + v\,u' = (5)(2) + (-1)(-3) = 13$ b) $\dfrac{d(u/v)}{dx} = \dfrac{u'v - u\,v'}{v^2} = \dfrac{(-3)(-1) - (5)(2)}{(-1)^2} = -7$

  c) $\dfrac{d(v/u)}{dx} = \dfrac{v'u - u'v}{u^2} = \dfrac{(2)(5) - (-3)(-1)}{5^2} = \dfrac{7}{25}$ d) $\dfrac{d(7v - 2u)}{dx} = 7v' - 2u' = 7(2) - 2(-3) = 20$

43. $y = x^3 - 3x + 1 \Rightarrow y' = 3x^2 - 3 \Rightarrow m = 3(2)^2 - 3 = 9$ the slope of the tangent line $\therefore \dfrac{-1}{9}$ is the slope of

the normal line; the desired equation is $y - 3 = \dfrac{-1}{9}(x-2) \Rightarrow y = \dfrac{-1}{9}x + \dfrac{29}{9}$

45. $y = 2x^3 - 3x^2 - 12x + 20 \Rightarrow y' = 6x^2 - 6x - 12 = 6(x^2 - x - 2) = 6(x-2)(x+1)$; since $m = y'$, solving
$6(x-2)(x+1) = 0 \Rightarrow x = 2$ or $-1 \Rightarrow (2, 0)$ and $(-1, 27)$ are the desired points.

47. $y = \dfrac{4x}{x^2 + 1}$, $y' = \dfrac{4(x^2 + 1) - 2x(4x)}{(x^2 + 1)^2} = \dfrac{4(1 - x^2)}{(x^2 + 1)^2} \Rightarrow m = \dfrac{4(1 - x^2)}{(x^2 + 1)^2}$; at $(0, 0)$ $m = 4$, $y - 0 = 4(x - 0) \Rightarrow$

$y = 4x$; at $(1, 2)$ $m = 0$, $y - 2 = 0(x - 1) \Rightarrow y = 2$

49. The curve $y = ax^2 + bx + c$ is tangent to $y = x$ at the origin $\Rightarrow (0,0)$ is a point on the curve.

$\therefore \ 0 = 0a + 0b + c$ or $c = 0$. If $y = ax^2 + bx$, then $\dfrac{dy}{dx} = 2ax + b$ measures the slope of the tangent

to the curve at any point. In particular, $y = x$ has slope 1 at $x = 0$. $\ \therefore \ 1 = 2a(0) + b \Rightarrow b = 1$ and

$y = ax^2 + 1$. Since the $(1,2)$ is also on the curve, $2 = (1)^2 a + 1$ or $a = 1$. $\ \therefore \ $ the equation is

$y = x^2 + x$.

51. $P = \dfrac{nRT}{V - nb} - \dfrac{an^2}{V^2} \Rightarrow \dfrac{dP}{dV} = \dfrac{0(V - nb) - 1(nRT)}{(V - nb)^2} - \dfrac{0V^2 - 2V(an^2)}{(V^2)^2} = -\dfrac{nRT}{(V - nb)^2} + \dfrac{2an^2}{V^3}$, $\dfrac{d^2P}{dV^2} =$

$-\dfrac{0(V - nb)^2 - 2(V - nb)(nRT)}{(V - nb)^4} + \dfrac{0(V^3) - 3V^2(2an^2)}{V^6} = \dfrac{2nRT}{(V - nb)^3} - \dfrac{6an^2}{V^4}$

53. $\dfrac{d}{dx}(cu) = \dfrac{d}{dx}(c) u + c \dfrac{d}{dx}(u) = 0 u + c \dfrac{du}{dx} = c \dfrac{du}{dx}$

2.2 VELOCITY AND OTHER RATES OF CHANGE

1. $s = 0.8t^2 \Rightarrow v = 1.6t$ and $a = 1.6$

   a) $v = 0.0$ m/sec and $a = 1.6$ m/sec^2 at $t = 0$; $v = 16$ m/sec and $a = 1.6$ m/sec^2 at $t = 10$

   b) $\Delta s = s(10) - s(0) = 80 - 0 = 80$ m; $v_{av} = \dfrac{\Delta s}{\Delta t} = 8.0$ m/sec

3. $s = t^3 - 3t^2 + 2t \Rightarrow v = 3t^2 - 6t + 2$ and $a = 6t - 6$

   a) $v = 2$ m/sec and $a = -6$ m/sec^2 at $t = 0$; $v = 2.0$ m/sec and $a = 6$ m/sec^2 at $t = 2$

   b) $\Delta s = s(2) - s(0) = 0 - 0 = 0.0$ m; $v_{av} = \dfrac{\Delta s}{\Delta t} = 0.0$ m/sec

5. $s = 125t^{-1} - 626t^{-3} \Rightarrow v = -125t^{-2} + 1875t^{-4}$ and $a = 250t^{-3} - 7500t^{-5}$

   a) $v = 1750$ m/sec amd $a = -7250$ m/sec^2 at $t = 1$; $v = -2$ m/sec and $a = -0.4$ m/sec^2 at $t = 5$

   b) $\Delta s = s(5) - s(1) = 25 - 5 = 20$ m; $v_{av} = \dfrac{\Delta s}{\Delta t} = 5$ m/sec

7. $s_m = 1.86t^2 \Rightarrow v_m = 3.72t$, solving $3.72t = 16.6 \Rightarrow t = 4.46$ sec on Mars; $s_j = 11.44t^2 \Rightarrow v_j = 22.88t$,

   solving $22.88t = 16.6 \Rightarrow t = .726$ sec on Jupiter

9. $s = 24t - 4.9t^2 \Rightarrow v = s' = 24 - 9.8t$, solving $v = 0 \Rightarrow t = 24/9.8$ sec is the time it takes to reach its

   maximum height. The maximum is $s(24/9.8) = 1440/49$ m ≈ 29.39 m.

11. $b(t) = 10^6 + 10^4 t - 10^3 t^2 \Rightarrow b'(t) = 10^4 - (2)(10^3 t) = 10^3[10 - 2t]$ a) $b'(0) = 10^4$ bacteria/hr

b) $b'(5) = 0$ bacteria/hr   c) $b'(10) = -10^4$ bacteria/hr

13. $c(x) = 2000 + 100x - .1x^2 \Rightarrow c'(x) = 100 - .2x$ a) $c_{av} = \dfrac{c(100)}{100} = \$110$ per machine

b)   Marginal cost $= c'(x) \Rightarrow$ the marginal cost of producing 100 machines is $c'(100) = \$80$.

c)   The cost of producing the 101st machine is $c(101) - c(100) = 100 - \dfrac{201}{10} = \$79.90$.

15. $s = t^3 - 6t^2 + 9t \Rightarrow v = s' = 3t^2 - 12t + 9 \Rightarrow a = v' = s'' = 6t - 12; \ v = 0 \Rightarrow 3t^2 - 12t + 9 = 0 \Rightarrow$

$3(t^2 - 4t + 3) = 0 \Rightarrow 3(t - 1)(t - 3) = 0 \Rightarrow$ when $t = 1$ or 3, the velocity is zero; $a(1) = -6$ m/sec^2 and

$a(3) = 6$ m/sec^2

17.   a)   190 ft/sec   b)   2 sec   c)   at 8 sec, 0 ft/sec

19. $s = 490t^2 \Rightarrow v = 980t \Rightarrow a = 980$. a)   Solving $160 = 490t^2 \Rightarrow t = \dfrac{4}{7}$ sec. The average velocity was

$\dfrac{s(4/7) - s(0)}{4/7} = 280$ cm/sec. b) At the 160 cm mark the balls are falling at $v(4/7) = 560$ cm/sec.

The acceleration at the 160 cm mark was 980 cm/sec^2. c)   The light was flashing at a rate

of $\dfrac{17}{4/7} = 29.75$ flashes per second.

21.   a) 0, 0   b)   largest 1700, smallest about 1400

23.   b   25.   d

27.   a)

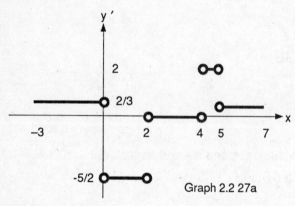

b)   0, 2, 4, and 5

Graph 2.2 27a

29.   a)

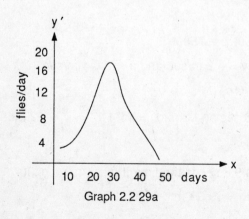

b)   The fastest is between the 20th and 30th days; slowest is between the 40th and 50th days.

Graph 2.2 29a

31.

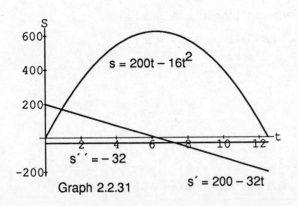

Graph 2.2.31

33.

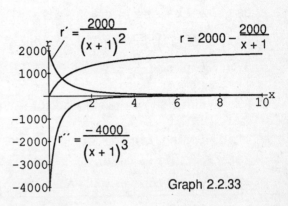

Graph 2.2.33

2.3 DERIVATIVES OF TRIGONOMETRIC FUNCTIONS

1. $y = 1 + x - \cos x \Rightarrow \dfrac{dy}{dx} = 1 + \sin x$

3. $y = \csc x - 5x + 7 \Rightarrow \dfrac{dy}{dx} = -\csc x \cot x - 5$

5. $y = x^2 \cot x \Rightarrow \dfrac{dy}{dx} = 2x \cot x - x^2 \csc^2 x$, product rule

7. $y = \sin x \sec x = \tan x \Rightarrow \dfrac{dy}{dx} = \sec^2 x$ 9. $y = \dfrac{4}{\cos x} = 4 \sec x \Rightarrow \dfrac{dy}{dx} = 4 \sec x \tan x$

11. $y = \dfrac{x}{1 + \cos x} \Rightarrow \dfrac{dy}{dx} = \dfrac{(1)(1 + \cos x) - (x)(- \sin x)}{(1 + \cos x)^2} = \dfrac{1 + \cos x + x \sin x}{(1 + \cos x)^2}$, quotient rule

13. $s = 2 \sin t - \tan t \Rightarrow \dfrac{ds}{dt} = 2 \cos t - \sec^2 t$ 15. $s = 2t + \cot t \Rightarrow \dfrac{ds}{dt} = 2 - \csc^2 t$

17. $r = 4 - \theta^2 \sin \theta \Rightarrow \dfrac{dr}{d\theta} = -2\theta \sin \theta - \theta^2 \cos \theta = -\theta[2 \sin \theta + \theta \cos \theta]$, product rule

19. $r = \sec \theta \csc \theta \Rightarrow \dfrac{dr}{d\theta} = (\sec \theta \tan \theta)(\csc \theta) + (\sec \theta)(- \csc \theta \cot \theta) = \sec^2 \theta - \csc^2 \theta$, product rule

21 $p = 5 + \dfrac{1}{\tan q} = 5 + \cot q = \dfrac{dp}{dq} = -\csc^2 q$

23. $p = \dfrac{\cot q}{1 + \cot q} \Rightarrow \dfrac{dp}{dq} = \dfrac{(- \csc^2 q)(1 + \cot q) - (- \csc^2 q)(\cot q)}{(1 + \cot q)^2} =$

$\dfrac{- \csc^2 q - \csc^2 q \cot q + \csc^2 q \cot q}{(1 + \cot q)^2} = -\dfrac{\csc^2 q}{(1 + \cot q)^2}$, quotient rule

25. $y = \csc x \Rightarrow y' = -\csc x \cot x \Rightarrow y'' = -[(-\csc x \cot x)(\cot x) + (\csc x)(- \csc^2 x)] =$
$\csc x \cot^2 x + \csc^3 x = \csc^3 x + \csc x[\csc^2 x - 1] = \csc^3 x + \csc^3 x - \csc x = 2\csc^3 x - \csc x$

27. $y = \sin x \Rightarrow y' = \cos x \Rightarrow m = 1$, tangent $y - 0 = 1(x - 0) \Rightarrow y = x$, normal $y - 0 = -1(x - 0) \Rightarrow y = -x$

29. $y = \cos x \Rightarrow y' = -\sin x \Rightarrow m = 0$, tangent $y + 1 = 0(x - \pi) \Rightarrow y = -1$, normal $x = \pi$

31. $y = \sec x \Rightarrow y' = \sec x \tan x \Rightarrow y'(0) = \sec(0)\tan(x) = 0 \Rightarrow$ a horizontal tangent when $x = 0$;

$y = \cos x \Rightarrow y' = -\sin x \Rightarrow y'(0) = -\sin(0) = 0 \Rightarrow$ a horizontal tangent when $x = 0$

33. $y = x + \sin x \Rightarrow y' = 1 + \cos x;\ y' = 0 \Rightarrow x = \pi \therefore$ a horizontal tangent at $x = \pi$

35. $y = x + \cos x \Rightarrow y' = 1 - \sin x;\ y' = 0 \Rightarrow x = \dfrac{\pi}{2} \therefore$ a horizontal tangent at $x = \pi/2$

37. $\displaystyle\lim_{x \to 2} \sin\left(\dfrac{1}{x} - \dfrac{1}{2}\right) = \sin(0) = 0$

39. $\displaystyle\lim_{x \to 0} (\sec x + \tan x) = \sec(0) + \tan(0) = 1$

41. Solving $\displaystyle\lim_{x \to 0^-} (x + b) = \lim_{x \to 0^+} \cos x \Rightarrow b = 1$

43. $y = \sqrt{2} \cos x \Rightarrow y' = -\sqrt{2} \sin x \Rightarrow m = -1,$ tangent $y - 1 = -1(x - \pi/4) \Rightarrow y = -x + \pi/4 + 1,$

normal $y - 1 = 1(x - \pi/4) \Rightarrow y = x + 1 - \pi/4$

45. a) When $y = 4 + \cot x - 2 \csc x \Rightarrow y' = -\left(\dfrac{1}{\sin x}\right)\left(\dfrac{1 - 2\cos x}{\sin x}\right)$. To find the location of the

horizontal tangent set $y' = 0 \Rightarrow x = \dfrac{\pi}{3}$ radians. When $x = \dfrac{\pi}{3}$, then $y = 4 - \sqrt{3}$, the horizontal

tangent.

b) When $x = \dfrac{\pi}{2}$, then $y' = -1$. The tangent line is $y = -x + \dfrac{\pi + 4}{2}$.

47.

h	1	0.1	0.01	0.001	
$\dfrac{1 - \cos h}{h^2}$	0.459697694	0.499583474	0.499958	0.5	The limit appears to be $\dfrac{1}{2}$.

49. a) $y = \sec x = \dfrac{1}{\cos x} \Rightarrow y' = \dfrac{(0)(\cos x) - (-\sin x)(1)}{\cos^2 x} = \left[\dfrac{1}{\cos x}\right]\left[\dfrac{\sin x}{\cos x}\right] = \sec x \tan x$

b) $y = \csc x = \dfrac{1}{\sin x} \Rightarrow y' = \dfrac{(0)(\sin x) - (\cos x)(1)}{\sin^2 x} = -\left[\dfrac{1}{\sin x}\right]\left[\dfrac{\cos x}{\sin x}\right] = -\csc x \cot x$

51.

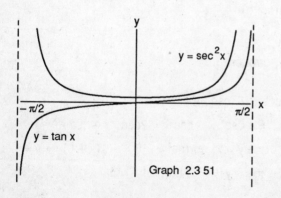

Graph 2.3 51

53.

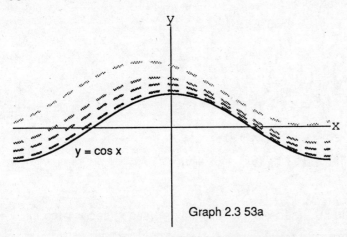

As h takes on the values of 1, 0.5, 0.3 and 0.1 the corresponding curves of $y = \dfrac{\sin(x + h) - \sin x}{h}$

move from grey to black, which is y = cos x.

Graph 2.3 53a

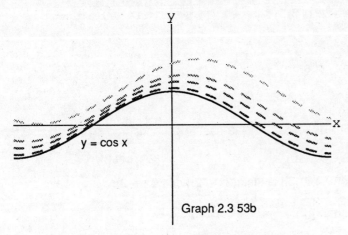

As h takes on the values of − 1, − 0.5, − 0.3 and − 0.1 the corresponding curves of $y = \dfrac{\sin(x + h) - \sin x}{h}$ move from grey to black,

which is y = cos x.

Graph 2.3 53b

If $h \rightarrow 0^+$ or $h \rightarrow 0^-$, then $y = \dfrac{\sin(x + h) - \sin x}{h} \rightarrow y = \cos x$.

2.4 THE CHAIN RULE

1. $y = \sin(x + 1) \Rightarrow \dfrac{dy}{dx} = \cos(x + 1)$

3. $y = x \cos(5x) \Rightarrow \dfrac{dy}{dx} = \cos(5x) - 5x \sin(5x)$

5. $y = \sin^3 x \Rightarrow \dfrac{dy}{dx} = 3(\sin^2 x)(\cos x)$

7. $y = x^2 \tan\left(\dfrac{1}{x}\right) = x^2 \tan(x^{-1}) \Rightarrow \dfrac{dy}{dx} = 2x \tan\left(x^{-1}\right) + x^2 \sec^2(x^{-1})(-x^{-2}) = 2x \tan\left(\dfrac{1}{x}\right) - \sec^2\left(\dfrac{1}{x}\right)$

9. $y = x + \sec\left(x^2 + 1\right) \Rightarrow \dfrac{dy}{dx} = 1 + \sec\left(x^2 + 1\right) \tan\left(x^2 + 1\right) (2x)$

11. $y = \cos(\sin x) \Rightarrow \dfrac{dy}{dx} = -[\sin(\sin x)][\cos x]$

13. $y = (x - 2)^4 \Rightarrow \dfrac{dy}{dx} = 4(x - 2)^3$

15. $y = (2 \sin x + 5)^{-5} \Rightarrow \dfrac{dy}{dx} = (-5)(2 \sin x + 5)^{-6}(2 \cos x)$

17. $y = \left(1 - \dfrac{x}{7}\right)^7 \Rightarrow \dfrac{dy}{dx} = 7\left(1 - \dfrac{x}{7}\right)^6\left(-\dfrac{1}{7}\right) = -\left(1 - \dfrac{x}{7}\right)^6$

19. $y = \left(1 + x - \dfrac{1}{x}\right)^{-4} \Rightarrow \dfrac{dy}{dx} = (-4)\left(1 + x - \dfrac{1}{x}\right)^{-5}\left(1 + \dfrac{1}{x^2}\right)$

21. $y = (2x - 5)^4(x + 1)^8 \Rightarrow \dfrac{dy}{dx} = 4(2x - 5)^3(2)(x + 1)^8 + (2x - 5)^4(8)(x + 1)^7 =$

    $8\,(2x - 5)^3(x + 1)^7[x + 1 + 2x - 5] = 8\,(2x - 5)^3(x + 1)^7(3x - 4)$

23. $y = (2x^3 + 3x^2 + 6x + 6)^6 \Rightarrow \dfrac{dy}{dx} = 6(2x^3 + 3x^2 + 6x + 6)^5(6x^2 + 6x + 6) =$

    $36(x^2 + x + 1)(2x^3 + 3x^2 + 6x + 6)^5$

25. $y = -(\sec x + \tan x)^{-1} \Rightarrow \dfrac{dy}{dx} = (\sec x + \tan x)^{-2}(\sec x \tan x + \sec^2 x) = (\sec x)(\sec x + \tan x)^{-1}$

27. $y = \sin\left[\dfrac{x - 2}{x + 3}\right] \Rightarrow \dfrac{dy}{dx} = \cos\left[\dfrac{x - 2}{x + 3}\right]\left[\dfrac{(1)(x + 3) - (1)(x - 2)}{(x + 3)^2}\right] = \cos\left[\dfrac{x - 2}{x + 3}\right]\left[\dfrac{5}{(x + 3)^2}\right]$

29. $s = (2t + 1)^{-4} \Rightarrow \dfrac{ds}{dt} = -8(2t + 1)^{-5}$

31. $s = \dfrac{4}{3\pi} \sin 3t + \dfrac{4}{3\pi} \sin 5t \Rightarrow \dfrac{ds}{dt} = \dfrac{4}{\pi} \cos 3t + \dfrac{4}{\pi} \cos 5t$

33. $r = \tan(2 - \theta) \Rightarrow \dfrac{dr}{d\theta} = -\sec^2(2 - \theta)$

35. $r = \dfrac{\cot 3\theta}{2 + \cot 3\theta} \Rightarrow \dfrac{dr}{d\theta} = \dfrac{-3 \csc^2 3\theta\left[2 + \cot 3\theta\right] - (\cot 3\theta)(-3 \csc^2 3\theta)}{(2 + \cot 3\theta)^2} = \dfrac{-6 \csc^2 3\theta}{(2 + \cot 3\theta)^2}$

37. $y = \sin^2(3x - 2) \Rightarrow \dfrac{dy}{dx} = [2 \sin(3x - 2)][\cos(3x - 2)][3] = 6[\sin(3x - 2)] \cos(3x - 2)$

39. $y = (1 + \cos 2x)^2 \Rightarrow \dfrac{dy}{dx} = 2(1 + \cos 2x)(-\sin 2x)(2) = -4(1 + \cos 2x)(\sin 2x)$

41. $y = \sin(\cos(2x - 5)) \Rightarrow \dfrac{dy}{dx} = [\cos(\cos(2x - 5))](-\sin(2x - 5))(2) = -2[\cos(\cos(2x - 5))][\sin(2x - 5)]$

43. $y = \tan x \Rightarrow y' = \sec^2 x \Rightarrow y'' = 2(\sec x)(\sec x \tan x) = 2(\sec^2 x)\tan x$

45. $y = \cot x \Rightarrow y' = -\csc^2 x \Rightarrow y'' = -2(\csc x)(-\csc x \cot x) = 2(\csc^2 x)(\cot x)$

47. $f(u) = u^5 + 1 \Rightarrow f'(u) = 5u^4,\ u = g(x) = \sqrt{x} \Rightarrow g'(x) = (1/2) x^{-1/2},\ x = 1 \Rightarrow u = 1 \Rightarrow (f \circ g)' =$

    $\left[5u^4 \Big|_{u = 1}\right]\left[(1/2)x^{-1/2} \Big|_{x = 1}\right] = (5)(1/2) = 5/2$

49. $f(u) = \cot \dfrac{\pi u}{10} \Rightarrow f'(u) = -\left[\csc^2 \dfrac{\pi u}{10}\right]\left[\dfrac{\pi}{10}\right],\ u = g(x) = 5\sqrt{x} \Rightarrow g'(x) = \dfrac{5}{2\sqrt{x}},\ x = 1 \Rightarrow u = 5,\ (f \circ g)' =$

    $\left[-\left[\csc^2 \dfrac{\pi u}{10}\right]\left[\dfrac{\pi}{10}\right]\Big|_{u = 5}\right]\left[\dfrac{5}{2\sqrt{x}}\Big|_{x = 1}\right] = \left[-\dfrac{\pi}{10}\right]\left[\dfrac{5}{2}\right] = -\dfrac{\pi}{4}$

51. $f(u) = \dfrac{2u}{u^2 + 1} \Rightarrow f'(u) = \dfrac{2(u^2 + 1) - 2u(2u)}{(u^2 + 1)^2} = \dfrac{2 - 2u^2}{(u^2 + 1)^2},\ u = g(x) = 10x^2 + x + 1 \Rightarrow g'(x) = 20x + 1,$

    $x = 0 \Rightarrow u = 1,\ (f \circ g)' = \left[\dfrac{2 - 2u^2}{(u^2 + 1)^2}\Big|_{u = 1}\right]\left[(20x + 1)\Big|_{x = 0}\right] = (0)(1) = 0$

53. a) $y = \cos u \Rightarrow \dfrac{dy}{du} = -\sin u$, $u = 6x + 2 \Rightarrow \dfrac{du}{dx} = 6$, $\dfrac{dy}{dx} = \dfrac{dy}{du}\dfrac{du}{dx} = \left[(-\sin u)\,|_{u\,=\,6x\,+\,2}\right][6] =$

$(-\sin(6x + 2))(6) = -6\sin(6x + 2)$

b) $y = \cos 2u \Rightarrow \dfrac{dy}{du} = -2\sin 2u$, $u = 3x + 1 \Rightarrow \dfrac{du}{dx} = 3$, $\dfrac{dy}{dx} = \dfrac{dy}{du}\dfrac{du}{dx} = \left[(-2\sin 2u)\,|_{u\,=\,3x\,+\,1}\right][3] =$

$[-2\sin(6x + 2)][3] = -6\sin(6x + 2)$

55. a) $y = u/5 + 7 \Rightarrow \dfrac{dy}{du} = 1/5$, $u = 5x - 35 \Rightarrow \dfrac{du}{dx} = 5$, $\dfrac{dy}{dx} = \dfrac{dy}{du}\dfrac{du}{dx} = \left[(1/5)\,|_{u\,=\,5x\,-\,35}\right][5] = (1/5)(5) = 1$

b) $y = 1 + 1/u = 1 + u^{-1} \Rightarrow \dfrac{dy}{du} = -u^{-2}$, $u = 1/(x - 1) = (x - 1)^{-1} \Rightarrow \dfrac{du}{dx} = -(x - 1)^{-2}$,

$\dfrac{dy}{dx} = \dfrac{dy}{du}\dfrac{du}{dx} = \left[-u^{-2}\big|_{u\,=\,1/(x\,-\,1)}\right]\left[-(x - 1)^{-2}\right] = \left[-(x - 1)^2\right]\left[-(x - 1)^{-2}\right] = 1$

57. $s = \cos\theta \Rightarrow \dfrac{ds}{d\theta} = -\sin\theta$, $\dfrac{d\theta}{dt} = 5 \Rightarrow \dfrac{ds}{dt}\bigg|_{\theta\,=\,3\pi/2} = \dfrac{ds}{d\theta}\dfrac{d\theta}{dt}\bigg|_{\theta\,=\,3\pi/2} = (-\sin 3\pi/2)(5) = -(-1)(5) = 5$

59. $y = \sin(x/2) \Rightarrow m = y' = (1/2)\cos(x/2)$ which has a maximum value of 1/2

61. $y = 2\tan(\pi x/4) \Rightarrow m\,|_{x\,=\,1} = y'\,|_{x\,=\,1} = (\pi/2)\sec^2(\pi x/4)\,|_{x\,=\,1} = \pi$, $y(1) = 2$ $\therefore$ the tangent line is

$y - 2 = \pi(x - 1) \Rightarrow y = \pi x + 2 - \pi$, while the normal line is $y - 2 = -\dfrac{1}{\pi}(x - 1) \Rightarrow y = -\dfrac{1}{\pi}x + 2 + \dfrac{1}{\pi}$

63. a) $\dfrac{d}{dx}\{2\,f(x)\}\bigg|_{x\,=\,2} = 2\,f'(2) = (2)(1/3) = 2/3$

b) $\dfrac{d}{dx}\{f(x) + g(x)\}\bigg|_{x\,=\,3} = f'(3) + g'(3) = 2\pi + 5$

c) $\dfrac{d}{dx}\{f(x) \cdot g(x)\}\bigg|_{x\,=\,3} = f'(3) \cdot g(3) + f(3) \cdot g'(3) = (2\pi)(-4) + (3)(5) = -8\pi + 15$

d) $\dfrac{d}{dx}\left\{\dfrac{f(x)}{g(x)}\right\}\bigg|_{x\,=\,2} = \dfrac{f'(2)\,g(2) - g'(2)\,f(2)}{(g(2))^2} = \dfrac{(1/3)(2) - (-3)(8)}{(2)^2} = \dfrac{2/3 + 24}{4} = \dfrac{37}{6}$

e) $\dfrac{d}{dx}\{f(g(x))\}\bigg|_{x\,=\,2} = f'(g(2))\,g'(2) = f'(2) \cdot (-3) = (1/3)(-3) = -1$

f) $\dfrac{d}{dx}\left\{\sqrt{f(x)}\right\}\bigg|_{x\,=\,2} = \dfrac{1}{2\sqrt{f(2)}}f'(2) = \dfrac{1/3}{2\sqrt{8}} = \dfrac{1}{12\sqrt{2}} = \dfrac{\sqrt{2}}{24}$

g) $\dfrac{d}{dx}\left\{\dfrac{1}{g(x)^2}\right\}\bigg|_{x\,=\,3} = \dfrac{-2}{g(3)^3}g'(3) = \dfrac{-10}{-64} = \dfrac{5}{32}$

h) $\dfrac{d}{dx}\left\{\sqrt{f(2)^2 + g(2)^2}\right\}\bigg|_{x\,=\,2} = \dfrac{1}{2\sqrt{f(x)^2 + g(x)^2}}[2\,f(2)\,f'(2) + 2\,g(2)\,g'(2)] =$

$\dfrac{(8)(1/3) + (2)(-3)}{\sqrt{8^2 + 2^2}} = \dfrac{-10}{3\sqrt{68}} = \dfrac{-5}{3\sqrt{17}}$

65. $s = A\cos(2\pi bt) \Rightarrow v = -2Ab\pi \sin(2\pi bt) \Rightarrow a = -4Ab^2\pi^2 \cos(2\pi bt)$, doubling the frequency would result in $s = A\cos(4\pi bt) \Rightarrow v = -4Ab\pi \sin(4\pi bt) \Rightarrow a = -16Ab^2\pi^2 \cos(4\pi bt)$ ∴ the range of the velocity doubles while the range of the acceleration quadruples

67. If $x = x(t)$ is the distance function, then $x'(t) = \dfrac{dx}{dt} = f(x(t)) \Rightarrow a(t) = v'(t) = \dfrac{d^2x}{dt^2} = f'(x(t))\, x'(t) =$

$f'(x(t))\, f(x(t)) = f(x(t))\, f'(x(t)) = f(x)\, f'(x)$.

69. a)

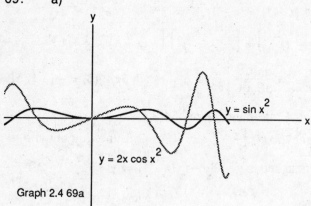

Graph 2.4 69a

b)

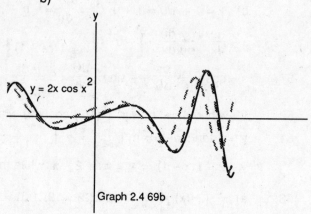

Graph 2.4 69b

As h takes on the values of 0.5 and 0.1 the corresponding curves of $y = \dfrac{\sin(x+h)^2 - \sin x^2}{h}$ in graph 2.4 69b move from grey to black, which is $y = 2x\cos x^2$. If $h \to 0^+$ or $h \to 0^-$, then $y = \dfrac{\sin(x+h)^2 - \sin x^2}{h} \to y = 2x\cos x^2$.

2.5 IMPLICIT DIFFERENTIATION AND FRACTIONAL POWERS

1. $y = x^{9/4} \Rightarrow \dfrac{dy}{dx} = (9/4)x^{5/4}$

3. $y = \sqrt[3]{x} = x^{1/3} = (1/3)x^{-2/3} = \dfrac{1}{3\sqrt[3]{x^2}}$

5. $y = (2x+5)^{-1/2} \Rightarrow y' = (-1/2)(2x+5)^{-3/2}(2) = -(2x+5)^{-3/2}$

7. $y = x\sqrt{x^2+1} = x(x^2+1)^{1/2} \Rightarrow y' = (x^2+)^{1/2} + (x/2)(x^2+1)^{-1/2}(2x) = \dfrac{2x^2+1}{\sqrt{x^2+1}}$

9. $x^2y + xy^2 = 6 \Rightarrow 2xy + x^2\dfrac{dy}{dx} + y^2 + 2xy\dfrac{dy}{dx} = 0 \Rightarrow (x^2+2xy)\dfrac{dy}{dx} = -2xy - y^2$ ∴ $\dfrac{dy}{dx} = \dfrac{-2xy-y^2}{x^2+2xy}$

11. $2xy + y^2 = x + y \Rightarrow 2y + 2x\dfrac{dy}{dx} + 2y\dfrac{dy}{dx} = 1 + \dfrac{dy}{dx} \Rightarrow (2x+2y-1)\dfrac{dy}{dx} = 1 - 2y$ ∴ $\dfrac{dy}{dx} = \dfrac{1-2y}{2x+2y-1}$

13. $x^2y^2 = x^2 + y^2 \Rightarrow 2xy^2 + 2x^2y\dfrac{dy}{dx} = 2x + 2y\dfrac{dy}{dx} \Rightarrow (2x^2y - 2y)\dfrac{dy}{dx} = 2x - 2xy^2$ ∴ $\dfrac{dy}{dx} = \dfrac{x(1-y^2)}{y(x^2-1)}$

15. $(x+y)^3 = 0 \Rightarrow \dfrac{dy}{dx} = 3(x+y)^2\left(1 + \dfrac{dy}{dx}\right) = 0 \Rightarrow \dfrac{dy}{dx} = -1$

17. $y = \sqrt{1 - \sqrt{x}} = (1 - x^{1/2})^{1/2} \Rightarrow \dfrac{dy}{dx} = (1/2)(1 - x^{1/2})^{-1/2}(-1/2)(x^{-1/2}) = \dfrac{-1}{4\sqrt{x}\sqrt{1 - \sqrt{x}}}$

19. $y = \sqrt{1 + \cos 2x} = (1 + \cos 2x)^{1/2} \Rightarrow \dfrac{dy}{dx} = (1/2)(1 + \cos 2x)^{-1/2}(-\sin 2x)(2) = \dfrac{-\sin 2x}{\sqrt{1 + \cos 2x}}$

21. $y = 3(\csc x)^{3/2} \Rightarrow \dfrac{dy}{dx} = (3)\left[(3/2)(\csc x)^{1/2}(-\csc x)(\cot x)\right] = (-9/2)(\csc^{3/2}x)(\cot x)$

23. $x = \tan y \Rightarrow 1 = \sec^2 y\dfrac{dy}{dx} \Rightarrow \dfrac{dy}{dx} = \dfrac{1}{\sec^2 y} = \cos^2 y$

25. $x + \tan(xy) = 0 \Rightarrow 1 + \left[\sec^2(xy)\right]\left[y + x\dfrac{dy}{dx}\right] = 0 \Rightarrow x\sec^2(xy)\dfrac{dy}{dx} = -1 - y\sec^2(xy) \Rightarrow$

$\dfrac{dy}{dx} = \dfrac{-1 - y\sec^2(xy)}{x\sec^2(xy)} = \dfrac{-1}{x\sec^2(xy)} - \dfrac{y}{x} = \dfrac{-\cos^2(xy)}{x} - \dfrac{y}{x} = \dfrac{-\cos^2(xy) - y}{x}$

27. $x^2 + y^2 = 1 \Rightarrow 2x + 2yy' = 0 \Rightarrow y' = -\dfrac{x}{y} \therefore \dfrac{dy}{dx} = -\dfrac{x}{y}$, also $2x + 2yy' = 0 \Rightarrow x + yy' = 0 \Rightarrow$

$1 + (y')^2 + yy'' = 0 \Rightarrow yy'' = -1 - (y')^2 \Rightarrow y'' = \dfrac{-1 - (y')^2}{y} \Rightarrow y'' = \dfrac{-1 - \left(-\dfrac{x}{y}\right)^2}{y} = -\dfrac{1}{y} - \dfrac{x^2}{y^3} = \dfrac{-y^2 - x^2}{y^3}$

$\therefore \dfrac{d^2y}{dx^2} = \dfrac{-y^2 - x^2}{y^3}$

29. $y^2 = x^2 + 2x \Rightarrow 2yy' = 2x + 2 \Rightarrow yy' = x + 1 \Rightarrow y' = \dfrac{x + 1}{y} \therefore \dfrac{dy}{dx} = \dfrac{x + 1}{y}$, also $yy' = x + 1 \Rightarrow$

$(y')^2 + yy'' = 1 \Rightarrow y'' = \dfrac{1}{y} - \dfrac{(y')^2}{y} \Rightarrow y'' = \dfrac{1}{y} - \left(\dfrac{1}{y}\right)\left(\dfrac{x + 1}{y}\right)^2 \Rightarrow y'' = \dfrac{1}{y} - \dfrac{(x + 1)^2}{y^3} \Rightarrow$

$y'' = \dfrac{y^2 - (x + 1)^2}{y^3} \therefore \dfrac{d^2y}{dx^2} = \dfrac{y^2 - (x + 1)^2}{y^3}$

31. $x^3 + y^3 = 16 \Rightarrow 3x^2 + 3y^2y' = 0 \Rightarrow y' = -\dfrac{x^2}{y^2} \Rightarrow y''\Big|_{(2,2)} = -\dfrac{2xy^2 - 2x^2yy'}{y^4}\Bigg|_{(2,2)} =$

$-\dfrac{(2)(2)(4) - (2)(4)(2)(-1)}{16} = -2$

33. $x^2 + xy - y^2 = 1 \Rightarrow 2x + y + xy' - 2yy' = 0 \Rightarrow (x - 2y)y' = -2x - y \Rightarrow y' = \dfrac{2x + y}{2y - x}$

a) the slope of the tangent line $m = y'\big|_{(2,3)} = \dfrac{2x + y}{2y - x}\bigg|_{(2,3)} = \dfrac{7}{4}$, the tangent line is

$y - 3 = \dfrac{7}{4}(x - 2) \Rightarrow y = \dfrac{7}{4}x - \dfrac{1}{2}$

b) the normal line is $y - 3 = -\dfrac{4}{7}(x - 2) \Rightarrow y = -\dfrac{4}{7}x + \dfrac{29}{7}$

35. $x^2y^2 = 9 \Rightarrow 2xy^2 + 2x^2yy' = 0 \Rightarrow x^2yy' = -xy^2 \Rightarrow y' = -\dfrac{y}{x}$

a) the slope of the tangent line , $= y'\big|_{(-1,3)} = -\dfrac{y}{x}\bigg|_{(-1,3)} = 3$, the tangent line is $y - 3 = 3(x + 1) \Rightarrow$

$y = 3x + 6$

b) the normal line is $y - 3 = -\dfrac{1}{3}(x + 1) \Rightarrow y = -\dfrac{1}{3}x + \dfrac{8}{3}$

37. Solving $x^2 + xy + y^2 = 7$ and $y = 0 \Rightarrow x^2 = 7 \Rightarrow x = \pm\sqrt{7} \Rightarrow (-\sqrt{7}, 0)$ and $(\sqrt{7}, 0)$ are the points where the curve crosses the x–axis. Now $x^2 + xy + y^2 = 7 \Rightarrow 2x + y + xy' + 2yy' = 0 \Rightarrow$

$(x + 2y)y' = -2x - y \Rightarrow y' = -\dfrac{2x + y}{x + 2y} \Rightarrow m = -\dfrac{2x + y}{x + 2y} \Rightarrow$ the slope at $(-\sqrt{7}, 0)$ is $m = -\dfrac{-2\sqrt{7}}{-\sqrt{7}} = -2$

and the slope at $(\sqrt{7}, 0)$ is $m = -\dfrac{2\sqrt{7}}{\sqrt{7}} = -2$ Since the slope is -2 in each case, the corresponding

tangents must be parallel.

39. $2xy + \pi \sin y = 2\pi \Rightarrow 2y + 2xy' + (\pi \cos y)y' = 0 \Rightarrow (2x + \pi \cos y)y' = -2y \Rightarrow y' = \dfrac{-2y}{2x + \pi \cos y} \Rightarrow$

$\dfrac{dy}{dx} = \dfrac{-2y}{2x + \pi \cos y} \Rightarrow \dfrac{dy}{dx}\bigg|_{(1,\pi/2)} = \dfrac{-2y}{2x + \pi \cos y}\bigg|_{(1,\pi/2)} = \dfrac{-\pi}{2 + \pi \cos(\pi/2)} = -\dfrac{\pi}{2}$

41. $y^4 = y^2 - x^2 \Rightarrow 4y^3 y' = 2yy' - 2x \Rightarrow (2)(2y^3 - y)y' = -2x \Rightarrow y' = \dfrac{x}{y - 2y^3}$; the slope of the tangent line

at $\left(\dfrac{\sqrt{3}}{4}, \dfrac{\sqrt{3}}{2}\right)$ is $\dfrac{x}{y - 2y^3}\bigg|_{\left(\frac{\sqrt{3}}{4}, \frac{\sqrt{3}}{2}\right)} = \dfrac{\frac{\sqrt{3}}{4}}{\frac{\sqrt{3}}{2} - \frac{6\sqrt{3}}{8}} = \dfrac{\frac{1}{4}}{\frac{1}{2} - \frac{3}{4}} = \dfrac{1}{2 - 3} = -1$; the slope of the tangent line

at $\left(\dfrac{\sqrt{3}}{4}, \dfrac{1}{2}\right)$ is $\dfrac{x}{y - 2y^3}\bigg|_{\left(\frac{\sqrt{3}}{4}, \frac{1}{2}\right)} = \dfrac{\frac{\sqrt{3}}{4}}{\frac{1}{2} - \frac{2}{8}} = \dfrac{2\sqrt{3}}{4 - 2} = \sqrt{3}$

43. a) $f(x) = \dfrac{3}{2}x^{2/3} - 3 \Rightarrow f'(x) = x^{-1/3} \Rightarrow f''(x) = -\dfrac{1}{3}x^{-4/3}$ ∴ a is false

b) $f(x) = \dfrac{9}{10}x^{5/3} - 7 \Rightarrow f'(x) = \dfrac{3}{2}x^{2/3} \Rightarrow f''(x) = x^{-1/3}$ ∴ b is true

c) $f''(x) = x^{-1/3} \Rightarrow f'''(x) = -\dfrac{1}{3}x^{-4/3}$ ∴ c is true

d) $f'(x) = \dfrac{3}{2}x^{2/3} + 6 \Rightarrow f''(x) + x^{-1/3}$ ∴ d is true

45. $y^2 = x \Rightarrow \dfrac{dy}{dx} = \dfrac{1}{2y}$. If a normal is drawn from $(a,0)$ to (x_1, y_1), $\dfrac{y_1 - 0}{x_1 - a} = -2y \Rightarrow y_1 = -2y_1(x_1 - a)$ or

$a = x_1 + \dfrac{1}{2}$. Since $x_1 \geq 0$, we must have that $a \geq \dfrac{1}{2}$. The two points on the parabola are $\left(x_1, \sqrt{x_1}\right)$

and $\left(x_1, -\sqrt{x_1}\right)$. To be perpendicular, $\left(\dfrac{\sqrt{x_1}}{x_1 - 0}\right)\left(\dfrac{\sqrt{x_1}}{a - x_1}\right) = -1 \Rightarrow x = \dfrac{1}{4}$ and $y = \pm\dfrac{1}{2}$. ∴ $\left(\dfrac{1}{4}, \pm\dfrac{1}{2}\right)$

and $a = \dfrac{3}{4}$.

47. $xy + 2x - y = 0 \Rightarrow x\dfrac{dy}{dx} + y + 2 - \dfrac{dy}{dx} = 0 \Rightarrow \dfrac{dy}{dx} = \dfrac{y + 2}{1 - x}$. The slope of line $2x + y = 0$ is -2. To

be parallel, the normal lines must also have slope of -2. Since a normal is perpendicular to a

tangent, the slope of the tangent is $\dfrac{1}{2}$. ∴ $\dfrac{y + 2}{1 - x} = \dfrac{1}{2} \Rightarrow 2y + 4 = 1 - x \Rightarrow x = -3 - 2y$.

Substituting, $y(-3 - 2y) + 2(-3 - 2y) - y = 0 \Rightarrow y^2 + 4y + 3 = 0 \Rightarrow y = -3, -1$. If $y = -3$, then

$x = 3$ and $y + 3 = -2(x - 3) \Rightarrow y + 2x = -3$. If $y = -1$, then $x = -1$ and $y + 1 = -2(x + 1) \Rightarrow$

$y + 2x = -3$.

49. $v = k\sqrt{s} = ks^{1/2} \Rightarrow a = v' = (1/2)ks^{-1/2}\dfrac{ds}{dt} = \dfrac{k}{2\sqrt{s}}\dfrac{ds}{dt} = \dfrac{k}{2\sqrt{s}}v = \dfrac{k}{2\sqrt{s}}k\sqrt{s} = \dfrac{k^2}{2}$, a constant

51. $T = 2\pi\sqrt{\dfrac{L}{g}} = 2\pi\left(Lg^{-1}\right)^{1/2}$ where g is constant $\Rightarrow \dfrac{dT}{du} = 2\pi\left(\dfrac{1}{2}\right)\left(Lg^{-1}\right)^{-1/2}\left[\left(g^{-1}\right)\left(\dfrac{dL}{du}\right)\right] =$

$\dfrac{\pi kL}{g}\sqrt{\dfrac{g}{L}} = \dfrac{\pi kL}{\sqrt{gL}} = \pi k\sqrt{\dfrac{L}{g}} = \left(\dfrac{k}{2}\right)\left(2\pi\sqrt{\dfrac{L}{g}}\right) = \dfrac{kT}{2}$

2.6 LINEARIZATION AND DIFFERENTIALS

1. $f(x) = x^4 \Rightarrow f'(x) = 4x^3 \Rightarrow L(x) = f'(1)(x-1) + f(1) = 4(x-1) + 1 \Rightarrow L(x) = 4x - 3$

3. $f(x) = x^3 - x \Rightarrow f'(x) = 3x^2 - 1 \Rightarrow L(x) = f'(1)(x-1) + f(1) = 2(x-1) + 0 \Rightarrow L(x) = 2x - 2$

5. $f(x) = \sqrt{x} = x^{1/2} \Rightarrow f'(x) = (1/2)x^{-1/2} = \dfrac{1}{2\sqrt{x}} \Rightarrow L(x) = f'(4)(x-4) + f(4) = \dfrac{1}{4}(x-4) + 2 \Rightarrow$

$L(x) = \dfrac{1}{4}x + 1$

7. $f(x) = x^2 + 2x \Rightarrow f'(x) = 2x + 2 \Rightarrow L(x) = f'(0)(x-0) + f(0) = 2(x-0) + 0 \Rightarrow L(x) = 2x$

9. $f(x) = 2x^2 + 4x - 3 \Rightarrow f'(x) = 4x + 4 \Rightarrow L(x) = f'(-1)(x+1) + f(-1) = 0(x+1) + (-5) \Rightarrow L(x) = -5$

11. $f(x) = \sqrt[3]{x} = x^{1/3} \Rightarrow f'(x) = (1/3)x^{-2/3} \Rightarrow L(x) = f'(8)(x-8) + f(8) = \dfrac{1}{12}(x-8) + 2 \Rightarrow L(x) = \dfrac{1}{12}x + \dfrac{4}{3}$

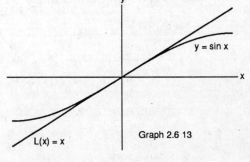

Graph 2.6 13

13. $f(x) = \sin x \Rightarrow f'(x) = \cos x \Rightarrow L(x) = f'(0)(x-0) + f(0) =$

$1(x-0) + 0 \Rightarrow L(x) = x$

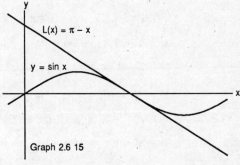

Graph 2.6 15

15. $f(x) = \sin x \Rightarrow f'(x) = \cos x \Rightarrow L(x) = f'(\pi)(x-\pi) + f(\pi) =$

$(-1)(x-\pi) + 0 \Rightarrow L(x) = -x + \pi$

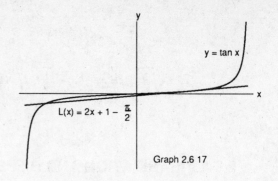

Graph 2.6 17

17. $f(x) = \tan x \Rightarrow f'(x) = \sec^2 x \Rightarrow$

   $L(x) = f'(\pi/4)(x - \pi/4) + f(\pi/4) = (2)(x - \pi/4) + 1 \Rightarrow$

   $L(x) = 2x + 1 - \pi/2$

19. $f(x)$ 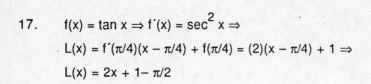$(1 + x)^k \approx 1 + kx$

a) $(1 + x)^2$ $1 + 2x$

b) $\dfrac{1}{(1 + x)^5} = (1 + x)^{-5}$ $1 + (-5)x = 1 - 5x$

c) $\dfrac{2}{1 - x} = 2(1 + (-x))^{-1}$ $2[1 + (-1)(-x)] = 2 + 2x$

d) $(1 - x)^6 = (1 + (-x))^6$ $1 + (6)(-x) = 1 - 6x$

e) $3(1 + x)^{1/3}$ $3[1 + (1/3)(x)] = 3 + x$

f) $\dfrac{1}{\sqrt{1 + x}} = (1 + x)^{-1/2}$ $1 + (-1/2)(x) = 1 - \dfrac{x}{2}$

21. $f(x) = \sqrt{x + 1} + \sin x = (x + 1)^{1/2} + \sin x \Rightarrow f'(x) = (1/2)(x + 1)^{-1/2} + \cos x \Rightarrow$

   $L_f(x) = f'(0)(x - 0) + f(0) = \dfrac{3}{2}(x - 0) + 1 \Rightarrow L_f(x) = \dfrac{3}{2}x + 1$, the linearization of f(x); $g(x) = \sqrt{x + 1} =$

   $(x + 1)^{1/2} \Rightarrow g'(x) = (1/2)(x + 1)^{-1/2} \Rightarrow L_g(x) = g'(0)(x - 0) + g(0) = \dfrac{1}{2}(x - 0) + 1 \Rightarrow L_g(x) = \dfrac{1}{2}x + 1$,

   the linearization of g(x); $h(x) = \sin x \Rightarrow h'(x) = \cos x \Rightarrow L_h(x) = h'(0)(x - 0) + h(0) =$

   $(1)(x - 0) + 0 \Rightarrow L_h(x) = x$, the linearization of h(x). $L_f(x) = L_g(x) + L_h(x)$ implies that the

   linearization of a sum is equal to the sum of the linearizations.

23. 1.414213562, 1.189207115, 1.090507733,1.044273782, 1.021897149, 1.010889286,

   1.005429901, ...

25. $f(x) = x^2 + 2x$, $x_o = 0$, $dx = .1 \Rightarrow f'(x) = 2x + 2$

   a) $\Delta f = f(x_o + dx) - f(x_o) = f(.1) - f(0) = .01 + .2 = .21$

   b) $df = f'(x_o)dx = [2(0) + 2][.1] = .2$

   c) $|\Delta f - df| = |.21 - .22| = .01$

27. $f(x) = x^3 - x$, $x_o = 1$, $dx = .1 \Rightarrow f'(x) = 3x^2 - 1$

   a) $\Delta f = f(x_o + dx) - f(x_o) = f(1.1) - f(1) = .231$

   b) $df = f'(x_o)dx = (3(1)^2 - 1)(.1) = .2$

   c) $|\Delta f - df| = |.231 - .2| = .031$

29. $f(x) = x^{-1}$, $x_o = .5$, $dx = .1 \Rightarrow f'(x) = -x^{-2}$

   a) $\Delta f = f(x_o + dx) - f(x_o) = f(.6) - f(.5) = -1/3$

   b) $df = f'(x_o)dx = (-4)(1/10) = -2/5$

   c) $|\Delta f - df| = |-1/3 + 2/5| = 1/15$

31. $V = \frac{4}{3}\pi r^3 \Rightarrow dV = 4\pi r_o^2 dr$ 33. $V = x^3 \Rightarrow dV = 3x_o^2 dx$

35. $V = \pi r^2 h$, height constant $\Rightarrow dV = 2\pi r_o h dr$

37. Given r = 2 m, dr = .02 m   a) $A = \pi r^2 \Rightarrow dA = 2\pi r dr = 2\pi(2)(.02) = .08\pi$ m^2
     b) $\left[\frac{.08\pi}{4\pi}\right][100\%] = 2\%$

39. The error in measurement dx = (1%)(10) = .1 cm, $V = x^3 \Rightarrow dV = 3x^2 dx = 3(10)^2(.1) = 30$ cm$^3 \Rightarrow$ the
     percentage error in the volume calculation is $\left[\frac{30}{1000}\right][100\%] = 3\%$

41. Given d = 100 cm, dd = 1 cm, $V = \frac{4}{3}\pi\left(\frac{d}{2}\right)^3 = \frac{\pi d^3}{6} \Rightarrow dV = \frac{\pi}{2}d^2(dd) = \frac{\pi}{2}(100)^2(1) = \frac{10^4\pi}{2}$,

     now $\frac{dV}{V}(10^2\%) = \left[\dfrac{\frac{10^4\pi}{2}}{\frac{10^6\pi}{6}}\right][10^2\%] = \left[\dfrac{\frac{10^6\pi}{2}}{\frac{10^6\pi}{6}}\right]\% = 3\%$

43. $V = \pi h^3 \Rightarrow dV = 3\pi h^2 dh$; recall that $\Delta V \approx dV$. $|\Delta V| \le (1\%)(V) = \left(\frac{(1)(\pi h^3)}{100}\right) \Rightarrow |dV| \le \left(\frac{(1)(\pi h^3)}{100}\right) \Rightarrow$

     $|3\pi h^2 dh| \le \left(\frac{(1)(\pi h^3)}{100}\right) \Rightarrow |dh| \le \left(\frac{1}{300}\right)(h) = \left(\frac{1}{3}\%\right)(h)$. $\therefore$ the greatest tolerated error in the
     measurement of h is $\frac{1}{3}\%$.

45. $V = \pi r^2 h$, h is constant $\Rightarrow dV = 2\pi r h dr$. Recall that $\Delta V \approx dV$. We want $|\Delta V| \le \frac{1}{1000} V \Rightarrow$

     $|dV| \le \frac{\pi r^2 h}{1000} \Rightarrow |2\pi r h dr| \le \frac{\pi r^2 h}{1000} \Rightarrow |dr| \le \frac{r}{2000} = (.05\%)(r)$. A .05% variation in the radius can
     be tolerated.

47. $V = kr^4 \Rightarrow dV = \frac{dV}{dr} dr = 4kr^3 dr \Rightarrow \frac{dV}{V} = \frac{4kr^3 dr}{kr^4} = 4\frac{dr}{r} \Rightarrow$ the relative change in V is four times

     the relative change in r, so a 10% increase in r will produce a 40% increase in flow.

49.

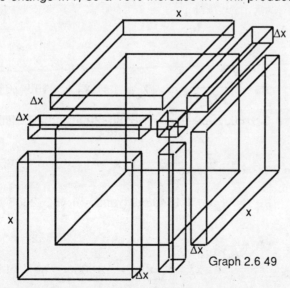

Volume $= (x + \Delta x)^3 =$
$x^3 + 3x^2(\Delta x) + 3x(\Delta x)^2 + (\Delta x)^3$

Graph 2.6 49

51. $y = x^3 - 3x \Rightarrow dy = (3x^2 - 3)dx$

53. $y = \dfrac{2x}{1+x} \Rightarrow dy = \dfrac{2(1+x) - 2x(1)}{(1+x)^2} dx = \dfrac{2}{(1+x)^2} dx$

55. $y + xy = x \Rightarrow y + xy - x = 0 \Rightarrow dy + ydx + xdy - dx = 0 \Rightarrow (1+x)dy = dx - ydx \Rightarrow dy = \dfrac{(1-y)}{1+x}dx$

57. $y = 4\tan^2\left(\dfrac{x}{2}\right) \Rightarrow dy = 8\tan\left(\dfrac{x}{2}\right)\sec^2\left(\dfrac{x}{2}\right)\left(\dfrac{1}{2}\right) dx = 4\tan\left(\dfrac{x}{2}\right)\sec^2\left(\dfrac{x}{2}\right) dx$

59. $y = x^2 - \cos\left(x^2\right) \Rightarrow dy = 2x + \left(\sin x^2\right)(2x)dx = 2x\left(1 + \sin x^2\right)dx$

61. a) $x = 1$ b) $x = 1$

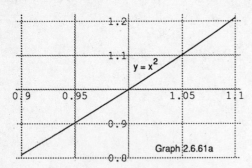

Graph 2.6.61a

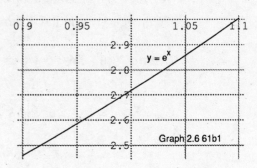

Graph 2.6 61b1

b) $x = 0$ b) $x = -1$

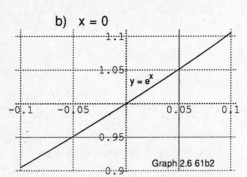

Graph 2.6 61b2

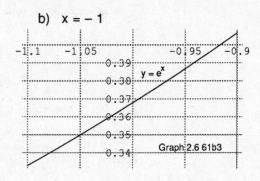

Graph 2.6 61b3

2.7 NEWTON'S METHOD

1. $y = x^2 + x - 1 \Rightarrow y' = x^2 + 1, x_{n+1} = x_n - \dfrac{x_n^2 + x_n - 1}{2x_n + 1} \therefore x_o = 1 \Rightarrow x_2 = .619047619$ and $x_o = -1 \Rightarrow$

   $x_2 = -1.66666667$ The roots are 0.618033989 and -1.61803399 by the Calculus Tool Kit.

3. $y = x^4 + x - 3 \Rightarrow y' = 4x^3 + 1, x_{n+1} = x_n - \dfrac{x_n^4 + x_n - 3}{4x_n^3 + 1} \therefore x_o = 1 \Rightarrow x_2 = 1.16541962$ and

   $x_o = -1 \Rightarrow x_2 = -1.64516129.$ The roots are 1.16403514 and -1.45262688 by the

   Calculus Tool Kit.

5. $y = x^4 - 2 \Rightarrow y' = 4x^3, x_{n+1} = x_n - \dfrac{x_n^4 - 2}{4x_n^3} \therefore x_o = 1 \Rightarrow x_2 = 1.1935.$ The root is 1.18920711.

7.   If $f'(x_0) \neq 0$ then $x_0 = x_1 = x_2 = x_3 =$

9.

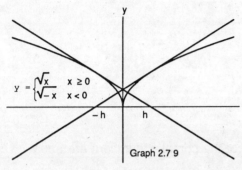

$y = \begin{cases} \sqrt{x} & x \geq 0 \\ \sqrt{-x} & x < 0 \end{cases}$

Graph 2.7 9

If $x_0 = h > 0 \Rightarrow x_1 = x_0 - \dfrac{f(x_0)}{f'(x_0)} = h - \dfrac{f(h)}{f'(h)} = h - \dfrac{\sqrt{h}}{\dfrac{1}{2\sqrt{h}}} =$

$h - (\sqrt{h})(2\sqrt{h}) = -h$ while if $x_0 = -h < 0 \Rightarrow x_1 = x_0 - \dfrac{f(x_0)}{f'(x_0)} =$

$-h - \dfrac{f(-h)}{f'(-h)} = -h - \dfrac{\sqrt{h}}{\dfrac{-1}{2\sqrt{h}}} = -h + (\sqrt{h})(2\sqrt{h}) = h.$

11.   If $f(x) = x^3 + 2x - 4$, then $f(1) = -1 < 0$; $f(2) = 8 > 0$; and by the Intermediate Value Theorem, the equation $x^3 + 2x - 4 = 0$ has a solution between 1 and 2. Consequently, $f'(x) = 3x^2 + 2$ and $x_{n+1} = x_n - \dfrac{x_n^3 + 2x_n - 4}{3x_n^2 + 2}$. The root is 1.17951.

13. $f(x) = \tan x \Rightarrow f'(x) = \sec^2 x \therefore x_{n+1} = x_n - \dfrac{\tan(x_n)}{\sec^2(x_n)}$. If $x_0 = 3$ we approximate π

to be 3.14159.

15. $f(x) = x - 1 - 0.5 \sin x \Rightarrow f'(x) = 1 - 0.5 \cos x \therefore x_{n+1} = x_n - \dfrac{x_n - 1 - 0.5 \sin x_n}{1 - 0.5 \cos x_n}$. If $x = 1.5$, then

$x_1 = 1.49870157$. The root is 1.49870 to five decimal places.

17.   Let $f(x) = \tan(x) - 2x \Rightarrow f'(x) = \sec^2(x) - 2$. $\therefore x_{n+1} = x_n - \dfrac{\tan(x_n) - 2x_n}{\sec^2(x_n) - 2}$. If $x_0 = 1$, then

$x_6 = x_7 = x_8 = 1.16556119$.

19. $f(x) = 2x^4 - 4x^2 + 1 \Rightarrow f'(x) = 8x^3 - 8x \therefore x_{n+1} = x_n - \dfrac{2x_n^4 - 4x_n^2 + 1}{8x_n^3 - 8x_n}$. If $x_0 = -2$, then

$x_6 = -1.30656296$; if $x_0 = -.5$, then $x_3 = -0.5411961$. The positive roots are 0.5411961 and

1.30656296 because $f(x)$ is an even function.

21. $f(x) = (x - 1)^{40} \Rightarrow f'(x) = 40(x - 1)^{39} \therefore x_{n+1} = x_n - \dfrac{(x_n - 1)^{40}}{40(x_n - 1)^{39}}$. With $x_0 = 2$ then $x_{87} = x_{88} =$

$x_{89} = \cdots = x_{200} = 1.11051$. The computer comes within .11051 of the root.

23.   We wish to solve $8x^4 - 14x^3 - 9x^2 + 11x - 1 = 0$. Let $f(x) = 8x^4 - 14x^3 - 9x^2 + 11x - 1$, then

$f'(x) = 32x^3 - 42x^2 - 18x + 11 \therefore x_{n+1} = x_n - \dfrac{8x_n^4 - 14x_n^3 - 9x_n^2 + 11x_n - 1}{32x_n^3 - 42x_n^2 - 18x_n + 11}$.

x_0	approximation of corresponding root
-1.0	-0.976823589
0.1	0.100363332
0.6	0.64274667
2.0	1.98371359

25.   In the following twenty points the real and imaginary parts of the complex numbers are equal.

$x_1 = 0.6875 + 0.6875\ i$ $x_2 = 0.323288 + 0.323288\ i$ $x_3 = -1.60727 - 1.60727\ i$

$x_4 = -1.1904 - 1.1904\ i$ $x_5 = -0.855748 - 0.855748\ i$ $x_6 = -0.542077 - 0.542077\ i$

$x_7 = -0.0141877 - 0.0141877\ i$ $x_8 = 13.1385 + 13.1385\ i$ $x_9 = 9.85385 + 9.85385\ i$

$x_{10} = 7.39032 + 7.39032\ i$ $x_{11} = 5.54259 + 5.54259\ i$ $x_{12} = 4.15657 + 4.15657\ i$

$x_{13} = 3.11656 + 3.11656\ i$ $x_{14} = 2.33536 + 2.33536\ i$ $x_{15} = 1.74661 + 1.74661\ i$

$x_{16} = 1.29823 + 1.29823\ i$ $x_{17} = 0.945106 + 0.945106\ i$ $x_{18} = 0.634794 + 0.634794\ i$

$x_{19} = 0.231763 + 0.231763\ i$ $x_{20} = -4.8467 - 4.8467\ i$

2.M MISCELLANEOUS EXERCISES

1. $y = (x + 1)^4 (x^2 + 2x) \Rightarrow \dfrac{dy}{dx} = 4(x + 1)^3 (x^2 + 2x) + (x + 1)^4 (2x + 2) =$

$(x + 1)^3 \left[4(x^2 + 2x) + (x + 1)(2x + 2) \right] = 2(x + 1)^3 (3x^2 + 6x + 1)$

3. $y = (x^2 + \sin x + 1)^3 \Rightarrow \dfrac{dy}{dx} = 3(x^2 + \sin x + 1)^2 (2x + \cos x)$

5. $y = (\sec x + \tan x)^5 \Rightarrow \dfrac{dy}{dx} = 5(\sec x + \tan x)^4 (\sec x \tan x + \sec^2 x) = 5(\sec x)(\sec x + \tan x)^5$

7. $x^2 + xy + y^2 - 5x = 2 \Rightarrow 2x + y + xy' + 2yy' - 5 = 0 \Rightarrow (x + 2y)y' = 5 - 2x - y \Rightarrow \dfrac{dy}{dx} = y' = \dfrac{5 - 2x - y}{x + 2y}$

9. $s = (2t - 5)(4 - t)^{-1} \Rightarrow \dfrac{ds}{dt} = (2)(4 - t)^{-1} + (2t - 5)(-1)(4 - t)^{-2}(-1) =$

$(4 - t)^{-2}[2(4 - t) + (2t - 5)] = 3(4 - t)^{-2}$

11. $s = (-1 - \dfrac{t}{2} - \dfrac{t^2}{4})^2 \Rightarrow \dfrac{ds}{dt} = 2(-1 - \dfrac{t}{2} - \dfrac{t^2}{4})(-\dfrac{1}{2} - \dfrac{t}{2}) = \left(1 + \dfrac{t}{2} + \dfrac{t^2}{4}\right)(1 + t)$

13. $s = (t^2 - 8t)^{-1/2} \Rightarrow \dfrac{ds}{dt} = (-1/2)(t^2 - 8t)^{-3/2}(2t - 8) = (4 - t)(t^2 - 8t)^{-3/2}$

15. $p^3 + 4pq - 3q^2 + 2 \Rightarrow 3p^2 dp + 4q dp + 4p dq - 6q dq = 0 \Rightarrow \left(3p^2 + 4q\right)dp = (6q - 4p)dq \Rightarrow$

$\dfrac{dp}{dq} = \dfrac{6q - 4p}{3p^2 + 4q}$

17. $r \cos 2s + \sin^2 s = \pi \Rightarrow \cos 2s\, dr - 2r \sin 2s\, ds + 2 \sin s \cos s\, ds = 0 \Rightarrow \cos 2s\, dr =$

$(2r \sin 2s - 2 \sin s \cos s)ds \Rightarrow \dfrac{dr}{ds} = \dfrac{2r \sin 2s - \sin 2s}{\cos 2s} = (2r - 1) \tan(2s)$

19. If (x_o, y_o) is on the graphs of both the line and parabola, then $2x_o = 1 \Rightarrow x_o = \frac{1}{2}$ and $y_o = \frac{1}{4} + c = 1. \therefore c = \frac{1}{4}$

21. $y = \cos(2x - A) \Rightarrow y' = -2\sin(2x - A)$ and $y'' = -4\cos(2x - A) = -4y$

23. a) $s(t) = 64t - 16t^2 \Rightarrow v(t) = s'(t) = 64 - 32t = 32(2 - t)$. The maximum height is reached when $v(t) = 0 \Rightarrow t = 2$ sec. The velocity when it leaves the hand is $v(0) = 64$ ft/sec.

    b) $s(t) = 64t - 2.6t^2 \Rightarrow v(t) = s'(t) = 64 - 5.2t$. The maximum height is reached when $v(t) = 0 \Rightarrow t \approx 12.31$ sec. The maximum height is about $s(12.31) = 393.85$ ft.

25.

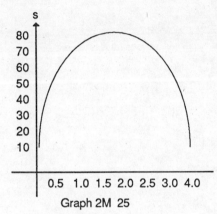

Graph 2M 25

a) $(1/2, 38), (3/2, 70), v = \dfrac{70 - 38}{3/2 - 1/2} = 32$ ft/sec

b) $(2, 74), (3, 58), v = \dfrac{74 - 58}{2 - 3} = -16$ ft/sec

c) $(3/2, 70), (5/2, 70), v = 0$ ft/sec

27. The total revenue, $r(x) = x\left(3 - \dfrac{x}{40}\right)^2$ where $0 \le x \le 60$. The marginal revenue $\dfrac{dr}{dx} = (3)\left(3 - \dfrac{x}{40}\right)\left(1 - \dfrac{x}{40}\right)$. When $\dfrac{dr}{dx} = 0 \Rightarrow x = 40$ or 120, but 120 is not in the domain. When 40 people are on the bus, the marginal revenue is equal to zero with a corresponding fare of \$4.00.

29. $m\left(v^2 - v_0^2\right) = k\left(x_0^2 - x^2\right) \Rightarrow m\left(2v\dfrac{dv}{dt}\right) = k\left(-2x\dfrac{dx}{dt}\right) \Rightarrow m\dfrac{dv}{dt} = k\left(-\dfrac{2x}{2v}\right)\dfrac{dx}{dt} \Rightarrow m\dfrac{dv}{dt} = -kx\left(\dfrac{1}{v}\right)\dfrac{dx}{dt}$. where $\dfrac{dx}{dt} = v \Rightarrow m\dfrac{dv}{dt} = -kx$

31. a) To be continuous at $x = \pi$ requires that $\underset{x \to \pi^-}{\text{Lim}} \sin x = \underset{x \to \pi^+}{\text{Lim}} mx + b \Rightarrow 0 = m\pi + b \Rightarrow m = -\dfrac{b}{\pi}$

    b) If $y' = \begin{cases} \cos x, & x < \pi \\ m, & x \ge \pi \end{cases}$ is differentiable at $x = \pi$, then $\underset{x \to \pi^-}{\text{Lim}} \cos x = m \Rightarrow m = -1$ and $b = \pi$.

33. If $y = \sin(x - \sin x)$ then $y' = [\cos(x - \sin x)][1 - \cos x]$. When $y' = 0 \Rightarrow x = 0 + n(2\pi)$ where n is an integer. The curve has an infinite number of horizontal tangents as indicated above.

35. The derivative of $\sin(x + a) = \sin x \cos a + \cos x \sin a$ is $\cos(x + a) = \cos x \cos a - \sin x \sin a$, an identity. This principle does not apply to the equation $x^2 - 2x - 8 = 0$ because it is not an identity.

37. When $s = t^2 + 5t$ and $t = (u^2 + 2u)^{1/3}$, then $\dfrac{ds}{dt} = 2t + 5$ and $\dfrac{dt}{du} = (1/3)(u^2 + 2u)^{-2/3}(2u + 2)$.

$\therefore \dfrac{ds}{du} = \left[\dfrac{ds}{dt}\Big|_{t = (u^2 + 2u)^{1/3}}\right]\left[\dfrac{dt}{du}\right] = \left[2(u^2 + 2u)^{1/3} + 5\right]\left[\left(\dfrac{2}{3}\right)(u^2 + 2u)^{-2/3}(u + 1)\right]$. Accordingly,

$\dfrac{ds}{du}\Big|_{u = 2} = \dfrac{9}{2}$.

39. If $y = \sqrt{x}$, then $y' = \dfrac{1}{2\sqrt{x}}$. The equation of the tangent line is $y - 2 = \dfrac{1}{4}(x - 4) \Rightarrow \dfrac{x}{-4} + \dfrac{y}{1} = 1$. The

intercepts are $(0, 1)$ and $(-4, 0)$.

41. $y = 2(x - 1)^{-1/2} \Rightarrow m = -(x - 1)^{-3/2}\Big|_{10} = -\dfrac{1}{27}$; $y - \dfrac{2}{3} = -\dfrac{1}{27}(x - 10) \Rightarrow y = -\dfrac{x}{27} + \dfrac{28}{27}$, the tangent

43. a) $y = (2x - 1)^{1/2} \Rightarrow \dfrac{dy}{dx} = \dfrac{1}{2}(2x - 1)^{-1/2}(2) = (2x - 1)^{-1/2} \Rightarrow \dfrac{d^2y}{dx^2} = -\dfrac{1}{2}(2x - 1)^{-3/2}(2) =$

$-(2x - 1)^{-3/2} \Rightarrow \dfrac{d^3y}{dx^3} = \dfrac{3}{2}(2x - 1)^{-5/2}(2) = 3(2x - 1)^{-5/2}$

b) $y = (3x + 2)^{-1} \Rightarrow \dfrac{dy}{dx} = -3(3x + 2)^{-2} \Rightarrow \dfrac{d^2y}{dx^2} = (-3)(-2)(3x + 2)^{-3}(3) = 18(3x + 2)^{-3} \Rightarrow$

$\dfrac{d^3y}{dx^3} = -54(3x + 2)^{-4} = -162(3x + 2)^{-4}$

c) $y = ax^3 + bx^2 + cx + d \Rightarrow \dfrac{dy}{dx} = 3ax^2 + 2bx + c \Rightarrow \dfrac{d^2y}{dx^2} = 6ax + 2b \Rightarrow \dfrac{d^3y}{dx^3} = 6a$

45. $x^2 + y^2 = a^2 \Rightarrow 2x + 2yy' = 0 \Rightarrow y' = -\dfrac{x}{y} \Rightarrow$ the normal line at (x_o, y_o) is $y =$

$\dfrac{y_o}{x_o}\left(x - x_o\right) + y_o$. If $x = 0$, then $y = \dfrac{y_o}{x_o}\left(0 - x_o\right) + y_o = -y_o + y_o = 0 \Rightarrow$ all normal lines to

the circle pass through the origin.

47. a) If $x^2 + 2y^2 = 9$, then $y' = -\dfrac{x}{2y}$. The tangent line is $y - 2 = \left(-\dfrac{1}{4}\right)(x - 1) \Rightarrow y = -\dfrac{1}{4}x + \dfrac{9}{4}$. The

normal line is $y - 2 = 4(x - 1) \Rightarrow y = 4x - 2$.

b) If $x^3 + y^2 = 2$, then $y' = -\dfrac{3x^2}{2y}$. The tangent line is $y - 1 = \left(-\dfrac{3}{2}\right)(x - 1) \Rightarrow y = -\dfrac{3}{2}x + \dfrac{5}{2}$. The

normal line is $y - 1 = \left(\dfrac{2}{3}\right)(x - 1) \Rightarrow y = \dfrac{2}{3}x + \dfrac{1}{3}$.

c) If $xy + 2x - 5y = 2$, then $y' = \dfrac{y + 2}{5 - x}$. The tangent line is $y - 2 = 2(x - 3) \Rightarrow y = 2x - 4$. The

normal line is $y - 2 = \left(-\dfrac{1}{2}\right)(x - 3) \Rightarrow y = -\dfrac{1}{2}x + \dfrac{7}{2}$.

49.

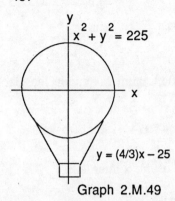

When $x^2 + y^2 = 225$, then $y' = -\dfrac{x}{y}$. The tangent line to the balloon at

$(12, -9)$ is $y + 9 = \dfrac{4}{3}(x - 12) \Rightarrow y = \dfrac{4}{3}x - 25$. The top of the gondola

is 23 ft below the center of the balloon. The intersection of $y = -23$ and

$y = \dfrac{4}{3}x - 25$ is at the far right edge of the gondola. The gondola is

3 ft wide.

$y = (4/3)x - 25$

Graph 2.M.49

51. $y^3 + y = 2\cos x \Rightarrow 3y^2 y' + y' = -2\sin x$, $y' = \dfrac{-2\sin x}{3y^2 + 1}$ and $y' \Big|_{(0,1)} = \dfrac{-2\sin x}{3y^2 + 1}\Big|_{(0,1)} = 0$;

$6y(y')^2 + 3y^2 y'' + y'' = -2\cos x$, $y'' = \dfrac{-2\cos x - 6y(y')^2}{3y^2 + 1}$ and $y'' \Big|_{(0,1)} =$

$\dfrac{-2\cos x - 6y(y')^2}{3y^2 + 1}\Big|_{(0,1)} = -\dfrac{1}{2}$

53.   a) $x^2 - y^2 = 1 \Rightarrow 2x - 2yy' = 0$ and $y' = \dfrac{x}{y}$

b) $y'' = \dfrac{y - xy'}{y^2} = \dfrac{y - x\left(\dfrac{x}{y}\right)}{y^2} = \dfrac{y - \dfrac{x^2}{y}}{y^2} = \dfrac{y^2 - x^2}{y^3} = -\dfrac{1}{y^3}$

55. $f(x) = \dfrac{1}{1 + \tan x} \Rightarrow f'(x) = \dfrac{-\sec^2 x}{(1 + \tan x)^2}$. The linearization at $x = 0$ is $L(x) = f'(0)(x - 0) + f(0) = 1 - x$.

57. $f(x) = \dfrac{2}{1 - x} + \sqrt{1 + x} - 3.1 = 2(1 - x)^{-1} + (1 + x)^{1/2} - 3.1 \Rightarrow f'(x) = -2(1 - x)^{-2}(-1) + \dfrac{1}{2}(1 + x)^{-1/2} =$

$\dfrac{2}{(1 - x)^2} + \dfrac{1}{2\sqrt{1 + x}} \Rightarrow L(x) = f'(0)(x - 0) + f(0) = 2.5x - 0.1$

59.   a) $S = 6r^2 \Rightarrow dS = 12r\,dr$. We want $|dS| \le (2\%)(S) \Rightarrow |12r\,dr| \le \dfrac{12r^2}{100} \Rightarrow |dr| \le \dfrac{r}{100}$. The

measurement must have an error less than 1%.

b) When $V = r^3$, then $dV = 3r^2 dr$. The accuracy of the volume is $\dfrac{dV}{V} 100\% = \dfrac{3r^2 dr}{r^3} 100\% =$

$\left(\dfrac{3}{r}\right) dr\ 100\% = \left(\dfrac{3}{r}\right)\left(\dfrac{r}{100}\right) 100\% = 3\%$

61.   With h representing the height of the tree, we have $h = 100\tan\alpha$ with α being the angle of elevation

of the tree top. Recall that $1° = \dfrac{\pi}{180}$ radians. Now $dh = 100\sec^2\alpha\,d\alpha \Rightarrow$

$dh = 100\sec^2\left(\dfrac{\pi}{6}\right)\left(\dfrac{\pi}{180}\right) = \dfrac{20\pi}{27}$. The error is $\pm\dfrac{20\pi}{27}$ ft $\approx \pm 2.3271$ ft.

63.   a) $(1)^2 = \dfrac{4\pi^2 L}{32.2} \Rightarrow L \approx 0.8156$ ft (we will use 3 significant digits 0.816 ft)

       b) $2TdT = \dfrac{4\pi^2}{g}\,dL \Rightarrow dT = \dfrac{2\pi^2}{Tg}\,dL = \dfrac{2\pi^2}{\dfrac{2\pi\sqrt{L}}{\sqrt{g}}\,g}\,dL = \dfrac{\pi}{\sqrt{gL}}\,dL \approx 0.006$ sec

       c) The original clock completed 86400 swings/day. The lengthened pendulum will cause the clock to gain about 8.64 min/day.

65.   We wish to solve $-x^2 + 3x + 4 = 0$. Let $f(x) = -x^2 + 3x + 4 \Rightarrow f'(x) = -2x + 3$ $\therefore$ $x_{n+1} =$

       $x_n - \dfrac{-x_n^2 + 3x_n + 4}{-2x_n + 3}$. If $x_0 = 2$, then $x_{10} = 2.19582$, an approximation of the x-intercept.

67.   Let $h(x) = (fg)(x) = f(x)\,g(x) \Rightarrow h'(x) = \underset{x \to x_0}{\text{Lim}} \dfrac{h(x) - h(x_0)}{x - x_0} = \underset{x \to x_0}{\text{Lim}} \dfrac{f(x)\,g(x) - f(x_0)\,g(x_0)}{x - x_0} =$

       $\underset{x \to x_0}{\text{Lim}} \dfrac{f(x)\,g(x) - f(x)\,g(x_0) + f(x)\,g(x_0) - f(x_0)\,g(x_0)}{x - x_0} = \underset{x \to x_0}{\text{Lim}} \left[\dfrac{f(x)\big[(g(x) - g(x_0)\big]}{x - x_0} \right] +$

       $\underset{x \to x_0}{\text{Lim}} \left[g(x_0) \left[\dfrac{f(x) - f(x_0)}{x - x_0} \right] \right] = f(x_0) \underset{x \to x_0}{\text{Lim}} \left[\dfrac{g(x) - g(x_0)}{x - x_0} \right] + g(x_0)\,f'(x_0) = 0 \; \underset{x \to x_0}{\text{Lim}} \left[\dfrac{g(x) - g(x_0)}{x - x_0} \right] +$

       $g(x_0)\,f'(x_0) = g(x_0)\,f'(x_0)$. $\therefore$ $(fg)(x)$ is differentable at x_0.

69.   If $y = u_1 u_2$, then $\dfrac{dy}{dx} = \dfrac{du_1}{dx}\,u_2 + u_1\,\dfrac{du_2}{dx}$ by the product rule. Assume the statement to be true for
       $n = k - 1$. If $y = u_1 u_2 \cdots u_{k-1} u_k = \big[u_1 u_2 \cdots u_{k-1}\big] u_k$, then

       $\dfrac{dy}{dx} = \dfrac{d\big[u_1 u_2 \cdots u_{k-1}\big]}{dx}\,u_k + \big[u_1 u_2 \cdots u_{k-1}\big] \dfrac{du_k}{dx} =$

       $\left[\dfrac{du_1}{dx}\,u_2 u_3 \cdots u_{k-1} + u_1\,\dfrac{du_2}{dx}\,u_3 u_4 \cdots u_{k-1} + \cdots + u_1 u_2 u_3 \cdots u_{k-2}\,\dfrac{du_{k-1}}{dx} \right] u_k +$

       $u_1 u_2 \cdots u_{k-1}\,\dfrac{du_k}{dx}$ and the generalized product rule follows.

71.   From the given conditions we have $f(x + h) = f(x)\,f(h)$, $f(h) - 1 = h\,g(h)$ and $\underset{h \to 0}{\text{Lim}}\,g(h) = 1$; $f'(x) =$

       $\underset{h \to 0}{\text{Lim}} \dfrac{f(x + h) - f(x)}{h} = \underset{h \to 0}{\text{Lim}} \dfrac{f(x)\,f(h) - f(x)}{h} = \underset{h \to 0}{\text{Lim}} \left[f(x)\,\dfrac{f(h) - 1}{h} \right] = f(x) \underset{h \to 0}{\text{Lim}}\,g(h) = f(x) \Rightarrow f'(x) = f(x)$.

CHAPTER 3

APPLICATIONS OF DERIVATIVES

3.1 RELATED RATES OF CHANGE

1. $A = \pi r^2 \Rightarrow \dfrac{dA}{dt} = 2\pi r \dfrac{dr}{dt}$

3. $V = x^3 \Rightarrow \dfrac{dV}{dt} = 3x^2 \dfrac{dx}{dt}$

5. $V = (1/3)\pi r^2 h \Rightarrow \dfrac{dV}{dt} = (1/3)\pi r^2 \dfrac{dh}{dt}$

7. Given $A = \pi r^2$, $\dfrac{dr}{dt} = 0.01$ cm/sec, and $r = 50$ cm. Since $\dfrac{dA}{dt} = 2\pi r \dfrac{dr}{dt}$, then $\dfrac{dA}{dt}\bigg|_{r\,=\,50} =$

 $2\pi(50)\left(\dfrac{1}{100}\right) = \pi$ cm^2/min.

9. Given $\dfrac{dl}{dt} = -2$ cm/sec, $\dfrac{dw}{dt} = 2$ cm/sec, $l = 12$ cm and $w = 5$ cm.

 a) $A = lw \Rightarrow \dfrac{dA}{dt} = l\dfrac{dw}{dt} + w\dfrac{dl}{dt} \Rightarrow \dfrac{dA}{dt} = 12(2) + 5(-2) = 14$ cm^2/sec, increasing

 b) $P = 2l + 2w \Rightarrow \dfrac{dP}{dt} = 2\dfrac{dl}{dt} + 2\dfrac{dw}{dt} = 2(-2) + 2(2) = 0$ cm/sec, constant

 c) $D = \sqrt{w^2 + l^2} = (w^2 + l^2)^{1/2} \Rightarrow \dfrac{dD}{dt} = (1/2)(w^2 + l^2)^{-1/2}\left[2w\dfrac{dw}{dt} + 2l\dfrac{dl}{dt}\right] \Rightarrow$

 $\dfrac{dD}{dt} = \dfrac{w\dfrac{dw}{dt} + l\dfrac{dl}{dt}}{\sqrt{w^2 + l^2}} = -\dfrac{14}{13}$ cm/sec, decreasing

11. Given: $\dfrac{dx}{dt} = 5$ ft/sec and the ladder is 13 ft. Since $x^2 + y^2 = 169 \Rightarrow \dfrac{dy}{dt} = -\dfrac{x}{y}\dfrac{dx}{dt}$, the

 ladder is sliding down the wall at -12 ft/sec. The area of the triangle formed by the ladder and walls

 is $A = \dfrac{1}{2}xy \Rightarrow \dfrac{dA}{dt} = \left(\dfrac{1}{2}\right)\left[x\dfrac{dy}{dt} + y\dfrac{dx}{dt}\right]$. The area is changing at $-\dfrac{119}{2}$ ft^2/sec.

13. Given: $r = \dfrac{h}{2}$, $h = 5$, and $\dfrac{dV}{dt} = 10$ ft^3/min. $V = \dfrac{1}{3}\pi r^2 h \Rightarrow V = \dfrac{\pi h^3}{12} \Rightarrow \dfrac{dV}{dt} = \dfrac{\pi h^2}{4}\dfrac{dh}{dt}$. The height is

 growing at ≈ 0.5093 ft/min.

15. If $V = \dfrac{4}{3}\pi r^3$, $r = 5$, and $\dfrac{dV}{dt} = 100\pi$ ft^3/min, then $\dfrac{dV}{dt} = 4\pi r^2 \dfrac{dr}{dt} \Rightarrow \dfrac{dr}{dt} = 1$ ft min. $S = 4\pi r^2 \Rightarrow$

 $\dfrac{dS}{dt} = 8\pi r\dfrac{dr}{dt} = 8\pi(5)(1) = 40\pi$ ft^2/min, the rate which the area is increasing.

17. Let s represent the distance between the bicycle and balloon, h the height of the balloon and x the

 horizontal distance between the balloon and bicycle. The relationship between the variables is

 $s^2 = h^2 + x^2 \Rightarrow \dfrac{ds}{dt} = \dfrac{1}{s}\left[h\dfrac{dh}{dt} + x\dfrac{dx}{dt}\right] \Rightarrow \dfrac{ds}{dt} = \dfrac{1}{85}[68(1) + 51(17)] = 11$ ft/sec.

19. $y = QD^{-1} \Rightarrow \dfrac{dy}{dt} = D^{-1}\dfrac{dQ}{dt} - QD^{-2}\dfrac{dD}{dt} = \dfrac{1}{41}(0) - \dfrac{233}{(41)^2}(-2) = \dfrac{466}{1681} \approx 0.2772$ L/min.

21.   Let $P(x, y)$ represent a point on the curve $y = x^2$ and θ the angle of inclination of a line containing P

and the origin. Consequently, $\text{Tan } \theta = \dfrac{y}{x} \Rightarrow \text{Tan } \theta = \dfrac{x^2}{x} = x \Rightarrow \sec^2\theta\dfrac{d\theta}{dt} = \dfrac{dx}{dt} \Rightarrow \dfrac{d\theta}{dt} = \cos^2\theta\dfrac{dx}{dt}$.

Recall that $\dfrac{dx}{dt} = 10$ m/sec $\Rightarrow \dfrac{d\theta}{dt}\Big|_{x=3} = 1$ rad/sec. $\underset{x \to \infty}{\text{Lim}} \left(\dfrac{x}{\sqrt{x^4 + x^2}}\right)^2 (10) =$

$(10)\underset{x \to \infty}{\text{Lim}} \ \dfrac{1}{x^2 + 1} = 0$ rad/sec

23.   a) $\dfrac{dr}{dt} = 9\dfrac{dx}{dt} = 9(0.1) = 0.9, \dfrac{dc}{dt} = (3x^2 - 12x + 15)\dfrac{dx}{dt} = \left(3(2)^2 - 12(2) + 15\right)(0.1) = 0.3,$

$\dfrac{dp}{dt} = 0.9 - 0.3 = 0.6$

b) $\dfrac{dr}{dt} = 70\dfrac{dx}{dt} = 70(0.05) = 3.5, \dfrac{dc}{dt} = 3x^2 - 12x - 45x^{-2} = \left(3(1.5)^2 - 12(1.5) - 45(1.5)^{-2}\right)(0.05) =$

$- 1.5625, \dfrac{dp}{dt} = 3.5 - (- 1.5625) = 5.0625$

25.   Let s represent the distance the ball has fallen, h the distance between the ball and the ground, and

l the distance between the shadow and a point directly beneath the ball. Accordingly, $s + h = 50$ and

$l = \dfrac{1500}{h - 50} \Rightarrow h = 50 - 16t^2$ and $\dfrac{dl}{dt} = -\dfrac{1500}{(h - 50)^2}\dfrac{dh}{dt} = -\dfrac{(1500)(32t)}{(16t^2)^2} = -\dfrac{1500}{16t^4} \Rightarrow$

$\dfrac{dl}{dt}\Big|_{t = 1/2} = - 1500$ ft/sec.

27.   Let s represent the horizontal distance between the car and plane while r is the distance between

the car and plane $\Rightarrow 9 + s^2 = r^2 \Rightarrow \dfrac{ds}{dt} = \dfrac{r}{\sqrt{r^2 - 9}}\dfrac{dr}{dt} \Rightarrow \dfrac{ds}{dt}\Big|_{r = 5} = \dfrac{5}{\sqrt{16}} (- 160) = - 200$ MPH $\Rightarrow$

speed of plane + speed of car $\Rightarrow$ the speed of the car is 80 MPH.

29.   When x represents the length of the shadow, then $\tan \theta = \dfrac{80}{x} \Rightarrow \sec^2\theta \dfrac{d\theta}{dt} = -\dfrac{80}{x^2}\dfrac{dx}{dt} \Rightarrow$

$\dfrac{dx}{dt} = \dfrac{-x^2 \sec^2\theta}{80}\dfrac{d\theta}{dt}$. Recall that $\dfrac{d\theta}{dt} = 0.27° = \dfrac{3\pi}{2000}$ rad/min. $\left|\dfrac{dx}{dt}\right|\Big|_{(\theta = 3\pi/2000 \text{ and } \sec \theta = 5/3)} =$

$\left|\dfrac{-x^2 \sec^2\theta}{80}\dfrac{d\theta}{dt}\right|\Big|_{(\theta = 3\pi/2000 \text{ and } \sec \theta = 5/3)} = \dfrac{3\pi}{16}$ ft/min ≈ 0.589 ft/min ≈ 7.1 in/min.

3.2 MAXIMA, MINIMA, AND THE MEAN VALUE THEOREM

1.   With $f(1) = - 1 < 0$, $f(3) = 7/3 > 0$, and the Intermediate Value Theorem, we may conclude that

$f(x) = 0$ has at least one root between 1 and 3 when $f(x) = x - 2x^{-1}$. Since $f'(x) = 1 + 2/x^2 \Rightarrow$

$f'(x) > 0$ for $1 < x < 3 \Rightarrow f(x)$ is increasing for all $1 < x < 3 \Rightarrow f(x) = 0$ has exactly one solution

when $1 < x < 3$.

3. With $f(-2) = 11 > 0$, $f(-1) = -1 < 0$, and the Intermediate Value Theorem, we may conclude that $f(x) = 0$ has at least one root between -2 and -1 when $f(x) = x^4 + 3x + 1$. When $-2 < x < -1 \Rightarrow$ $-8 < x^3 < -1 \Rightarrow -32 < 4x^3 < -4 \Rightarrow -29 < 4x^3 + 3 < -1 \Rightarrow f'(x) < 0$ for $-2 < x < -1 \Rightarrow f(x)$ is decreasing for all $-2 < x < -1 \Rightarrow f(x) = 0$ has exactly one solution when $-2 < x < -1$.

5. When $f(x) = x^2 + 2x - 1$ for $0 \le x \le 1$, then $\dfrac{f(1) - f(0)}{1 - 0} = f'(c) \Rightarrow 3 = 2c + 2 \Rightarrow c = 1/2$.

6. When $f(x) = x^{2/3}$ for $0 \le x \le 1$, then $\dfrac{f(1) - f(0)}{1 - 0} = f'(c) \Rightarrow 1 = (2/3)c^{-1/3} \Rightarrow c = 8/27$.

7. When $f(x) = x + 1/x$ for $1/2 \le x \le 2$, then $\dfrac{f(2) - f(1/2)}{2 - 1/2} = f'(c) \Rightarrow 0 = 1 - 1/c^2 \Rightarrow c = 1$.

9. Let $d(t)$ represent the distance the automobile traveled in time t. The average speed is $\dfrac{d(2) - d(0)}{2 - 0}$. The Mean Value Theorem indicates that for some $0 \le t_o \le 2$, $d'(t_o) = \dfrac{d(2) - d(0)}{2 - 0}$, where $d'(t_o)$ is the speed of the automobile at time t_o.

11. a)

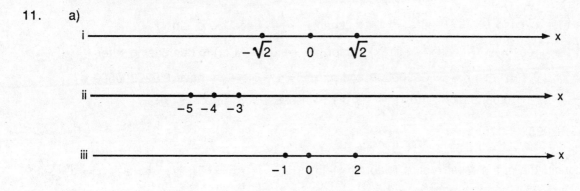

b) Let r_1 and r_2 be zeros of the polynomial $P(x) = x^n + a_{n-1}x^{n-1} + \dots + \alpha_1 x + a_0$. Then $P(r_1) = P(r_2) = 0$. Since polynomials are everywhere continuous and differentiable by Rolle's Theorem, $P'(r) = 0$ for some $r_1 < r < r_2$. $P'(x) = nx^{n-1} + (n-1)a_{n-1}x^{n-2} + \dots + a_1$.

13. Since $f(x)$ is continuous at $x = 0$ and $x = 1$ we have $\lim\limits_{x \to 0^+} f(x) = a = f(0) \Rightarrow a = 3$ and $\lim\limits_{x \to 1^-} f(x) =$ $\lim\limits_{x \to 1^+} f(x) \Rightarrow -1 + 3 + a = m + b \Rightarrow 5 = m + b$. Since $f(x)$ is differentiable at $x = 1$ we have $\lim\limits_{x \to 1^-} f'(x) = \lim\limits_{x \to 1^+} f'(x) \Rightarrow 1 = m$. $\therefore a = 3$, $m = 1$ and $b = 4$.

15. $f(x)$ is not continuous on $[0,1]$ because $\lim\limits_{x \to 1^-} f(x) = 1 \ne 0 = f(1)$.

17. Let $f(x) = \sin x$ for $a \le x \le b$. From the Mean Value Theorem there exists a c between a and b such that $\dfrac{\sin b - \sin a}{b - a} = \cos c \Rightarrow -1 \le \dfrac{\sin b - \sin a}{b - a} \le 1 \Rightarrow \left| \dfrac{\sin b - \sin a}{b - a} \right| \le 1 \Rightarrow$

$|\sin b - \sin a| \le |b - a|$.

19.   a)   When $f(x) = 1/x \Rightarrow f'(x) = -1/x^2 \Rightarrow f'(x) < 0$ for all non-zero numbers $\Rightarrow f(x) = 1/x$ is always decreasing on any interval on which it is defined.

b) $f(x) = 1/x^2 \Rightarrow f'(x) = -2/x^3 \Rightarrow f'(x) > 0$ for all negative numbers and $f'(x) < 0$ for all positive numbers. Consequently, $f(x)$ is increasing on any interval to the left of the origin and decreasing on any interval to the right of the origin.

21.   By the Mean Value Theorem there exists a c between a and b such that $\dfrac{f(b) - f(a)}{b - a} = f'(c)$ and $\dfrac{g(b) - g(a)}{b - a} = f'(c) \Rightarrow$ at x = c the tangents to the graphs f and g are parallel.

23.   From the Mean Value Theorem we have $\dfrac{f(b) - f(a)}{b - a} = f'(c)$ where c is between a and b, but $f'(c) = 2pc + q \Rightarrow$ only one solution between a and b.

25.   If $f(0) = 3$ and $f'(x) = 0$ for all x. From the Mean Value Theorem we have that $\dfrac{f(x) - f(0)}{x - 0} = f'(c)$ where c is between x and 0. Now $\dfrac{f(x) - 3}{x} = 0 \Rightarrow f(x) - 3 = 0 \Rightarrow f(x) = 3$.

27. $f'(x) = \left(1 + x^4 \cos x\right)^{-1} \Rightarrow f''(x) = -\left(1 + x^4 \cos x\right)^{-2}\left(4x^3 \cos x - x^4 \sin x\right) = -x^3\left(1 + x^4 \cos x\right)^{-2}(4 \cos x - x \sin x) < 0$ for $0 < x < 0.1 \Rightarrow f'(x)$ is decreasing when $0 < x < 0.1 \Rightarrow$ min $f' \approx 0.999900509$ and max $f' = 1$. Now we have $0.999900509 \leq \dfrac{f(0.1) - 1}{0.1} \leq 1 \Rightarrow 0.09999005 \leq f(0.1) - 1 \leq 0.1 \Rightarrow 1.09999005 \leq f(0.1) \leq 1.1 \Rightarrow$ $f(0.1) \approx 1.099995025$.

29.   The conclusion of the Mean Value Theorem yields $\dfrac{\frac{1}{b} - \frac{1}{a}}{b - a} = -\dfrac{1}{c^2} \Rightarrow c^2\left(\dfrac{a - b}{ab}\right) = a - b \Rightarrow c = \sqrt{ab}$.

3.3 GRAPHING WITH Y´ AND Y´´

1.   When $y = \dfrac{x^3}{3} - \dfrac{x^2}{2} - 2x + \dfrac{1}{3} \Rightarrow y' = x^2 - x - 2 = (x - 2)(x + 1) \Rightarrow y'' = 2x - 1 = 2(x - 1/2)$. The graph is rising on $(-\infty, -1) \cup (2, \infty)$, falling on $(-1, 2)$, concave upward on $(1/2, \infty)$, and concave downward on $(-\infty, 1/2)$. Consequently, the local maximum is 3/2 at x = -1, the local minimum is -3 at x = 2, and $(1/2, -3/4)$ is a point of inflection.

3.   When $y = \sin |x|$, then the graph is rising on $(-3\pi/2, -\pi/2) \cup (0, \pi/2) \cup (3\pi/2, 2\pi)$, falling on $(-2\pi, -3\pi/2) \cup (-\pi/2, 0) \cup (\pi/2, 3\pi/2)$, concave upward on $(-2\pi, -\pi) \cup (\pi, 2\pi)$, and concave downward on $(-\pi, 0) \cup (0, \pi)$. Consequently, the maximum is 1 at x = $\pm \pi/2$, the minimum is -1 at x = $\pm 3\pi/2$, and the points of inflection are $(-\pi, 0)$ and $(\pi, 0)$.

5.   When $y = x^2 - 4x + 3$, then $y´ = 2x - 4 = 2(x - 2)$ and $y´´ = 2$.
      The curve rises on $(2, \infty)$ and falls on $(-\infty, 2)$. At $x = 2$ there
      is a minimum. Since $y´´ > 0$, the curve is concave upward
      for all x.

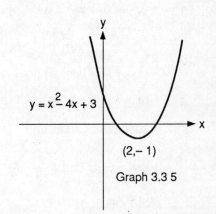

7.   When $y = 2x^3 - 3x^2$, then $y´ = 6x^2 - 6x = 6x(x - 1)$ and
      $y´´ = 12x - 6 = 6(2x - 1)$. The curve rises on $(-\infty, 0) \cup$
      $(1, \infty)$ and falls on $(0,1)$. At $x = 0$ there is a local
      maximum and at $x = 1$ a local minimum. The curve is
      concave downward on $(-\infty, 1/2)$ and concave upward on
      $(1/2, \infty)$. At $x = 1/2$ there is a point of inflection.

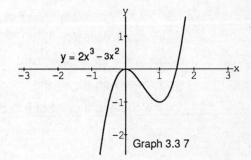

9.   When $y = 4 + 3x - x^3$, then $y´ = 3 - 3x^2 = 3(1 - x)(1 + x)$
      and $y´´ = -6x$. The curve rises on $(-1, 1)$ and falls on
      $(-\infty, -1) \cup (1, \infty)$. At $x = -1$ there is a local minimum
      and at $x = 1$ a local maximum. The curve is concave
      upward on $(-\infty, 0)$ and concave downward on $(0, \infty)$.
      At $x = 0$ there is a point of inflection.

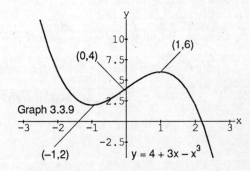

11.   When $y = (x - 2)^3 + 1$, then $y´ = 3(x - 2)^2$ and $y´´ =$
      $6(x - 2)$. The curve never falls. The curve is concave
      downward on $(-\infty, 2)$ and concave upward on $(2, \infty)$.
      At $x = 2$ there is a point of inflection.

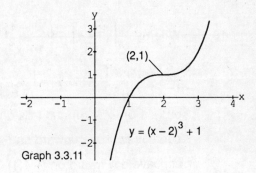

13. When $y = 2x^4 - 4x^2 + 1$, then $y' = 8x\,(x - 1)(x + 1)$ and $y'' = 8\left(x - \dfrac{1}{\sqrt{3}}\right)\left(x + \dfrac{1}{\sqrt{3}}\right)$. The curve rises on $(-1,0) \cup (1,\infty)$ and falls on $(-\infty,-1) \cup (0,1)$. At $x = \pm 1$ there are minimums and at $x = 0$ a local maximum. The curve is concave upward on $\left(-\infty,-\dfrac{\sqrt{3}}{3}\right) \cup \left(\dfrac{\sqrt{3}}{3},\infty\right)$ and concave downward on $\left(-\dfrac{\sqrt{3}}{3}, \dfrac{\sqrt{3}}{3}\right)$. At $x = \pm\dfrac{\sqrt{3}}{3}$ there are points of inflection.

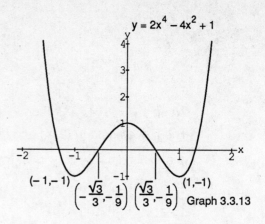

Graph 3.3.13

15.. When $y = x + \sin x$, then $y' = 1 + \cos x$ and $y'' = -\sin x$. The curve rises on $(0, 2\pi)$. At $x = 0$ there is a minimum and at $x = 2\pi$ a maximum. The curve is concave downward on $(0, \pi)$ and concave upward on $(\pi, 2\pi)$. At $x = \pi$ there is a point of inflection.

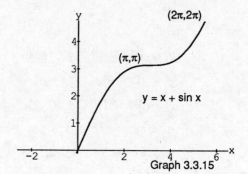

Graph 3.3.15

17. When $y = (x + 1)^{3/2}$, then $y' = \dfrac{3}{2}(x + 1)^{1/2}$ and $y'' = \dfrac{3}{4}(x + 1)^{-1/2}$. The curve rises on $(-1,\infty)$ and never falls. The curve is concave upward on $(-1,\infty)$. There are no points of inflection.

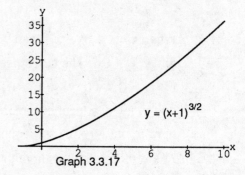

Graph 3.3.17

19. When $y = \left|x^2 - 1\right| = \begin{cases} x^2 - 1, & |x| \geq 1 \\ 1 - x^2, & |x| < 1 \end{cases}$, then $y' = \begin{cases} 2x, & |x| \geq 1 \\ -2x, & |x| < 1 \end{cases}$ and $y'' = \begin{cases} 2, & |x| \geq 1 \\ -2, & |x| < 1 \end{cases}$. The curve rises on $(-1,0) \cup (1,\infty)$ and falls on $(-\infty,-1) \cup (0,1)$. The curve is concave upward on $(-\infty,-1) \cup (1,\infty)$ and concave downward on $(-1,1)$. At $x = \pm 1$ there are points of inflection.

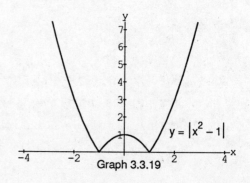

Graph 3.3.19

21.   If $x = y^2 - y + 1$, then $\dfrac{dx}{dy} = 2y - 1$ and $\dfrac{dx^2}{dy^2} = 2$. The curve rises when $y > \dfrac{1}{2}$ and falls when $y < \dfrac{1}{2}$. There is no point of inflection.

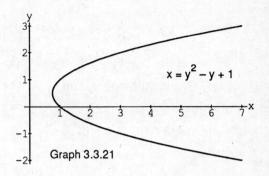

Graph 3.3.21

23.   If $x = y - y^3$, then $\dfrac{dx}{dy} = 1 - 3y^2$ and $\dfrac{dx^2}{dy^2} = -6y$. The curve rises when $-\sqrt{\dfrac{1}{3}} < y < \sqrt{\dfrac{1}{3}}$ and falls when $y < -\sqrt{\dfrac{1}{3}}$ or $y > \sqrt{\dfrac{1}{3}}$. At $y = 0$ there is a point of inflection.

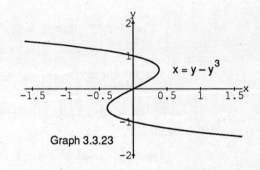

Graph 3.3.23

25.   If $y = 5x^{1/5}$, then $y′ = x^{-4/5}$. There is a vertical tangent at $x = 0$.

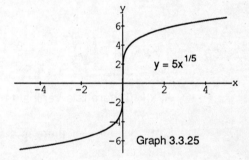

Graph 3.3.25

27.   If $y = \sqrt{x + 1}$, then $y′ = \dfrac{1}{2\sqrt{x + 1}}$. There is a vertical tangent at $x = -1$.

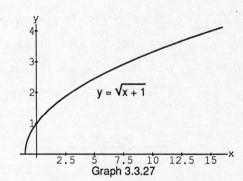

Graph 3.3.27

29.   If $y = (x - 8)^{2/3}$, then $y' = \frac{2}{3}(x - 8)^{-1/3}$. At $x = 8$ there is a

vertical tangent, $\displaystyle\lim_{x \to 8^-} y' = -\infty$ and $\displaystyle\lim_{x \to 8^+} y' = \infty$. ∴ there

is a cusp at $x = 8$.

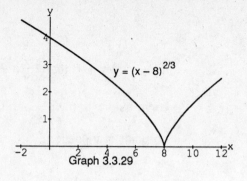
Graph 3.3.29

31.   If $y = (x^2 - 1)^{2/3}$, then $y' = \frac{2}{3}(x^2 - 1)^{-1/3}(2x) = \dfrac{4x}{3\sqrt[3]{x^2 - 1}}$.

At $x = \pm 1$ there are vertical tangents; $\displaystyle\lim_{x \to -1^-} y' = \infty$,

$\displaystyle\lim_{x \to -1^+} y' = -\infty$, $\displaystyle\lim_{x \to 1^-} y' = -\infty$ and $\displaystyle\lim_{x \to 1^+} y' = \infty$.

∴ there are cusps at $x = \pm 1$.

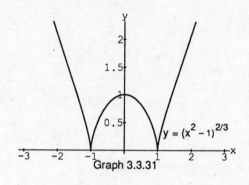
Graph 3.3.31

33.   If $y = \sqrt{|x|} = \begin{cases} \sqrt{x}, & x \geq 0 \\ \sqrt{-x}, & x < 0 \end{cases}$, then $y' = \begin{cases} \dfrac{1}{2\sqrt{x}}, & x > 0 \\ \dfrac{-1}{2\sqrt{-x}}, & x < 0 \end{cases}$

At $x = 0$ there is a vertical tangent, $\displaystyle\lim_{x \to 0^-} y' = -\infty$ and

$\displaystyle\lim_{x \to 0^+} y' = \infty$. ∴ there is a cusp at $x = 0$.

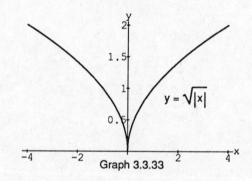
Graph 3.3.33

35.   The velocity will be zero when the slope of the tangent line for $y = s(t)$ is horizontal. The velocity is

zero when t is approximately 2, 6 ,or 9.5 sec. The acceleration will be zero at those values of t

where the curve $y = s(t)$ has points of inflection. The acceleration is zero when t is approximately 4,

8, or 12.5 sec.

37.

Point	y′	y″
P	−	+
Q	+	0
R	+	−
S	0	−
T	−	−

39.

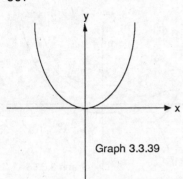

Graph 3.3.39

41.

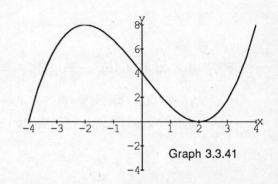

Graph 3.3.41

43.   a)   b)   c)

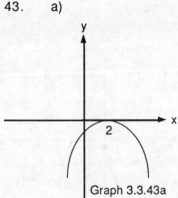

Graph 3.3.43a

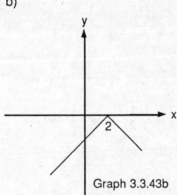

Graph 3.3.43b

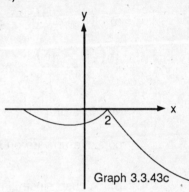

Graph 3.3.43c

45. When $y´ = (x - 1)^2(x - 2)$, then $y´´ = 2(x - 1)(x - 2) + (x - 1)^2$. The curve falls on $(-\infty, 2)$ and rises on $(2, \infty)$. At $x = 2$ there is a local minimum. The curve is concave upward on $(-\infty, 1) \cup (5/3, \infty)$ and concave downward on $(1, 5/3)$. At $x = 1$ or $x = 5/3$ there are inflection points.

47. No. When $f(x) = x^3$, then $f´(x) = 3x^2$ and $f´(0) = 0$ but $x = 0$ is neither a local minimum nor maximum.

49. a) True. If $y = ax^2 + bx + c$ where $a \neq 0$, then $y´ = 2ax + b$ and $y´´ = 2a$. Since $2a$ is a constant, it is not possible for $y´´$ to change signs.

b) True. When $y = ax^3 + bx^2 + cx + d$ where $a \neq 0$, then $y´ = 3ax^2 + 2bx + c$ and $y´´ = 6ax + 2b$. Since $-b/3a$ is a solution of $y´´ = 0$, we have that $y´´$ changes its sign at $x = -b/3a$ an inflection point.

51. If $c \to -\infty$, then $y \to mx$ where $m \to \infty$. If $c \to \infty$, then $y \to mx$ where $m \to -\infty$.

53. If $y = x^5 - 5x^4 - 240$, then $y´ = 5x^3(x - 4)$ and $y´´ = 20x^2(x - 3)$. The zeros of $y´ = 0$ and $y´´ = 0$ are extrema and points of inflection respectively.

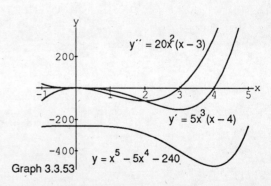

Graph 3.3.53

55.   If $y = \frac{4}{5}x^5 + 16x^2 - 25$, then $y' = 4x(x^3 + 8)$ and

$y'' = 16(x^3 + 2)$. The zeros of $y' = 0$ and $y'' = 0$ are

extrema and points of inflection respectively.

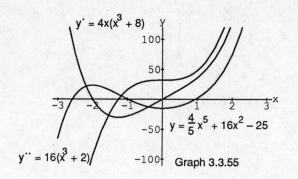

Graph 3.3.55

57.   a) If $y = x^{2/3}\left(x^2 - 2\right)$, then $y' = \dfrac{8\left(x^2 - 1/2\right)}{3x^{1/3}}$ and $y'' =$

$\dfrac{40\left(x^2 + 1/10\right)}{9x^{4/3}}$. The curve rises on $\left(-\sqrt{\dfrac{1}{2}},0\right) \cup \left(\sqrt{\dfrac{1}{2}},\infty\right)$

and falls on $\left(-\infty,-\sqrt{\dfrac{1}{2}}\right) \cup \left(0,\sqrt{\dfrac{1}{2}}\right)$. The curve is

concave upward on $(-\infty,0) \cup (0,\infty)$.

b) A cusp since $\lim\limits_{x \to 0^-} y' = \infty$ and $\lim\limits_{x \to 0^+} y' = -\infty$.

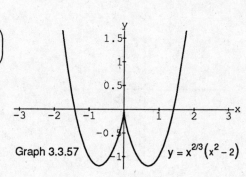

Graph 3.3.57

59.   a) computer generated   b) the correct graph

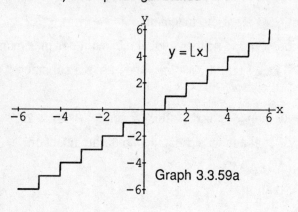

Graph 3.3.59a

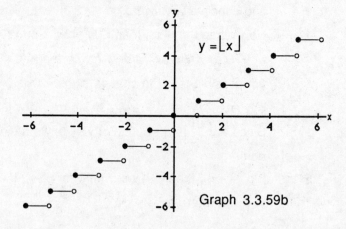

Graph 3.3.59b

61.   a) computer generated   b) the correct graph

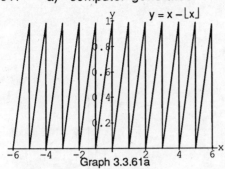

Graph 3.3.61a

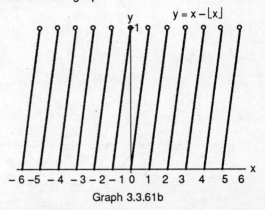

Graph 3.3.61b

63.   a) computer generated

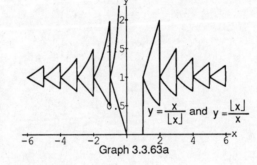

Graph 3.3.63a

        b) the correct graphs

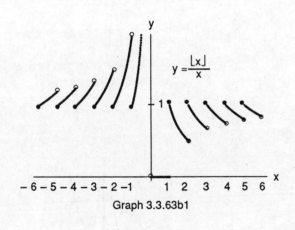

Graph 3.3.63b1

Graph 3.3.63b2

65.   No difference

Graph 3.3.65

3.4 GRAPHING RATIONAL FUNCTIONS – ASYMPTOTES AND DOMINANT TERMS

1.

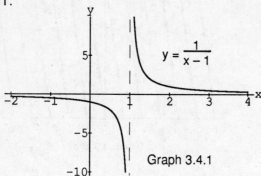

$y = \dfrac{1}{x - 1}$

Graph 3.4.1

3.

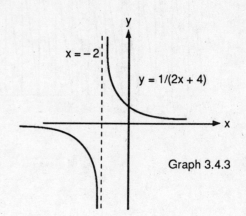

$x = -2$

$y = 1/(2x + 4)$

Graph 3.4.3

5.

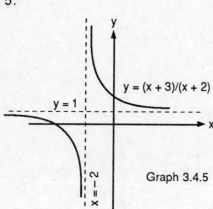

$y = (x + 3)/(x + 2)$

$y = 1$

$x = -2$

Graph 3.4.5

7.

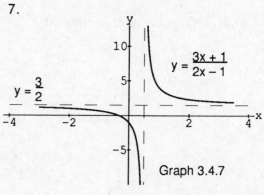

$y = \dfrac{3}{2}$

$y = \dfrac{3x + 1}{2x - 1}$

Graph 3.4.7

9.

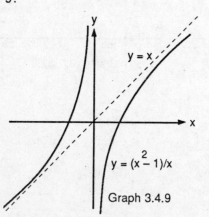

$y = x$

$y = (x^2 - 1)/x$

Graph 3.4.9

11.

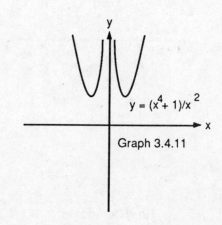

$y = (x^4 + 1)/x^2$

Graph 3.4.11

13.

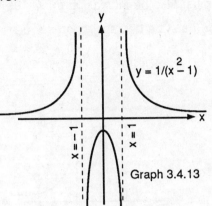

Graph 3.4.13

$y = 1/(x^2 - 1)$

$x = -1$ $x = 1$

15.

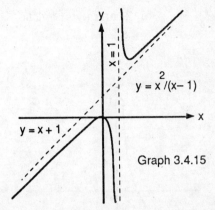

Graph 3.4.15

$x = 1$

$y = x^2/(x - 1)$

$y = x + 1$

17.

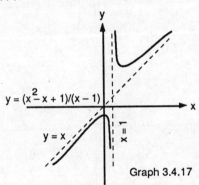

Graph 3.4.17

$y = (x^2 - x + 1)/(x - 1)$

$y = x$

$x = 1$

19.

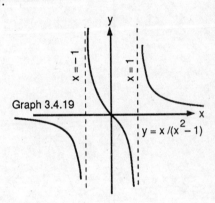

Graph 3.4.19

$x = -1$ $x = 1$

$y = x/(x^2 - 1)$

21.   a)   If $f(x) = \begin{cases} 0, & x = \pm 1 \\ \dfrac{x^{2/3}}{x^2 - 1}, & x \neq \pm 1 \end{cases}$, then $f'(x) = \begin{cases} 0, & x = \pm 1 \\ \dfrac{-2(2x^2 + 1)}{3x^{1/3}(x^2 - 1)^2}, & x \neq \pm 1 \end{cases}$ and the critcal numbers

are 0 and ± 1. At $x = 0$ there is a local minimum since $f'(x) > 0$ for $x \, \varepsilon \, (-1, 0)$ and $f'(x) < 0$ when $x \, \varepsilon \, (0, 1)$. At $x = \pm 1$ we have $\lim\limits_{x \to -1^-} f(x) = \infty$, $\lim\limits_{x \to -1^+} f(x) = -\infty$, $\lim\limits_{x \to 1^-} f(x) = -\infty$ and

$\lim\limits_{x \to 1^+} f(x) = \infty$.

b1)   The window is $-5 \leq x \leq 5$.

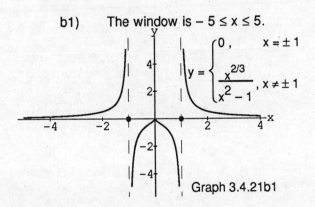

$y = \begin{cases} 0, & x = \pm 1 \\ \dfrac{x^{2/3}}{x^2 - 1}, & x \neq \pm 1 \end{cases}$

Graph 3.4.21b1

b2)   The window is $-0.25, x \leq 0.25$.

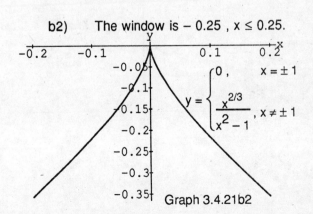

$y = \begin{cases} 0, & x = \pm 1 \\ \dfrac{x^{2/3}}{x^2 - 1}, & x \neq \pm 1 \end{cases}$

Graph 3.4.21b2

23.   a)   If an odd function is increasing on x > 0, then the function must also be increasing on x < 0.

       b)   If and even function is increasing on x < 0, then the function must be decreasing on x > 0.

25.

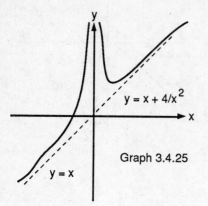

Graph 3.4.25

27.

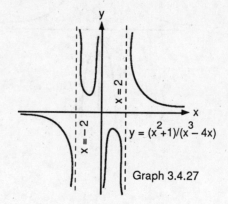

Graph 3.4.27

29.

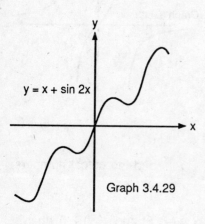

Graph 3.4.29

31.

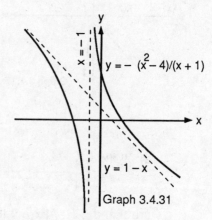

Graph 3.4.31

33.   a)

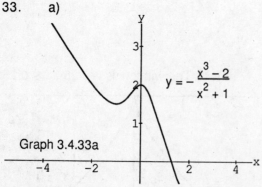

Graph 3.4.33a

       b)

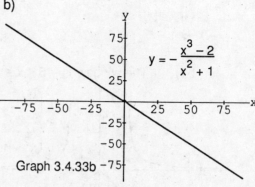

Graph 3.4.33b

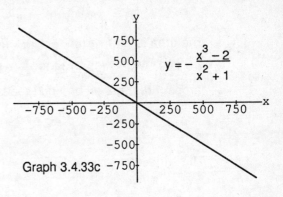

Graph 3.4.33c

c) The distance in part c is so great that small
    movements are not visible.

35. See exercise 21. The reason that the cusp did not appear in the first graph is because of the size
     if the interval was too large to show such detail.

3.5 OPTIMIZATION

1. If $A = \frac{1}{2}$ rs and $P = s + 2r = 100$, then $A(r) = \frac{1}{2}r(100 - 2r) = 50r - r^2$. $A'(r) = 50 - 2r \Rightarrow 25$ is a critical

    point. At r = 25 there is a maximum because $A''(r) < 0$ for all possible r. The values of r = 25 ft and
    s = 50 ft will give the sector its greatest area.

3. Let l and w represent the length and width of the rectangle respectively. With an area of 16 in^2, we
    have that $(l)(w) = 16 \Rightarrow w = 16l^{-1}$ and the perimeter, $P = 2l + 2w = 2l + 32l^{-1}$. Solving
    $P'(l) = 0 \Rightarrow \frac{2(l + 4)(l - 4)}{l^2} = 0 \Rightarrow l = -4, 4,$ or 0. Due to the fact that -4 and 0 are not possible

    lengths of a rectangle, l must be 4, w = 4, and the perimeter is 16 in.

5. a) The line containing point P also contains the points (0, 1) and (1, 0). $\therefore$ the line containing P is
        $y = 1 - x$ and a general point on that line is $(x, 1 - x)$.
    b) The area A(x) is $2x(1 - x)$, where $0 \le x \le 1$.
    c) When $A(x) = 2x - 2x^2$, then $A'(x) = 0 \Rightarrow x = 1/2$. Due to A(0) = 0, and A(1) = 0, we may
        conclude that A(1/2) = 1/2 sq units, the largest area.

7. The volume of the box is $V(x) = x(15 - 2x)(8 - 2x)$, where $0 \le x \le 4$.
    Solving $V'(x) = 0 \Rightarrow x = 5/3$ or 6, but 6 is not in the domain. Since V(0) =
    V(4) = 0, V(5/3) must be the maximum volume. The dimensions are:
    14/3 x 35/3 x 5/3 inches.

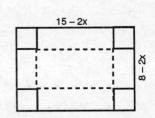

9.   The area is A(x) = x(800 − 2x), where $0 \le x \le 400$. Solving A′(x) = 0 $\Rightarrow$
      x = 200. With A(0) = A(400) = 0, the maximum area is A(200). The
      largest area that an be enclosed is A(200) = 80000 m^2.

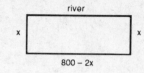

11.   The surface area is S(x) = $x^2 + 4x\left(\dfrac{500}{x^2}\right)$, where $0 < x$. The critical points

      are 0 or 10, but 0 is not in the domain. S″(10) > 0 $\Rightarrow$ at x = 10
      there is a minimum. ∴ the dimensions should be 10 ft on the base
      edge and 5 ft for the height.

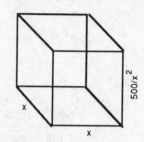

13.   The area of the printing is (y − 4)(x − 8) = 50. Consequently, y = (50/(x − 8)) + 4.
      The area of the paper is A(x) = $x\left(\dfrac{50}{x-8} + 4\right)$, where 8 < x. The critical points are

      − 2, or 18, but − 2 is not in the domain. Now A″(18) > 0 $\Rightarrow$ that at x = 18 we
      have a minimum.   ∴ the dimensions which will minimize the amount of paper are
      18 by 9 inches.

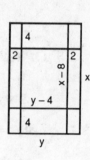

15.   The area of the triangle is A(θ) = $\dfrac{ab\sin\theta}{2}$, where $0 < \theta < \pi$. Solving
      A′(θ) = 0 $\Rightarrow \theta = \dfrac{\pi}{2}$. A″$\left(\dfrac{\pi}{2}\right)$ < 0 $\Rightarrow$ there is a maximum at $\theta = \dfrac{\pi}{2}$.

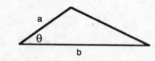

17.   With a volume of 1000 cm and V = $\pi r^2 h$, then h = $\dfrac{1000}{\pi r^2}$. The surface area is

      S = $2\pi rh + \pi r^2 = 2000r^{-1} + \pi r^2$, where $0 < r$. The critical points are 0 and $\dfrac{10}{\sqrt[3]{\pi}}$,

      but 0 is not in the domain. At r = $\dfrac{10}{\sqrt[3]{\pi}}$ we have a minimum because

      s″$\left(\dfrac{10}{\sqrt[3]{\pi}}\right)$ > 0. ∴ r = h = $\dfrac{10}{\sqrt[3]{\pi}}$

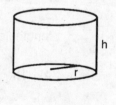

19.   a)   From the diagram we have 4x + l = 108 and V = x^2l. The volume of the
      box is V(x) = x^2(108 − 4x), where $0 \le x \le 27$. The critical points are 0 and
      18, but x = 0 would result in no box. At x = 18 we have V″(x) < 0 $\Rightarrow$
      a maximum. The dimensions of the box are 18 x 18 x 36 in.

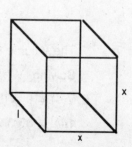

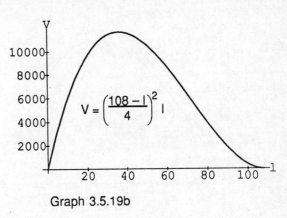

Graph 3.5.19b

b) The graph indicates that the maximum volume occurs near $l = 36$, which is compatible with the result of part a.

21. a) From figure 3.57 we have $P = 2x + 2y \Rightarrow y = P/2 - x$. If $P = 36$, then $y = 18 - x$. When the cylinder is formed, $x = 2\pi r \Rightarrow r = x/2\pi$ and $h = y \Rightarrow h = 18 - x$. The volume of the cylinder is $V = \pi r^2 h \Rightarrow V(x) = \dfrac{18x^2 - x^3}{4\pi}$. Solving $V'(x) = 0 \Rightarrow x = 0$ or 12; but when $x = 0$, there is no cylinder. Since $V''(12) < 0$ at $x = 12$, there is a maximum. The values of $x = 12$ and $y = 6$ will give the largest volume.

b) From part a) we have $V(x) = \pi x^2 (18 - x)$. Solving $V'(x) = 0 \Rightarrow x = 0$ or 12, but $x = 0$ would result in no cylinder. Since $V''(12) < 0$ at $x = 12$, there is a maximum. The values of $x = 12$ and $y = 6$ will give the largest volume.

23. a) $f(x) = x^2 + \dfrac{a}{x} \Rightarrow f'(x) = x^{-2}\left(2x^3 - a\right)$. Solving $f'(x) = 0$ at $x = 2$ implies that $a = 16$.

b) $f(x) = x^2 + \dfrac{a}{x} \Rightarrow f''(x) = 2x^{-3}\left[x^3 + a\right]$. Solving $f''(x) = 0$ at $x = 2$ implies that $a = -1$.

25. If $f(x) = x^3 + ax^2 + bx$, then $f'(x) = 3x^2 + 2ax + b$ and $f''(x) = 6x + 2a$.

a) A local maximum at $x = -1$ and local minimum at $x = 3 \Rightarrow f'(-1) = 0$ and $f'(3) = 0$. $\therefore 3 - 2a + b = 0$ and $27 + 6a + b = 0 \Rightarrow a = -3$ and $b = -9$.

b) A local maximum at $x = -1$ and a point of inflection at $x = 1 \Rightarrow f'(-1) = 0$ and $f''(1) = 0$. $\therefore 48 + 8a + b = 0$ and $6 + 2a = 0 \Rightarrow a = -3$ and $b = -24$.

27. a) From the diagram we have $d^2 = 4r^2 - w^2$. The strength of the beam is $S = kwd^2 = kw(4r^2 - w^2)$. When $r = 6$, then $S = 144kw - kw^3$. $S'(w) = 144k - 3kw^2 = 3k(48 - w^2)$. $S'(w) = 0 \Rightarrow w = \pm 4\sqrt{3}$. $S''(4\sqrt{3}) < 0$ and $-4\sqrt{3}$ is not acceptable. $\therefore S(4\sqrt{3})$ is the maximum strength. The dimensions of the strongest beam are $4\sqrt{3}$ by $4\sqrt{6}$ inches.

b)

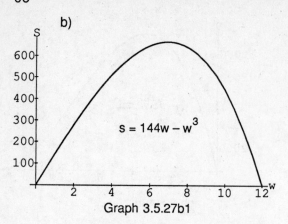

Graph 3.5.27b1

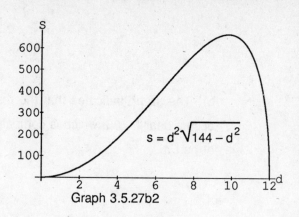

Graph 3.5.27b2

Both graphs indicate the same maximum and are compatible with each other. The changing of k has no effect.

29.   a)   If $y = \cot x - \sqrt{2} \csc x$ where $0 < x < \pi$, then $y' = [\csc x][\sqrt{2} \cot x - \csc x]$. Solving $y' = 0 \Rightarrow$
$\cos x = \dfrac{1}{\sqrt{2}} \Rightarrow x = \dfrac{\pi}{4}$. For $0 < x < \dfrac{\pi}{4}$ we have $y' > 0$ and $y' < 0$ when $\dfrac{\pi}{4} < x < \pi$ $\therefore$ at $x = \dfrac{\pi}{4}$ there is a

maximum. The maximum value of y is -1.

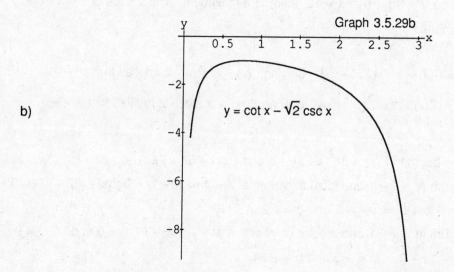

Graph 3.5.29b

b)

$y = \cot x - \sqrt{2} \csc x$

31.   The distance between a point on the curve $y = \sqrt{x}$ and $(3/2, 0)$ is represented by $D(x) =$
$\sqrt{(3/2 - x)^2 + (0 - \sqrt{x})^2}$. $D(x)$ is minimized when $I(x) = (3/2 - x)^2 + (0 - \sqrt{x})^2 = (3/2 - x)^2 + x$ is
minimized. $I'(x) = 2(3/2 - x)(-1) + 1 = 2(x - 1)$. The solution of $I'(x) = 0$ is $x = 1$ and $I''(1) = 2 > 0$.
$\therefore$ the minimum value of $I(x)$ is $I(1) = \dfrac{5}{4} \Rightarrow$ the minimum distance is $D(1) = \sqrt{\dfrac{5}{4}} = \dfrac{\sqrt{5}}{2}$.

33. From the diagram the area of the cross section is $A(\theta) = \cos \theta + \sin \theta \cos \theta$, where $0 < \theta < \pi/2$.

$A'(\theta) = -\sin \theta + \cos^2 \theta - \sin^2 \theta = -\left(2 \sin^2 \theta + \sin \theta - 1\right) = -(2 \sin \theta - 1)(\sin \theta + 1)$. $A'(\theta) = 0 \Rightarrow$

$\sin \theta = 1/2$ or $\sin \theta = -1 \Rightarrow \theta = \pi/6$ because $\sin \theta \neq -1$ when $0 < \theta < \pi/2$. $A'(\theta) > 0$ for $0 < \theta < \pi/6$

and $A'(\theta) < 0$ for $\pi/6 < \theta < \pi/2$ ∴ at $\theta = \pi/6$ there is a maximum. The volume of the trough is

maximized when the area of the cross section is maximized.

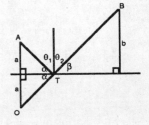

35. The distance $\overline{OT} + \overline{TB}$ is minimized when $\overline{OB}$ is a straight line.

Hence, $\angle \alpha = \angle \beta \Rightarrow \theta_1 = \theta_2$.

37. If $v = kax - kx^2$, then $v' = ka - 2kx$ and $v'' = -2k$. $v' = 0 \Rightarrow x = a/2$. At $x = a/2$ there is a maximum

since $v''(a/2) = -2k < 0$. The maximum value of v is $ka^2/4$.

39. a) Let $4 - x$ be the perimeter of the square and x the circumference of the circle. Now $\dfrac{x}{2\pi}$ is the

radius of the circle while $1 - \dfrac{x}{4}$ the side of the square. The enclosed area is $A = \left(1 - \dfrac{x}{4}\right)^2 +$

$\pi \left(\dfrac{x}{2\pi}\right)^2 = \left(1 - \dfrac{x}{4}\right)^2 + \dfrac{x^2}{4\pi}$ where $0 \leq x \leq 4$. $\dfrac{dA}{dx} = 0 \Rightarrow \dfrac{x}{8} - \dfrac{1}{2} + \dfrac{x}{2\pi} + 0 \Rightarrow x = \dfrac{4\pi}{\pi + 4}$. Since $A(0) = 1$,

$A\left(\dfrac{4\pi}{\pi + 4}\right) = \dfrac{4}{\pi + 4}$ and $A(4) = \dfrac{4}{\pi}$, then at $x = 4$ there is a maximum.

b) This graph indicates the same maximum and is compatible with part a.

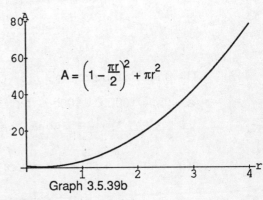

$$A = \left(1 - \dfrac{\pi r}{2}\right)^2 + \pi r^2$$

Graph 3.5.39b

c) This graph indicates the same maximum and is compatible with part a.

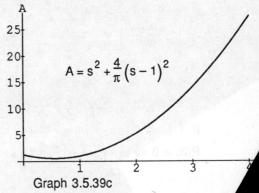

$$A = s^2 + \dfrac{4}{\pi}(s - 1)^2$$

Graph 3.5.39c

41. a) The distance is given by $d(x) = \sqrt{(x-c)^2 + x}$ where $x \geq 0$. $d'(x) = \dfrac{2x - 2c + 1}{2\sqrt{(x-c)^2 + x}} \Rightarrow$

at $x = c - \dfrac{1}{2}$ there is a minimum and $\left(c - \dfrac{1}{2}, \sqrt{c - \dfrac{1}{2}}\right)$ is nearest $(c,0)$.

b) If $c < \dfrac{1}{2}$, then $d'(x)$ has no critical numbers and the minimum must occur at the endpoint $x = 0$.

∴ $(0,0)$ is nearest $(c,0)$

43. If $x > 0$, then $(x - 1)^2 \geq 0 \Rightarrow x^2 + 1 \geq 2x \Rightarrow \dfrac{x^2 + 1}{x} \geq 2$. In particular if a, b, c and d are positive

integers, then $\left(\dfrac{a^2 + 1}{a}\right)\left(\dfrac{b^2 + 1}{b}\right)\left(\dfrac{c^2 + 1}{c}\right)\left(\dfrac{d^2 + 1}{d}\right) \geq 16$.

45. a) If $v = cr_0 r^2 - cr^3$, then $v' = 2cr_0 r - 3cr^2 = cr[2r_0 - 3r]$ and $v'' = 2cr_0 - 6cr = 2c[r_0 - 3r]$. The solution
of $v' = 0$ is $r = 0$ or $2r_0/3$, but 0 is not in the domain. $v' > 0$ for $r < 2r_0/3$ and $v' < 0$ for $r > 2r_0/3 \Rightarrow$ at
$r = 2r_0/3$ there is a maximum.

b)

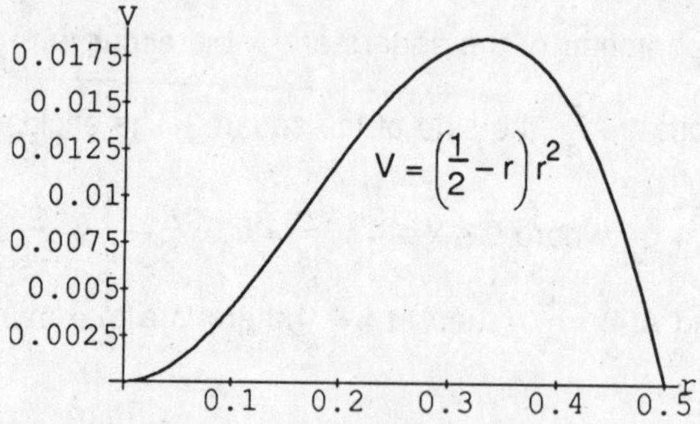

Graph 3.5.45b

47. The profit is $p = nx - nc = n(x - c) = \left[a(x-c)^{-1} + b(100 - x)\right](x - c) = a + b(100 - x)(x - c) =$
$a + (bc + 100b)x - 100bc - bx^2$. $p'(x) = bc + 100b - 2bx \Rightarrow p''(x) = -2b$. Solving $p'(x) = 0 \Rightarrow$
$x = \dfrac{c}{2} + 50$. At $x = \dfrac{c}{2} + 50$ there is a maximum since $p''\left(\dfrac{c}{2} + 50\right) < 0$.

49. $A(q) = kmq^{-1} + cm + \dfrac{h}{2} q$, where $q > 0 \Rightarrow A'(q) = -kmq^{-2} + \dfrac{h}{2} = \dfrac{hq^2 - 2km}{2q^2}$ and $A''(q) = 2kmq^{-3}$.

The critical points are $-\sqrt{\dfrac{2km}{h}}$, 0, and $\sqrt{\dfrac{2km}{h}}$, but only $\sqrt{\dfrac{2km}{h}}$ is in the domain.

$A''\left(\sqrt{\dfrac{2km}{h}}\right) > 0 \Rightarrow q = \sqrt{\dfrac{2km}{h}}$ is a minimum.

51. The profit $p(x) = r(x) - c(x) = (6x) - (x^3 - 6x^2 + 15x) = -x^3 + 6x^2 - 9x$, where $x \geq 0$.
$p'(x) = -3x^2 + 12x - 9 = -3(x - 3)(x - 1)$ and $p''(x) = -6x + 12$. The critical points are 1 and 3.
$''(1) = 6 > 0 \Rightarrow$ at $x = 1$ there is a local minimum and $P''(3) = -6 < 0 \Rightarrow$ at $x = 3$ there is a maximum.
the best you can do is break even.

3.6 INDETERMINATE FORMS AND L'HOPITAL'S RULE

1. $\displaystyle\lim_{x \to 2} \frac{x-2}{x^2-4} = \lim_{x \to 2} \frac{1}{2x} = \frac{1}{4}$

3. $\displaystyle\lim_{x \to 1} \frac{x^3-1}{4x^3-x-3} = \lim_{x \to 1} \frac{3x^2}{12x^2-1} = \frac{3}{11}$

5. $\displaystyle\lim_{t \to 0} \frac{\sin t^2}{t} = \lim_{t \to 0} \frac{(\cos t^2)(2t)}{1} = 0$

7. $\displaystyle\lim_{x \to 0} \frac{\sqrt{ax+a^2}-a}{x} = \lim_{x \to 0} \frac{\frac{1}{2}\left(ax+a^2\right)^{-1/2} a}{1} = \lim_{x \to 0} \frac{a}{2\sqrt{ax+a^2}} = \frac{a}{2\,|a|} = \frac{1}{2}$ when $a > 0$

9. $\displaystyle\lim_{x \to 0} \frac{x \sin x}{1 - \cos x} = \lim_{x \to 0} \frac{\sin x + x \cos x}{\sin x} = \lim_{x \to 0} \frac{\cos x + \cos x - x \sin x}{\cos x} = 2$

11. $\displaystyle\lim_{x \to \pi/2} \left(x - \frac{\pi}{2}\right)^2 \sec x = \lim_{x \to \pi/2} \frac{\left(x - \frac{\pi}{2}\right)^2}{\cos x} = \lim_{x \to \pi/2} \frac{2\left(x - \frac{\pi}{2}\right)}{-\sin x} = 0$

13. $\displaystyle\lim_{x \to 0^+} \frac{2x}{x + 7\sqrt{x}} = \lim_{x \to 0^+} \frac{2\sqrt{x}}{\sqrt{x}+7} = 0$

15. $\displaystyle\lim_{t \to 0} \frac{10(\sin t - t)}{t^3} = \lim_{t \to 0} \frac{10(\cos t - 1)}{3t^2} = \lim_{t \to 0} \frac{10(-\sin t)}{6t} = \lim_{t \to 0} -\frac{10\cos t}{6} = -\frac{5}{3}$

17. $\displaystyle\lim_{x \to 0^+} \left(\frac{1}{x} - \frac{1}{\sqrt{x}}\right) = \lim_{x \to 0^+} \frac{1-\sqrt{x}}{x} = -\lim_{x \to 0^+} \frac{\sqrt{x}-1}{x} = -\lim_{x \to 0^+} \frac{(1/2)x^{-1/2}}{1} = -\frac{1}{2}\lim_{x \to 0^+} \frac{1}{\sqrt{x}} = \infty$

19. $\displaystyle\lim_{x \to -\pi/2^+} (\sec x + \tan x) = \lim_{x \to -\pi/2^+} \frac{1+\sin x}{\cos x} = \lim_{x \to -\pi/2^+} \frac{\cos x}{-\sin x} = \frac{0}{1} = 0$

21. $\displaystyle\lim_{x \to \infty} \frac{\sqrt{9x+1}}{\sqrt{x+1}} = \sqrt{\lim_{x \to \infty} \frac{9x+1}{x+1}} = \sqrt{9} = 3$

23. a) $\displaystyle\lim_{\theta \to \pi/2^-} (r-y) = \lim_{\theta \to \pi/2^-} \left(r - r\sin\theta\right) = \lim_{\theta \to \pi/2^-} \frac{1-\sin\theta}{\cos\theta} = \lim_{\theta \to \pi/2^-} \frac{-\cos\theta}{-\sin\theta} = 0$

    b) $\displaystyle\lim_{\theta \to \pi/2^-} \left(r^2-y^2\right) = \lim_{\theta \to \pi/2^-} \left(r^2 - r^2\sin^2\theta\right) = \lim_{\theta \to \pi/2^-} \frac{1-\sin^2\theta}{\cos^2\theta} = \lim_{\theta \to \pi/2^-} \frac{\cos^2\theta}{\cos^2\theta} = 1$

    c) $\displaystyle\lim_{\theta \to \pi/2^-} \left(r^3-y^3\right) = \lim_{\theta \to \pi/2^-} \left(r^3 - r^3\sin^3\theta\right) = \lim_{\theta \to \pi/2^-} \frac{1-\sin^3\theta}{\cos^3\theta} = \lim_{\theta \to \pi/2^-} \frac{-3\sin^2\theta\cos\theta}{3\cos^2\theta} =$

    $\displaystyle -\lim_{\theta \to \pi/2^-} \frac{\sin^2\theta}{\cos\theta} = -\lim_{\theta \to \pi/2^-} \frac{1-\cos^2\theta}{\cos\theta} = \lim_{\theta \to \pi/2^-} \left(\cos\theta - \sec\theta\right) = \infty$

25. $\displaystyle\lim_{x \to 0} \frac{f(x)}{g(x)} = \lim_{x \to 0} \frac{x+2}{x+1} = 2$, but $\displaystyle\lim_{x \to 0} \frac{f'(t)}{g'(t)} = 1$ is not a contradiction to L'Hopital's Rule since f and

g are not differentiable at $x = 0$.

27. $f'(c)$ must equal $\displaystyle\lim_{x \to 0} \frac{2 \tan x \sec^2 x}{\frac{8x}{\pi} \cos\left(\frac{4x^2}{\pi}\right)} = \frac{\pi}{4} \lim_{x \to 0} \frac{\sec^4 x + 2 \tan^2 x \sec^2 x}{\cos\left(\frac{4x^2}{\pi}\right) - \frac{8x^2}{\pi} \sin\left(\frac{4x^2}{\pi}\right)} = 1$ and $f'(c)$ must equal

$\displaystyle\frac{f(\pi/4) - f(0)}{\pi/4 - 0} = \frac{\sqrt{2} - c}{\pi/4}$ $\therefore \frac{\sqrt{2} - c}{\pi/4} = 1 \Rightarrow c = \frac{4\sqrt{2} - \pi}{4} \approx 0.6288$

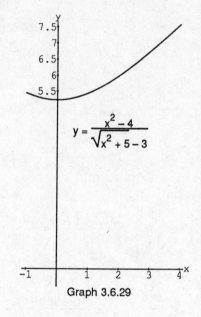

$y = \dfrac{x^2 - 4}{\sqrt{x^2 + 5} - 3}$

29. $\displaystyle\lim_{x \to 2} \frac{x^2 - 4}{\sqrt{x^2 + 5} - 3} = \lim_{x \to 2} \frac{2x}{(1/2)\left(x^2 + 5\right)^{-1/2}(2x)} =$

$2 \displaystyle\lim_{x \to 2} \sqrt{x^2 + 5} = 6$

Graph 3.6.29

3.7 APPROXIMATION ERRORS, QUADRATIC APPROXIMATIONS, AND THE MEAN VALUE THEOREM

1. $\sin x \approx x \Rightarrow x \sin x \approx x^2 \therefore Q(x) = x^2$

3. $\cos x \approx 1 - \dfrac{x^2}{2}$ and $\sqrt{1 + x} \approx 0$ at $x \approx -1 \Rightarrow \cos\left(\sqrt{1 + x}\right) \approx 1 - \dfrac{1 + x}{2} = \dfrac{1}{2} - \dfrac{x}{2}$ $\therefore Q(x) = \dfrac{1}{2} - \dfrac{x}{2}$

5. $\sec x = (\cos x)^{-1} \approx \dfrac{2}{2 - x^2} \approx 1 + \dfrac{x^2}{2} \Rightarrow \dfrac{\sec x}{1 - x} \approx \left(1 + \dfrac{x^2}{2}\right)\left(1 + x + x^2\right) = 1 + x + x^2 + \dfrac{x^2}{2} + \dfrac{x^3}{2} + \dfrac{x^4}{2} \approx$

$1 + x + \dfrac{3 x^2}{2}$ $\therefore Q(x) = 1 + x + \dfrac{3 x^2}{2}$

7. $\sqrt{1 + x} \approx 1 + \dfrac{x}{2} \Rightarrow f''(x) = -\dfrac{1}{4}(1 + x)^{-3/2}$ which is maximum when $x = -0.1$; $\therefore \left|e_1(x)\right| \le$

$\dfrac{1}{2}\left(\dfrac{1}{4(0.9)^{3/2}}\right)(0.1)^2 < 0.00146$

9.   a) $f(x) = \sqrt{1+x}$, $f(0) = 1$; $f'(x) = \dfrac{1}{2\sqrt{1+x}}$, $f'(0) = \dfrac{1}{2}$; $f''(x) = -\dfrac{1}{4\sqrt{(1+x)^3}}$ and $f''(0) = -\dfrac{1}{4}$ $\Rightarrow$

$\sqrt{1+x} \approx 1 + \dfrac{x}{2} - \dfrac{x^2}{8}$ $\therefore Q(x) = 1 + \dfrac{x}{2} - \dfrac{x^2}{8}$

    b) since $f'''(x) = \dfrac{3}{8\sqrt{(1+x)^5}}$ we have $\left|e_2(x)\right| \le \dfrac{1}{6}\left(\dfrac{3}{8}\right)(1-0.1)^{-5/2}(0.1)^3 \approx 0.0000813$

11.   a) $f(x) = \sec x$, $f(0) = 1$; $f'(x) = \sec x \tan x$, $f'(0) = 0$; $f''(x) = \sec x \tan^2 x + \sec^3 x$, $f''(0) = 1$; $\Rightarrow$

$\sec \approx 1 + \dfrac{x^2}{2}$. $\therefore Q(x) = 1 + \dfrac{x^2}{2}$

    b) If $x = \dfrac{\pi}{6}$ (does not require a calculator) and $f'''(x) = \sec x \tan^3 x + 2\sec^3 x \tan x + 3\sec^3 x \tan x$

which is bounded by $f'''\left(\dfrac{\pi}{6}\right) = \dfrac{14}{3}$ on $[-\dfrac{\pi}{6},\dfrac{\pi}{6}]$, then $\left|e_2(x)\right| \le \dfrac{1}{6}\left(\dfrac{14}{3}\right)(0.1)^3 \approx 0.00078$

13.   a) $f(x) = \sin x$, $f(\pi/2) = 1$; $f'(x) = \cos x$, $f'(\pi/2) = 0$; $f''(x) = -\sin x$, $f''(\pi/2) = -1 \Rightarrow \sin x \approx$

$1 - \dfrac{1}{2}(x - \pi/2)^2$ when $a = \pi/2$. $\therefore Q(x) = 1 - \dfrac{1}{2}(x - \pi/2)^2$

    b) $\left|e_2(x)\right| \le \dfrac{1}{6}(0.1)^3 \approx 0.000167$

15.   a) $f(x) = \cos x$, $f(\pi/2) = 0$; $f'(x) = -\sin x$, $f'(\pi/2) = -1$; $f''(x) = -\cos x$, $f''(\pi/2) = 0$; $f'''(x) = \sin x \Rightarrow$

$\cos x \approx -1(x - \pi/2) = \pi/2 - x$ $\therefore Q(x) = \pi/2 - x$

    b) $\left|e_2(x)\right| \le \dfrac{1}{6}(1)(0.1)^3 \approx 0.000167$

17. $\left|e_2(x)\right| \le \max\left\{\dfrac{f'''(x)}{6}\right\} x^3 = \dfrac{x^3}{6}$

    a) $\dfrac{x^3}{6} \le 0.01 \Rightarrow |x| \le \left(\dfrac{6}{100}\right)^{1/3} \approx 0.391$

    b) $\dfrac{x^3}{6} \le 0.01|x| \Rightarrow |x| \le \left(\dfrac{6}{100}\right)^{1/2} \approx 0.245$

19.   a) $f(x) = 2x + 3$, $f(0) = 3$; $f'(x) = 2$, $f'(0) = 2$; $f''(x) = 0 \Rightarrow Q(x) = 3 + 2x$

    b) $f(x) = mx + b$, $f(0) = b$; $f'(x) = m$, $f'(0) = m$; $f''(x) = 0 \Rightarrow Q(x) = b + mx$

21. $f(x) = (1 + x)^k$, $f(0) = 1$; $f'(x) = k(1 + x)^{k-1}$, $f'(0) = k$; $f''(x) = k(k-1)(1+x)^{k-2}$, $f''(0) = k(k-1) \Rightarrow$
$Q(x) = 1 + kx + \dfrac{k(k-1)}{2}x^2$

23.   a) If $f(x)$ is twice differentiable and at $x = a$ there is a point of inflection, then $f''(a) = 0$.

    $\therefore L(x) = Q(x) = f(a) + f'(a)(x - a)$.

3.M MISCELLANEOUS EXERCISES

1. From the diagram we have $\triangle OXP \cong$
$\triangle OPA$ and $\triangle PYO \cong \triangle BPO \Rightarrow$

$\overline{PA} = \dfrac{ar}{b}$ and $\overline{BP} = \dfrac{br}{a}$; $a^2 + b^2 =$

$\overline{BA} = \left(\overline{BP} + \overline{PA} \right)^2 = \left(\dfrac{br}{a} + \dfrac{ar}{b} \right)^2 \Rightarrow$

$\dfrac{1}{a^2} + \dfrac{1}{b^2} = \dfrac{1}{r^2} \Rightarrow a \bigg|_{b} = 2r = \dfrac{2r}{\sqrt{3}}$ and

$\dfrac{da}{dt} = -\dfrac{a^3}{b^3}\dfrac{db}{dt} = \dfrac{r}{10\sqrt{3}}$ m/sec

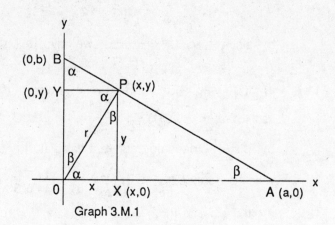

Graph 3.M.1

3. If $V = kr^4$, then $\dfrac{dV/dt}{V} = \dfrac{4kr^3}{kr^4}\dfrac{dr}{dt} = \dfrac{4}{r}\dfrac{dr}{dt}$. $\dfrac{dV/dt}{V} = 4\left(\dfrac{1}{100}\right) = 0.04$ at $\dfrac{dr/dt}{r} = \dfrac{1}{100}$. $\therefore$ the blood

flow will increase 4%/min.

5. a) $f(x) = x^4 + 2x^2 - 2 \Rightarrow f'(x) = 4x^3 + 4x$. Since $f(0) = -2 < 0$, $f(1) = 1 > 0$ and $f'(x) \geq 0$ for

$0 \leq x \leq 1$, we may conclude that $f(x)$ has exactly one solution when $0 \leq x \leq 1$.

b) 0.8555996772

7. a) $y = \dfrac{x}{x + 1} \Rightarrow y' = \dfrac{1}{(x + 1)^2} > 0$, for all x in the domain of $\dfrac{x}{x + 1}$. $\therefore$ $y = \dfrac{x}{x + 1}$ is increasing.

b) $y = x^3 + 2x \Rightarrow y' = 3x^2 + 2 > 0$ for all x. $\therefore$ the graph of $y = x^3 + 2x$ is always increasing and can

not have a maximum nor minimum.

9. If f is continuous on [a,c) and $f'(x) \leq 0$ on (a,c), then for all $x \in [a,c)$ we have $\dfrac{f(c) - f(x)}{c - x} \leq 0 \Rightarrow$

$f(c) - f(x) \leq 0 \Rightarrow f(x) \geq f(c)$. Also if f is continuous on (c,b] and $f'(x) \geq 0$ on (c,b], then for all

$x \in (c,b]$ we have $\dfrac{f(x) - f(c)}{x - c} \geq 0 \Rightarrow f(x) - f(c) \geq 0 \Rightarrow f(x) \geq f(c)$. $\therefore$ $f(x) \geq f(c)$ for all $x \in [a,b]$.

11.

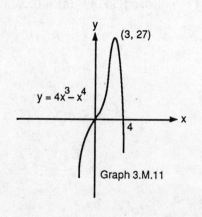

Graph 3.M.11

13.

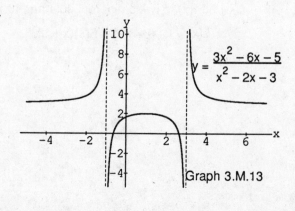

Graph 3.M.13

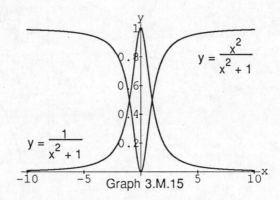

Graph 3.M.15

15.

17. The sign pattern indicates a local maximum at x = 1 and a local minimum at x = 3.

19. If $f(x) = -\left|x^2 - 2x - 3\right| = -\left(\left(x^2 - 2x - 3\right)^2\right)^{1/2}$, then
$f'(x) = -\dfrac{2(x-3)(x+1)(x-1)}{|x-3|\,|x+1|}$. $\lim\limits_{x \to -1^-} f'(x) = 4$,
$\lim\limits_{x \to -1^+} f'(x) = -4$, $\lim\limits_{x \to 1^-} f'(x) = 0$, $\lim\limits_{x \to 1^+} f'(x) = 0$,
$\lim\limits_{x \to 3^-} f'(x) = 4$ and $\lim\limits_{x \to 3^+} f'(x) = -4$

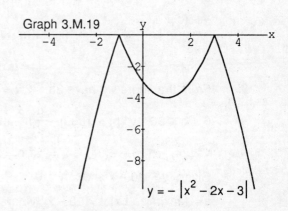

Graph 3.M.19

$y = -\left|x^2 - 2x - 3\right|$

21. a) The graph does not indicate any local extremum.

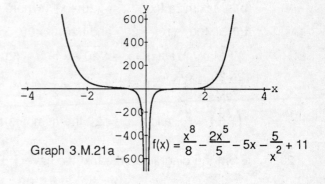

Graph 3.M.21a

$f(x) = \dfrac{x^8}{8} - \dfrac{2x^5}{5} - 5x - \dfrac{5}{x^2} + 11$

b) $f'(x) = x^7 - 2x^4 - 5 + \dfrac{10}{x^3} = x^{-3}(x^3 - 2)(x^7 - 5)$.

The sign pattern indicates a local maximum at $x = \sqrt[7]{5}$ and a local minimum at $x = \sqrt[3]{2}$.

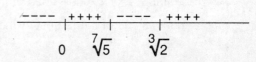

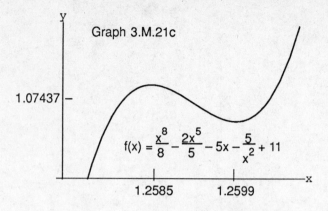

Graph 3.M.21c

c)

23.   a)   If $f'' \leq 0$, $f'(a) = 0$ and a is contained in the interval I, then $f(x) + \dfrac{-f''(c_2)}{2}(x-a)^2 = f(a)$ where

$\dfrac{-f''(c_2)}{2}(x-a)^2 \geq 0 \Rightarrow f(a) \geq f(x)$ for all $x \, \varepsilon \, I \Rightarrow$ f has a local maximum at $x = a$.

b)   If $f'' \geq 0$, $f'(a) = 0$ and a is contained in the interval I, then $f(x) + \dfrac{-f''(c_2)}{2}(x-a)^2 = f(a)$ where

$\dfrac{-f''(c_2)}{2}(x-a)^2 \leq 0 \Rightarrow f(x) \geq f(a)$ for all $x \, \varepsilon \, I \Rightarrow$ f has a local minimum at $x = a$.

25.   From the figure we have $3h + 2w = 108$ and $V = h^2 w$. $V(h) = h^2 \left(\dfrac{108 - 3h}{2} \right) = 54h^2 - \dfrac{3}{2}h^3$, where

$0 \leq h \leq 36$. $V'(h) = 108\,h - \dfrac{9}{2}h^2$ and $V'(h) = 0 \Rightarrow \dfrac{9}{2}h\,[24 - h] = 0 \Rightarrow h = 0$, no box, or $h = 24$.

$V''(24) = 108 - 9(24) < 0 \Rightarrow$ at $h = 24$ is a maximum.

27.   Given $y = 20 - x$ and $x, y \geq 0$.

a)   Maximize $f(x) = x(20 - x)$, where $0 < x < 20$. $f'(x) = 0 \Rightarrow 2(10 - x) = 0 \Rightarrow x = 10$. $f''(10) \leq 0 \Rightarrow$
when each number is 10 the product is maximized.

b)   Maximize $f(x) = x^2 + (20 - x)^2$, where $0 \leq x \leq 20$. $f'(x) = 0 \Rightarrow 4(x - 10) = 0 \Rightarrow x = 10$.

$f(0) = 400$, $f(10) = 200$ and $f(20) = 400$. $\therefore$ the numbers are 0 and 20.

c)   Maximize $f(x) = x + \sqrt{20 - x} = x + (20 - x)^{1/2}$, where $0 \leq x \leq 20$. $f'(x) = 0 \Rightarrow$

$\dfrac{2\sqrt{20 - x} - 1}{2\sqrt{20 - x}} = 0 \Rightarrow \sqrt{20 - x} = \dfrac{1}{2} \Rightarrow x = \dfrac{79}{4}$. The critical numbers are $x = \dfrac{79}{4}$ and $x = 20$. Since

$f\left(\dfrac{79}{4} \right) = \dfrac{81}{4}$ and $f(20) = 20$ the numbers must be $\dfrac{79}{4}$ and $\dfrac{1}{4}$.

29.   From the diagram in the text we have $\left(\dfrac{h}{2} \right)^2 + r^2 = (\sqrt{3})^2 \Rightarrow r^2 = \dfrac{12 - h^2}{4}$. The volume of the cylinder

is $V = \pi r^2 h = \pi \left(\dfrac{12 - h^2}{4} \right) h = \dfrac{\pi}{4}(12h - h^3)$, where $0 \leq h \leq 2\sqrt{3}$. $V'(h) = \dfrac{3\pi}{4}(2 + h)(2 - h) \Rightarrow$ the critical

points -2 or 2, but -2 is not in the domain. At $h = 2$ there is a maximum for $V''(2) = -3\pi < 0$. The
dimensions of the largest cylinder are radius $= \sqrt{2}$ and height $= 2$.

31. From the diagram in the text we have the cost $c = (50000)x + (30000)(20 - y)$ and $x = \sqrt{y^2 + 144}$,

where $0 \le y \le 20$. $c(y) = 50000\sqrt{y^2 + 144} + 30000(20 - y) \Rightarrow c'(y) = \dfrac{10000\left(5y - 3\sqrt{y^2 + 144}\right)}{\sqrt{y^2 + 144}}$.

The critical points are -9 or 9, but -9 is not in the domain. $c(0) = \$1,200,000$, $c(9) = \$1,080,000$,

and $c(20) = \$1,166,190$. The mimimum cost occurs when $x = 15$ miles and $y = 9$ miles.

33. The profit $P = 2px + py = 2px + p\left[\dfrac{40 - 10x}{5 - x}\right]$, where p is the profit on grade B tires and $0 \le x \le 4$.

$P'(x) = \dfrac{2p}{(5 - x)^2}\left(x^2 - 10x + 20\right) \Rightarrow$ the critical points $(5 - \sqrt{5})$, 5 or $(5 + \sqrt{5})$, but only $(5 - \sqrt{5})$ is in

the domain. $P'(x) > 0$ for $0 < x < (5 - \sqrt{5})$ and $P'(x) < 0$ for $(5 - \sqrt{5}) < x < 4 \Rightarrow$ at $x = (5 - \sqrt{5})$ there

is a local maximum. $P(0) = 8p$, $P(5 - \sqrt{5}) = 4p(5 - \sqrt{5}) \approx 11p$, and $P(4) = 8p \Rightarrow$ at $x = (5 - \sqrt{5})$ there

is absolute maximum. The maximum occurs when $x = (5 - \sqrt{5})$ and $y = 2(5 - \sqrt{5})$ i.e. $x \approx 276$ tires

and $y \approx 553$ tires.

35. The length of the ladder is $d_1 + d_2 = 8 \sec \theta + 6 \csc \theta$. We wish

to maximize $l(\theta) = 8 \sec \theta + 6 \csc \theta \Rightarrow l'(\theta) = 8 \sec \theta \tan \theta -$

$6 \csc \theta \cot \theta$. $l'(\theta) = 0 \Rightarrow 8 \sin^3 \theta - 6 \cos^3 \theta = 0 \Rightarrow \tan\theta = \dfrac{\sqrt[3]{6}}{2} \Rightarrow$

$d_1 = 4\sqrt{4 + \sqrt[3]{36}}$ and $d_2 = \sqrt[3]{36}\sqrt{4 + \sqrt[3]{36}} \Rightarrow$ the length of

the ladder is about $\left(4 + \sqrt[3]{36}\right)\sqrt{4 + \sqrt[3]{36}} = \left(4 + \sqrt[3]{36}\right)^{3/2} \approx 20$ ft.

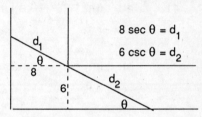

Graph 3.M 35

37. The area of the $\triangle ABC$, $A(x) = \dfrac{1}{2}(2)\sqrt{1 - x^2} = \left(1 - x^2\right)^{1/2}$,

where $0 \le x \le 1$. $A'(x) = \dfrac{-x}{\sqrt{1 - x^2}} \Rightarrow 0$ and ± 1 are critical

numbers. $A(\pm 1) = 0$, while $A(0) = 1$ the maximum. When

$x = 0$ the $\triangle ABC$ is isosceles.

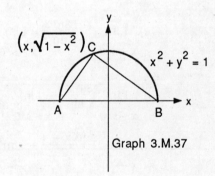

Graph 3.M.37

39.   a) By completing the square we have $f(x) = a\left(x - \dfrac{b}{a}\right)^2 + \dfrac{ac - b^2}{a} \geq 0$. If $a > 0$ and $f(x) \geq 0$, then

$\dfrac{ac - b^2}{a} \geq 0 \Rightarrow ac - b^2 > 0 \Rightarrow ac > b^2$. If $ac > b^2$ and $a > 0$, then $\dfrac{ac - b^2}{a} > 0 \Rightarrow f(x) \geq 0$. $\therefore$ the

equivalence follows.

   b) If $f(x) = \left(a_1 x + b_1\right)^2 + \ldots + \left(a_n x + b_n\right)^2$, then let $g(x) = Ax^2 + 2Bx + C$, where $A = \displaystyle\sum_{i=1}^{n} a_i^2$,

$B = \displaystyle\sum_{i=1}^{n} a_i b_i$ and $C = \displaystyle\sum_{i=1}^{n} b_i^2$. Part a) $\Rightarrow$ $B^2 \leq AC$ or $\left(\displaystyle\sum_{i=1}^{n} a_i b_i\right)^2 \leq \left(\displaystyle\sum_{i=1}^{n} a_i^2\right)\left(\displaystyle\sum_{i=1}^{n} b_i^2\right)$.

   c) $B^2 = AC$ only if $g(x_o) = A\left(x_o - \dfrac{B}{A}\right)^2 + \dfrac{AC - B^2}{A} = 0$, from part b) $\therefore f(x_o) = 0 \Rightarrow$ that each

$a_i x_o + b_i = 0 \Rightarrow a_i x_o = -b_i$ for $i = 1, 2, \ldots, n$.

41.   a) $\displaystyle\lim_{x \to 0} \frac{2 \sin 5x}{3x} = \lim_{x \to 0} \frac{10 \cos 5x}{3} = \frac{10}{3}$

   b) $\displaystyle\lim_{x \to 0} \sin 5x \cot 3x = \lim_{x \to 0} \frac{\sin 5x}{\tan 3x} = \lim_{x \to 0} \frac{5 \cos 5x}{3 \sec^2 3x} = \frac{5}{3}$

   c) $\displaystyle\lim_{x \to 0^+} x \csc^2 \sqrt{2x} = \lim_{x \to 0^+} \frac{x}{\sin^2 \sqrt{2x}} = \lim_{x \to 0^+} \frac{\sqrt{2x}}{2 \sin\sqrt{2x} \cos\sqrt{2x}} = \lim_{x \to 0^+} \frac{1}{2 \cos\left(2\sqrt{2x}\right)} = \frac{1}{2}$

   d) $\displaystyle\lim_{x \to \pi/2} \sec x - \tan x = \lim_{x \to \pi/2} \frac{1 - \sin x}{\cos x} = \lim_{x \to \pi/2} \frac{-\cos x}{-\sin x} = 0$

   e) $\displaystyle\lim_{x \to 0} \frac{x - \sin x}{x - \tan x} = \lim_{x \to 0} \frac{1 - \cos x}{1 - \sec^2 x} = \lim_{x \to 0} \frac{\sin x}{-2 \sec^2 x \tan x} = \lim_{x \to 0} \frac{\cos x}{-4 \sec^2 x \tan x - 2 \sec^4 x} = -\frac{1}{2}$

   f) $\displaystyle\lim_{x \to 0} \frac{\sin x^2}{x \sin x} = \lim_{x \to 0} \frac{2x \cos x^2}{\sin x + x \cos x} = \lim_{x \to 0} \frac{2 \cos x^2 - 4x^2 \sin x^2}{-x \sin x + \cos x + \cos x} = 1$

   g) $\displaystyle\lim_{x \to 0} \frac{\sec x - 1}{x^2} = \lim_{x \to 0} \frac{\sec x \tan x}{2x} = \lim_{x \to 0} \frac{\sec^3 x + \sec x \tan^2 x}{2} = \frac{1}{2}$

   h) $\displaystyle\lim_{x \to 2} \frac{x^3 - 8}{x^2 - 4} = \lim_{x \to 2} \frac{3x^2}{2x} = 3$

   i) $\displaystyle\lim_{x \to \pi/2} \frac{1 - \sin x}{1 + \cos x} = \lim_{x \to \pi/2} \frac{-\cos x}{-2 \sin 2x} = \lim_{x \to \pi/2} \frac{1}{4 \sin x} = \frac{1}{4}$

   j) $\displaystyle\lim_{x \to \pi/3} \frac{\cos x - 0.5}{x - \pi/3} = \lim_{x \to \pi/3} \frac{-\sin x}{1} = -\frac{\sqrt{3}}{2}$

43. a) $F(a) = F(b) = 0$ since $F(a) = f(a) - f(a) - (a - a)f'(a) - \ldots - k(a - a)^{n+1} = 0$ and

$F(b) = 0$ by the definition of k.

b) $F^{(k)}(a) = f^{(k)}(a) - f^{(k)}(a) - f^{(k+1)}(a - a) - \ldots - k(n + 1)(n)(n - 1) \ldots (n + 1 - k)(a - a)^{n+1-k} = 0$,

$k = 1, 2, 3, \ldots, n$.

c) Since, $F(a) = F(b) = 0$, Rolle's Theorem there exists a $c_1 \varepsilon (a,b)$ and $F'(c_1) = 0$. Since $F'(a) =$

$F'(c_1) = 0$, there exists a c_2 such that $c_2 \varepsilon (a,c_1)$ and $F''(c_2) = 0$. Also by Rolle's Theorem we have

$a < c_{n+1} < c_n < c_{n-1} < \ldots < c_1 < b$ and $F''(c_1) = F''(c_2) = \ldots F^{(n+1)}(c_{n+1}) = 0$.

d) Since $F^{(n+1)}(c_{n+1}) = 0$ and $F^{(n+1)}(x) = f^{(n+1)}(x) - k(n + 1)!$ we have $f^{(n+1)}(c_{n+1}) - k(n + 1)! = 0 \Rightarrow$

$k = \dfrac{f^{(n+1)}(c_{n+1})}{(n + 1)!}$.

CHAPTER 4

INTEGRATION

SECTION 4.1 INDEFINITE INTEGRALS

1. a) $3x^2 + C$ b) $-2x + C$ c) $3x^2 - 2x + C$

3. a) $x^{-3} + C$ b) $-\frac{1}{3}x^{-3} + C$ c) $-\frac{1}{3}x^{-3} + x^2 + 3x + C$

5. a) $x^{3/2} + C$ b) $\frac{8}{3}x^{3/2} + C$ c) $\frac{x^3}{3} - \frac{8}{3}x^{3/2} + C$

7. a) $x^{2/3} + C$ b) $x^{1/3} + C$ c) $x^{-1/3} + C$

9. a) $\frac{\cos 3x}{3} + C$ b) $-3\cos(x) + C$ c) $-3\cos x + \frac{\cos 3x}{3} + C$

11. a) $\tan(x) + C$ b) $\tan(5x) + C$ c) $\frac{\tan 5x}{5} + C$

13. a) $\sec(x) + C$ b) $\sec(2x) + C$ c) $2\sec(2x) + C$

15. $(\sin x - \cos x)^2 = \sin^2 x - 2\sin x \cos x + \cos^2 x = \sin^2 x + \cos^2 x - \sin 2x = 1 - \sin 2x.$ The antiderivative is
$x + \frac{\cos 2x}{2} + C.$

17. $\displaystyle\int 7\,dx = 7x + C$ 19. $\displaystyle\int 3\sqrt{x}\,dx = 2x^{3/2} + C$ 21. $\displaystyle\int t^{-1/3}\,dt = \frac{3}{2}t^{2/3} + C$

23. $\displaystyle\int (5v^2 + 2v)\,dv = \frac{5v^3}{3} + v^2 + C$ 25. $\displaystyle\int (2s^3 - 5s + 7)\,ds = \frac{1}{2}s^4 - \frac{5}{2}s^2 + 7s + C$

27. $\displaystyle\int (x-3)(4x^2 - 7)\,dx = \int (4x^3 - 12x^2 - 7x + 21)\,dx = x^4 - 4x^3 - \frac{7x^2}{2} + 21x + C$

29. $\displaystyle\int \frac{t^3 - 1}{t-1}\,dt = \int \frac{(t-1)(t^2 + t + 1)}{t-1}\,dt = \int (t^2 + t + 1)\,dt = \frac{t^3}{3} + \frac{t^2}{2} + t + C$

31. $\displaystyle\int (x + 2\cos x)\,dx = \frac{x^2}{2} + 2\sin x + C$

33. $\int \sin\frac{t}{3}\,dt = -3\cos\frac{t}{3} + C$ 35. $\int 3\csc^2 x\,dx = -3\cot x + C$ 37. $\int \frac{\csc x \cot x}{2}\,dx = -\frac{1}{2}\csc x + C$

39. $\int \frac{\sec^3\theta + \tan\theta}{\sec\theta}\,d\theta = \int \left(\sec^2\theta + \frac{\tan\theta}{\sec\theta}\right)d\theta = \int (\sec^2\theta + \sin\theta)\,d\theta = \tan\theta - \cos\theta + C$

41. $\int (\sin 2y - \csc^2 y)\,dy =$

$-\frac{1}{2}\cos 2y + \cot y + C$

43. $\int 4\sin^2 v\,dv = \int 4\left(\frac{1}{2} - \frac{1}{2}\cos 2v\right)dv =$

$\int (2 - 2\cos 2v)\,dv = 2v - \sin 2v + C$

45. $\int \sin x \cos x\,dx = \int \frac{1}{2}\sin 2x\,dx =$

$-\frac{1}{4}\cos 2x + C$

47. $\int (1 + \tan^2\theta)\,d\theta = \int \sec^2\theta\,d\theta =$

$\tan\theta + C$

49. $\int \cos\theta(\tan\theta + \sec\theta)\,d\theta = \int (\sin\theta + 1)\,d\theta = -\cos\theta + \theta + C$

51. $\frac{d}{dx}\left(\frac{(7x-2)^4}{28} + C\right) = \frac{4(7x-2)^3(7)}{28} = (7x-2)^3$ 53. $\frac{d}{dx}\left(\frac{-1}{x+1} + C\right) = (-1)(-1)(x+1)^{-2} = \frac{1}{(x+1)^2}$

55. a) $\frac{d}{dx}\left(\frac{x^2}{2}\sin x + C\right) = \frac{2x}{2}\sin x + \frac{x^2}{2}\cos x =$ b) $\frac{d}{dx}(-x\cos x + C) = -\cos x + x\sin x$, Wrong

$x\sin x + \frac{x^2}{2}\cos x$, Wrong

c) $\frac{d}{dx}(-x\cos x + \sin x + C) = -\cos x + x\sin x + \cos x = x\sin x$, Right

57. a) $\int f(x)\,dx = 1 - \sqrt{x} + C = -\sqrt{x} + C'$ b) $\int g(x)\,dx = x + 2 + C = x + C'$

c) $\int -f(x)\,dx = -(1 - \sqrt{x}) + C = \sqrt{x} + C'$ d) $\int -g(x)\,dx = -(x + 2) + C = -x + C'$

e) $\int (f(x) + g(x))\,dx = (1 - \sqrt{x}) + (x + 2) + C =$ f) $\int (f(x) - g(x))\,dx = (1 - \sqrt{x}) - (x + 2) + C =$

$x - \sqrt{x} + C'$ $-x - \sqrt{x} + C'$

g) $\int (x + f(x))\,dx = \frac{x^2}{2} + 1 - \sqrt{x} + C =$ h) $\int (g(x) - 4)\,dx = x + 2 - 4x + C =$

$\frac{x^2}{2} - \sqrt{x} + C'$ $-3x + C'$

59. a)

$f(x) = \sin x \sin(x + 2) - \sin^2(x + 1)$

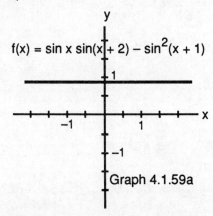

Graph 4.1.59a

b)

$f'(x) = \cos x \sin(x + 2) + \cos(x + 2) \sin x - 2 \sin(x + 1) \cos(x + 1)$

$= \sin(x + (x + 2)) - \sin 2(x + 1) = \sin(2x + 2) - \sin(2x + 2) = 0$

Since $f'(x) = 0$, f must be a constant function $\Rightarrow f(x) = c$ for all x.

Let $x = 0 \Rightarrow f(0) = \cos 0 \sin(0 + 2) + \cos(0 + 2) \sin 0 = \sin 2$

$\therefore f(x) = \sin 2$.

SECTION 4.2 INITIAL VALUE PROBLEMS AND MATHEMATICAL MODELING

1. $y = \int (2x - 7)\, dx = x^2 - 7x + C.$ $y(2) = 0 \Rightarrow 0 = 2^2 - 7(2) + C \Rightarrow C = 10.$ $\therefore y = x^2 - 7x + 10.$

3. $y = \int \left(\dfrac{1}{x^2} + x\right) dx = \int (x^{-2} + x)\, dx = -x^{-1} + \dfrac{x^2}{2} + C.$ $y(2) = 1 \Rightarrow 2^{-1} + \dfrac{2^2}{2} + C = 1 \Rightarrow C = -\dfrac{1}{2}.$

   $\therefore y = -x^{-1} + \dfrac{x^2}{2} - \dfrac{1}{2}$ or $y = -\dfrac{1}{x} + \dfrac{x^2}{2} - \dfrac{1}{2}.$

5. $y = \int 3\sqrt{x}\, dx = \int 3x^{1/2}\, dx = 2x^{3/2} + C.$ $y(9) = 4 \Rightarrow 4 = 2(9)^{3/2} + C \Rightarrow 4 = 54 + C \Rightarrow -50 = C$

   $\therefore y = 2x^{3/2} - 50$

7. $s = \int (1 + \cos t)\, dt = t + \sin t + C.$ $s(0) = 4 \Rightarrow 0 + \sin 0 + C = 4 \Rightarrow C = 4.$ $\therefore s = t + \sin t + 4.$

9. $v = \int -\pi \sin \pi t\, dt = (\pi)\left(\dfrac{1}{\pi}\right) \cos \pi t + C = \cos \pi t + C;$ $v(0) = 0 \Rightarrow 0 = \cos \pi(0) + C \Rightarrow 0 = 1 + C \Rightarrow$

   $C = -1$ $\therefore v = \cos(\pi t) - 1$

11. $\dfrac{d^2 y}{dx^2} = 2 - 6x \Rightarrow \dfrac{dy}{dx} = \int (2 - 6x)\, dx = 2x - 3x^2 + C_1.$ $\dfrac{dy}{dx} = 4$ when $x = 0 \Rightarrow 2(0) - 3(0)^2 + C_1 = 4$

   $\Rightarrow C_1 = 4 \Rightarrow \dfrac{dy}{dx} = 2x - 3x^2 + 4.$ $y = \int \left(2x - 3x^2 + 4\right) dx = x^2 - x^3 + 4x + C_2.$ $y(0) = 1 \Rightarrow$

   $0^2 - 0^3 + 4(0) + C_2 = 1 \Rightarrow C_2 = 1.$ $\therefore y = x^2 - x^3 + 4x + 1.$

13. $\frac{dr}{dt} = \int \frac{2}{t^3}\,dt = \int 2t^{-3}\,dt = -t^{-2} + C_1 = -\frac{1}{t^2} + C_1;\quad \frac{dr}{dt}(1) = 1 \Rightarrow 1 = -\frac{1}{(1)^2} + C_1 \Rightarrow 1 = -1 + C_1 \Rightarrow$

$C_1 = 2 \quad \therefore \frac{dr}{dt} = -\frac{1}{t^2} + 2 \Rightarrow r = \int\left(-\frac{1}{t^2} + 2\right)dt = \int\left(-t^{-2} + 2\right)dt = t^{-1} + 2t + C_2;\quad r(1) = 1 \Rightarrow$

$1 = 1^{-1} + 2(1) + C_2 \Rightarrow C_2 = -2 \quad \therefore r = t^{-1} + 2t - 2 = \frac{1}{t} + 2t - 2$

15. $\frac{d^2y}{dx^2} = \int 6\,dx = 6x + C_1;\quad \frac{d^2y}{dx^2}(0) = -8 \Rightarrow -8 = 6(0) + C_1 \Rightarrow C_1 = -8 \quad \therefore \frac{d^2y}{dx^2} = 6x - 8$

$\frac{dy}{dx} = \int (6x - 8)\,dx = 3x^2 - 8x + C_2;\quad \frac{dy}{dx}(0) = 0 \Rightarrow 0 = 3(0)^2 - 8(0) + C_2 \Rightarrow C_2 = 0 \quad \therefore \frac{dy}{dx} = 3x^2 - 8x$

$y = \int\left(3x^2 - 8x\right)dx = x^3 - 4x^2 + C_3;\quad y(0) = 5 \Rightarrow 5 = 0^3 - 4(0)^2 + C_3 \Rightarrow C_3 = 5 \quad \therefore y = x^3 - 4x^2 + 5$

17. $v = s'(t) = 9.8t,\ s = 10$ when $t = 0 \Rightarrow s = \int (9.8\,t)\,dt = 4.9t^2 + C.\ s(0) = 10 \Rightarrow 4.98(0)^2 + C = 10 \Rightarrow$

$C = 10 \Rightarrow s = 4.9t^2 + 10.$

19. $a = s''(t) = 32,\ v = 20$ and $s = 0$ when $t = 0 \Rightarrow v = \int 32\,dt = 32t + C_1.\ v(0) = 20 \Rightarrow 32(0) + C_1 = 20$

$\Rightarrow C_1 = 20 \Rightarrow v = 32t + 20.\ s = \int (32t + 20)\,dt = 16t^2 + 20t + C_2.\ s(0) = 0 \Rightarrow 16(0)^2 + 20(0) + C_2 = 0$

$\Rightarrow C_2 = 0 \Rightarrow s = 16t^2 + 20t.$

21. $m = y' = 3\sqrt{x} = 3x^{1/2} \Rightarrow y = \int 3x^{1/2}\,dx = 2x^{3/2} + C.\ y(9) = 4 \Rightarrow 2(9)^{3/2} + C = 4 \Rightarrow C = -50.$

$\therefore y = 2x^{3/2} - 50$ or $y = 2\sqrt{x^3} - 50.$

23. $a(t) = v'(t) = 1.6 \Rightarrow v(t) = \int 1.6\,dt = 1.6t + C.\ v(0) = 0 \Rightarrow 1.6(0) + C = 0 \Rightarrow C = 0 \Rightarrow v(t) = 1.6t.$ When $t =$

30, then $v(30) = 48$ m/sec.

25. $a(t) = v'(t) = 9.8 \Rightarrow v(t) = \int 9.8\,dt = 9.8t + C_1.\ v(0) = 0 \Rightarrow 9.8(0) + C_1 = 0 \Rightarrow C_1 = 0 \Rightarrow v(t) = s'(t) = 9.8t \Rightarrow$

$s(t) = \int 9.8t\,dt = 4.9t^2 + C_2.\ s(0) = 0 \Rightarrow 4.9(0)^2 + C_2 = 0 \Rightarrow C_2 = 0 \Rightarrow s(t) = 4.9t^2.$

Solving $s(t) = 10 \Rightarrow t^2 = \frac{10}{4.9} \Rightarrow t = \sqrt{\frac{10}{4.9}}.\ v\left(\sqrt{\frac{10}{4.9}}\right) = 9.8\sqrt{\frac{10}{4.9}} = \frac{(2)(4.9)\sqrt{10}}{\sqrt{4.9}} = (2)\sqrt{4.9}\sqrt{10} = 14$ m/sec.

27. $\frac{dy}{dt} = -k\sqrt{y} \Rightarrow y^{-1/2}\,dy = -k\,dt \Rightarrow \int y^{-1/2}\,dy = \int -k\,dt \Rightarrow 2y^{1/2} + C_1 = -kt + C_2 \Rightarrow$

$2y^{1/2} = -kt + C_2 - C_1 = -kt + C \Rightarrow y^{1/2} = \frac{C - kt}{2} \Rightarrow y = \frac{(C - kt)^2}{4}.$

29. Step 1: $\dfrac{ds}{dt} = \displaystyle\int -k\,dt = -kt + C_1$; $\dfrac{ds}{dt}(0) = 88 \Rightarrow 88 = -k(0) + C_1 \Rightarrow C_1 = 88$ $\therefore \dfrac{ds}{dt} = -kt + 88$

$s = \displaystyle\int (-kt + 88)\,dt = \dfrac{-kt^2}{2} + 88t + C_2$; $s(0) = 0 \Rightarrow 0 = \dfrac{-k(0)^2}{2} + 88(0) + C_2 \Rightarrow C_2 = 0$ $\therefore s = \dfrac{-kt^2}{2} + 88t$

Step 2: If $\dfrac{ds}{dt} = 0$, then $0 = -kt + 88 \Rightarrow t = \dfrac{88}{k}$

Step 3: $242 = \dfrac{-k\left(\dfrac{88}{k}\right)^2}{2} + 88\left(\dfrac{88}{k}\right) \Rightarrow 242 = \dfrac{-k(88)^2}{2k^2} + \dfrac{(88)^2}{k} \Rightarrow 242 = \dfrac{(88)^2}{k}\left(-\dfrac{1}{2} + 1\right) \Rightarrow 242 = \dfrac{(88)^2}{2k}$

$\Rightarrow 484\,k = (88)^2 \Rightarrow k = 16$

31. $\dfrac{ds}{dt} = \displaystyle\int -5.2\,dt = -5.2t + C_1$ $\dfrac{ds}{dt}(0) = 0 \Rightarrow 0 = -5.2(0) + C_1 \Rightarrow C_1 = 0$ $\therefore \dfrac{ds}{dt} = -5.2t$

$s = \displaystyle\int -5.2t\,dt = -2.6t^2 + C_2$ $s(0) = 4 \Rightarrow 4 = -2.6(0)^2 + C_2 \Rightarrow C_2 = 4$ $\therefore s = -2.6t^2 + 4$

If $s = 0$, $0 = -2.6t^2 + 4 \Rightarrow t^2 = \dfrac{4}{2.6} \Rightarrow t = \pm\sqrt{\dfrac{4}{2.6}} \Rightarrow t = \sqrt{\dfrac{4}{2.6}} \approx 1.24$ sec since $t > 0$

33. a) Force = Mass times Acceleration (Newton's Second Law) or F = ma. Let $a = \dfrac{dv}{dt} = \dfrac{dv}{ds}\dfrac{ds}{dt} = v\dfrac{dv}{ds}$. Then

$ma = -mgR^2s^{-2} \Rightarrow a = -gR^2s^{-2} \Rightarrow v\dfrac{dv}{ds} = -gR^2s^{-2} \Rightarrow v\,dv = -gR^2s^{-2}\,ds \Rightarrow \displaystyle\int v\,dv = \displaystyle\int -gR^2s^{-2}\,ds$

$\Rightarrow \dfrac{v^2}{2} = \dfrac{gR^2}{s} + C \Rightarrow v^2 = \dfrac{2gR^2}{s} + 2C = \dfrac{2gR^2}{s} + C'$. When $t = 0$, $v = v_0$, $s = R \Rightarrow v_0{}^2 = \dfrac{2gR^2}{R} + C' \Rightarrow$

$C' = v_0{}^2 - 2gR$. $\therefore v^2 = \dfrac{2gR^2}{s} + v_0{}^2 - 2gR$.

b) If $v_0 = \sqrt{2gR}$, then $v^2 = \dfrac{2gR^2}{s} \Rightarrow v = \sqrt{\dfrac{2gR^2}{s}}$ since $v \geq 0$ if $v_0 \geq \sqrt{2gR}$. Then $\dfrac{ds}{dt} = \dfrac{\sqrt{2gR^2}}{\sqrt{s}} \Rightarrow$

$\sqrt{s}\,ds = \sqrt{2gR^2}\,dt \Rightarrow \displaystyle\int s^{1/2}\,ds = \displaystyle\int \sqrt{2gR^2}\,dt \Rightarrow \dfrac{2}{3}s^{3/2} = \sqrt{2gR^2}\,t + C \Rightarrow s^{3/2} = \dfrac{3}{2}\sqrt{2gR^2}\,t + C'$.

$t = 0 \Rightarrow s = R \Rightarrow R^{3/2} = \dfrac{3}{2}\sqrt{2gR^2}(0) + C' \Rightarrow C' = R^{3/2}$. Then $s^{3/2} = \dfrac{3}{2}\sqrt{2gR^2}\ t + R^{3/2}$

$= \dfrac{3}{2}R\sqrt{2g}\ t + R^{3/2} \Rightarrow s^{3/2} = R^{3/2}\left[\dfrac{3}{2}R^{-1/2}\sqrt{2g}\,t + 1\right] \Rightarrow s^{3/2} = R^{3/2}\left[\dfrac{3\sqrt{2gR}}{2R}t + 1\right] = R^{3/2}\left[\dfrac{3v_0}{2R}t + 1\right]$

$\Rightarrow s = R\left[1 + \dfrac{3v_0}{2R}t\right]^{2/3}$

SECTION 4.3 DEFINITE INTEGRALS

1. $\displaystyle\sum_{k=1}^{2}\frac{6k}{k+1}=\frac{6(1)}{1+1}+\frac{6(2)}{2+1}=7$

3. $\displaystyle\sum_{k=1}^{4}(-1)^k\cos k\pi=-\cos\pi+\cos 2\pi-$

   $\cos 3\pi+\cos 4\pi=4$

5. $\displaystyle\sum_{k=1}^{6}k$

7. $\displaystyle\sum_{k=1}^{4}\frac{1}{2^k}$

9. $\displaystyle\sum_{k=1}^{5}(-1)^{k+1}\frac{1}{k}$

11. a) $\displaystyle\sum_{k=1}^{6}2^{k-1}=2^{1-1}+2^{2-1}+2^{3-1}+2^{4-1}+2^{5-1}+2^{6-1}=1+2+4+8+16+32$

b) $\displaystyle\sum_{k=0}^{5}2^{k}=2^{0}+2^{1}+2^{2}+2^{3}+2^{4}+2^{5}=1+2+4+8+16+32$

c) $\displaystyle\sum_{k=-1}^{4}2^{k+1}=2^{-1+1}+2^{0+1}+2^{1+1}+2^{2+1}+2^{3+1}+2^{4+1}=1+2+4+8+16+32$

All of them represent $1+2+4+8+16+32$

13. a) $\displaystyle\sum_{k=1}^{n}3a_k=3\left(\sum_{k=1}^{n}a_k\right)=3(-5)=-15$

b) $\displaystyle\sum_{k=1}^{n}\frac{b_k}{6}=\frac{1}{6}\left(\sum_{k=1}^{n}b_k\right)=\frac{1}{6}(6)=1$

c) $\displaystyle\sum_{k=1}^{n}(a_k+b_k)=\sum_{k=1}^{n}a_k+\sum_{k=1}^{n}b_k=$

   $-5+6=1$

d) $\displaystyle\sum_{k=1}^{n}(a_k-b_k)=\sum_{k=1}^{n}a_k-\sum_{k=1}^{n}b_k$

   $=-5-6=-11$

e) $\displaystyle\sum_{k=1}^{n}(b_k-2a_k)=\sum_{k=1}^{n}b_k-\sum_{k=1}^{n}2a_k=\sum_{k=1}^{n}b_k-2\left(\sum_{k=1}^{n}a_k\right)=6-2(-5)=16$

15.

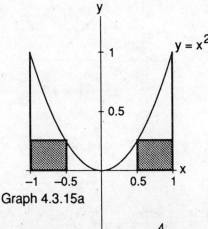

Graph 4.3.15a

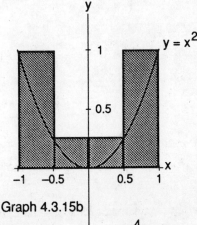

Graph 4.3.15b

Rectangles associated with $\displaystyle\sum_{k=1}^{4} \min_k \Delta x$

Rectangles associated with $\displaystyle\sum_{k=1}^{4} \max_k \Delta x$

$\Delta x = \dfrac{1}{2}$ $\qquad \displaystyle\sum_{k=1}^{4} \min_k \Delta x = \left(\dfrac{1}{4}\right)\left(\dfrac{1}{2}\right) +$

$\qquad (0)\left(\dfrac{1}{2}\right) + + 0\left(\dfrac{1}{2}\right) + \left(\dfrac{1}{4}\right)\left(\dfrac{1}{2}\right) = \dfrac{1}{4}$

$\Delta x = \dfrac{1}{2}$ $\qquad \displaystyle\sum_{k=1}^{4} \max_k \Delta x = (1)\left(\dfrac{1}{2}\right) + \left(\dfrac{1}{4}\right)\left(\dfrac{1}{2}\right) +$

$\qquad \left(\dfrac{1}{4}\right)\left(\dfrac{1}{2}\right) + (1)\left(\dfrac{1}{2}\right) = \dfrac{5}{4}$

17. $\displaystyle\int_{0}^{2} x^2\,dx$

19. $\displaystyle\int_{-7}^{5} (x^2 - 3x)\,dx$

21. $\displaystyle\int_{2}^{3} \dfrac{1}{1-x}\,dx$

23. $\displaystyle\int_{0}^{4} \cos x\,dx$

25. $\displaystyle\int_{-\pi/4}^{0} \sec x\,dx$

27. $\displaystyle\int_{0}^{5} \sqrt{25 - x^2}\,dx$

29. $-\displaystyle\int_{-2}^{0} (x^2 - 4)\,dx$ or $\displaystyle\int_{0}^{-2} (x^2 - 4)\,dx$

31. $\displaystyle\int_{1}^{\sqrt{2}} (x^2 - 1)\,dx - \displaystyle\int_{0}^{1} (x^2 - 1)\,dx$

33. $y = \sqrt{1 - x^2}$ is the upper half of a circle whose center is

$(0,0)$ and radius is 1. $\quad \therefore \quad \displaystyle\int_{-1}^{1} \sqrt{1 - x^2}\,dx = \dfrac{1}{2}\pi(1)^2 = \dfrac{1}{2}\pi$

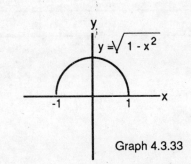

Graph 4.3.33

35. $y = 1 - |x| \Rightarrow y = \begin{cases} 1 + x, x < 0 \\ 1 - x, x \geq 0 \end{cases} \Rightarrow$ the region is a triangle with

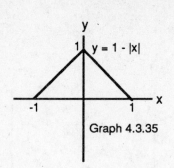

Graph 4.3.35

base = 2 and height = 1. $\therefore \displaystyle\int_{-1}^{1} (1 - |x|) \, dx = \frac{1}{2}(2)(1) = 1$

37. To find where $x - x^2 \geq 0$, let $x - x^2 = 0 \Rightarrow x(1 - x) = 0 \Rightarrow x = 0$ or $x = 1$. If $0 < x < 1$, then $x^2 < x \Rightarrow$
    $0 < x - x^2$. $\therefore x = 0$ and 1 maximize the integral.

39. a) $U = \max_1 \Delta x + \max_2 \Delta x + \cdots + \max_n \Delta x$ where $\max_1 = f(x_1), \max_2 = f(x_2), \cdots, \max_n = f(x_n)$ since f is
    increasing on [a,b]. $L = \min_1 \Delta x + \min_2 \Delta x + \cdots + \min_n \Delta x$ where $\min_1 = f(x_0), \min_2 = f(x_1), \cdots, \min_n =$
    $f(x_{n-1})$ since f is increasing on [a,b]. $\therefore U - L = (\max_1 - \min_1)\Delta x + (\max_2 - \min_2)\Delta x + \cdots +$
    $(\max_n - \min_n)\Delta x = (f(x_1) - f(x_0))\Delta x + (f(x_2) - f(x_1))\Delta x + \cdots + (f(x_n) - f(x_{n-1}))\Delta x = (f(x_n) - f(x_0))\Delta x =$
    $(f(b) - f(a))\Delta x$.

    b) $U = \max_1 \Delta x_1 + \max_2 \Delta x_2 + \cdots + \max_n \Delta x_n$ where $\max_1 = f(x_1), \max_2 = f(x_2), \cdots, \max_n = f(x_n)$ since f is
    increasing on [a,b]. $L = \min_1 \Delta x_1 + \min_2 \Delta x_2 + \cdots + \min_n \Delta x_n$ where $\min_1 = f(x_0), \min_2 = f(x_1), \cdots, \min_n =$
    $f(x_{n-1})$ since f is increasing on [a,b]. $\therefore U - L = (\max_1 - \min_1)\Delta x_1 + (\max_2 - \min_2)\Delta x_2 + \cdots +$
    $(\max_n - \min_n)\Delta x_n = (f(x_1) - f(x_0))\Delta x_1 + (f(x_2) - f(x_1))\Delta x_2 + \cdots + (f(x_n) - f(x_{n-1}))\Delta x_n \leq (f(x_1) - f(x_0))\Delta x_{max}$
    $+ (f(x_2) - f(x_1))\Delta x_{max} + \cdots + (f(x_n) - f(x_{n-1}))\Delta x_{max}$ where $\Delta x_{max} = \|P\|$. Then $U - L \leq (f(x_n) - f(x_0))\Delta x_{max}$
    $= (f(b) - f(a))\Delta x_{max} = |f(b) - f(a)| \Delta x_{max}$ since $f(b) \geq f(a)$. Thus $\displaystyle\lim_{\|P\| \to 0}(U - L) = \lim_{\|P\| \to 0}(f(b) - f(a)) \Delta x_{max}$
    $= 0$ since $\Delta x_{max} = \|P\|$.

41. $\displaystyle\int_0^1 (1 - x) \, dx$

n	upper endpoint sum	lower endpoint sum
4	0.625	0.375
10	0.55	0.45
20	0.525	0.475
50	0.51	0.49

43. $\displaystyle\int_{-\pi}^{\pi} \cos x \, dx$

n	upper endpoint sum	lower endpoint sum
4	3.14159551	−3.14157979
10	1.25664474	−1.25662538
20	0.628327454	−0.628307606
50	0.251337007	−0.251317017

45. $\displaystyle\int_{-1}^{1} |x|\, dx$

n	upper endpoint sum	lower endpoint sum
4	1.5	0.5
10	1.2	0.8
20	1.1	0.9
50	1.04	0.96

SECTION 4.4 EVALUATING DEFINITE INTEGRALS

1. a) $\displaystyle\int_{2}^{2} g(x)\, dx = 0$

   b) $\displaystyle\int_{5}^{1} g(x)\, dx = -\int_{1}^{5} g(x)\, dx = -8$

   c) $\displaystyle\int_{1}^{2} 3f(x)\, dx = 3\int_{1}^{2} f(x)\, dx = 3(-4)$

   $= -12$

   d) $\displaystyle\int_{2}^{5} f(x)\, dx = \int_{1}^{5} f(x)\, dx - \int_{1}^{2} f(x)\, dx =$

   $6 - (-4) = 10$

   e) $\displaystyle\int_{1}^{5} (f(x) - g(x))\, dx =$

   $\displaystyle\int_{1}^{5} f(x)\, dx - \int_{1}^{5} g(x)\, dx$

   $= 6 - 8 = -2$

   f) $\displaystyle\int_{1}^{5} (4f(x) - g(x))\, dx =$

   $4\displaystyle\int_{1}^{5} f(x)\, dx - \int_{1}^{5} g(x)\, dx =$

   $4(6) - 8 = 16$

3. a) $\displaystyle\int_{1}^{2} f(u)\, du = 5$

   b) $\displaystyle\int_{1}^{2} f(z)\, dz = 5$

   c) $\displaystyle\int_{2}^{1} f(t)\, dt = -5$

5. Since f is decreasing on [0,1]; the maximum of f occurs at 0 and is 1, the minimum of f occurs at 1 and is $\frac{1}{2}$.

   $\therefore$ the upper bound is $1(1 - 0) = 1$; the lower bound is $\frac{1}{2}(1 - 0) = \frac{1}{2}$.

7. $\displaystyle\int_{-2}^{2} 4.4\, dx = 4.4(2-(-2)) = 17.6$ 9. $\displaystyle\int_{3}^{1} 7\, dx = 7(1-3) = -14$ 11. $\displaystyle\int_{0}^{2} 5x\, dx = 5\left(\frac{2^2}{2}-\frac{0^2}{2}\right) = 10$

13. $\displaystyle\int_{0}^{2} (3-2x)\, dx = \int_{0}^{2} 3\, dx - 2\int_{0}^{2} x\, dx = 3(2-0) - 2\left(\frac{2^2}{2}-\frac{0^2}{2}\right) = 6-4 = 2$

15. $\displaystyle\int_{2}^{1} (1-x)\, dx = \int_{2}^{1} dx - \int_{2}^{1} x\, dx = (1-2) - \left(\frac{1^2}{2}-\frac{2^2}{2}\right) = \frac{1}{2}$

17. $\displaystyle\int_{1}^{2} x^2\, dx = \int_{0}^{2} x^2\, dx - \int_{0}^{1} x^2\, dx = \frac{8}{3} - \frac{1}{3} = \frac{7}{3}$

19. $\displaystyle\int_{0}^{1} (x-1)^2\, dx = \int_{0}^{1} (x^2-2x+1)\, dx = \int_{0}^{1} x^2\, dx - 2\int_{0}^{1} x\, dx + \int_{0}^{1} dx = \frac{1^3}{3} - 2\left(\frac{1^2}{2}-\frac{0^2}{2}\right) + (1-0) = \frac{1}{3}$

21. Area $= \displaystyle\int_{0}^{3} (x^2+1)\, dx = \int_{0}^{3} x^2\, dx + \int_{0}^{3} dx = \frac{3^3}{3} + (3-0) = 12$

23. Average value $= \displaystyle\frac{1}{1-(-1)}\int_{-1}^{1} 2|x|\, dx = \int_{-1}^{1} |x|\, dx = \int_{-1}^{0} -x\, dx + \int_{0}^{1} x\, dx = -\left(0-\frac{(-1)^2}{2}\right) + \left(\frac{1^2}{2}-0\right) = 1$

25. Average value $= \displaystyle\frac{1}{1-(-1)}\int_{-1}^{1} \sqrt{1-x^2}\, dx = \frac{1}{2}\left(\frac{\pi}{2}\right) = \frac{\pi}{4}$ since the integral is the area of the semicircle with

center $(0,0)$ and $r = 1 \Rightarrow$ Area $= \dfrac{1}{2}\pi(1)^2 = \dfrac{\pi}{2}$

27. Average value $= \displaystyle\frac{1}{2\sqrt{2}-0}\int_{0}^{2\sqrt{2}} (2x+1)\, dx = \frac{1}{2\sqrt{2}}\left(2\int_{0}^{2\sqrt{2}} x\, dx + \int_{0}^{2\sqrt{2}} dx\right) = \frac{1}{2\sqrt{2}}\left(2\left(\frac{(2\sqrt{2})^2}{2}-\frac{0^2}{2}\right) + \left(2\sqrt{2}-0\right)\right)$

$= 2\sqrt{2} + 1$

29. Since f is continuous on [1,2], there exists at least one $c \in (1,2)$ so that $f(c)(2-1) = \int_{1}^{2} f(x)\, dx \Rightarrow$

$f(c)(1) = 4 \Rightarrow f(c) = 4$.

31. Let $F(x) = f(x) - g(x)$. Then F is continuous on [a,b] since f and g are continuous on [a,b]. $\int_{a}^{b} F(x)\, dx =$

$\int_{a}^{b} (f(x) - g(x))\, dx = 0$. By the Mean Value Theorem, there exists a c in [a,b] so that $F(c) = \dfrac{1}{b-a} \int_{a}^{b} F(x)\, dx$

$= \dfrac{1}{b-a}(0) = 0$. $F(c) = f(c) - g(c) \Rightarrow 0 = f(c) - g(c) \Rightarrow f(c) = g(c)$.

33. $\displaystyle\int_{0}^{\sqrt{2}} 3x^3\, dx = 3 \int_{0}^{\sqrt{2}} x^3\, dx = 3\left(\dfrac{(\sqrt{2})^4}{4}\right) = 3$

35. $\displaystyle\int_{0}^{1/2} \left(4x^3 - \dfrac{3}{2}x^2 + 5x - 7\right) dx = 4 \int_{0}^{1/2} x^3\, dx - \dfrac{3}{2} \int_{0}^{1/2} x^2\, dx + 5 \int_{0}^{1/2} x\, dx - 7 \int_{0}^{1/2} dx = 4\left(\dfrac{(1/2)^4}{4}\right) - \dfrac{3}{2}\left(\dfrac{(1/2)^3}{3}\right) +$

$5\left(\dfrac{(1/2)^2}{2} - \dfrac{0^2}{2}\right) - 7\left(\dfrac{1}{2} - 0\right) = -\dfrac{23}{8}$

37. Partition [a,b] into n subintervals, each of length $\Delta x = \dfrac{b-a}{n}$ with the points $x_0 = a$, $x_1 = a + \Delta x$, $x_2 = a + 2\Delta x$,

$\cdots$, $x_{n-1} = a + (n-1)\Delta x$, $x_n = a + n\Delta x = b$. The circumscribed rectangles so defined have areas $f(x_1)\Delta x =$

$(a + \Delta x)\Delta x$, $f(x_2)\Delta x = (a + 2\Delta x)\Delta x$, $\cdots$, $f(x_n)\Delta x = (a + n\Delta x)\Delta x$. The sum of these areas is $S_n =$

$\left((a + \Delta x) + (a + 2\Delta x) + \cdots + (a + n\Delta x)\right)\Delta x = \left(na + (1 + 2 + \cdots + n)\Delta x\right)\Delta x = \left(na + \dfrac{n(n+1)}{2}\Delta x\right)\Delta x =$

$\left(a + \dfrac{n+1}{n}\Delta x\right)n\Delta x = \left(a + \dfrac{n+1}{2}\left(\dfrac{b-a}{n}\right)\right)(b-a)$ since $\Delta x = \dfrac{b-a}{n}$. Then $\displaystyle\int_{a}^{b} x\, dx = \lim_{n \to \infty} S_n =$

$\displaystyle\lim_{n \to \infty}\left(a + \dfrac{n+1}{2}\left(\dfrac{b-a}{n}\right)\right)(b-a) = \left(a + \dfrac{b-a}{2}\right)(b-a) = \left(\dfrac{b+a}{2}\right)(b-a) = \dfrac{b^2}{2} - \dfrac{a^2}{2}$

39. Partition [0,1] into n subintervals, each of length $\Delta x = \dfrac{1}{n}$ with the points $x_0 = 0$, $x_1 = \dfrac{1}{n}$, $x_2 = \dfrac{2}{n}$, $\cdots$, $x_n = \dfrac{n}{n} = 1$.

The inscribed rectangles so determined have areas $f(x_0)\Delta x = (0)\Delta x$, $f(x_1)\Delta x = \left(\dfrac{1}{n}\right)\Delta x$, $f(x_2)\Delta x = \left(\dfrac{2}{n}\right)\Delta x$, $\cdots$,

$f(x_{n-1}) = \left(\dfrac{n-1}{n}\right)\Delta x$. The sum of these areas is $S_n = \left(0 + \dfrac{1}{n} + \dfrac{2}{n} + \cdots + \dfrac{n-1}{n}\right)\Delta x =$

39. (Continued)

$\left(\dfrac{1}{n} + \dfrac{2}{n} + \cdots + \dfrac{n-1}{n}\right)\dfrac{1}{n}$. Then $\displaystyle\int_0^1 x \, dx = \lim_{n \to \infty} S_n = \lim_{n \to \infty} \left(\dfrac{1}{n} + \dfrac{2}{n} + \cdots + \dfrac{n-1}{n}\right)\dfrac{1}{n} = \dfrac{1}{2}$ since $\displaystyle\int_0^1 x \, dx =$

$\dfrac{1^2}{2} - \dfrac{0^2}{2} = \dfrac{1}{2}$.

41. a) Partition $\left[0, \dfrac{\pi}{2}\right]$ into n subintervals, each of length $\Delta x = \dfrac{\pi}{2n}$ with points $x_0 = 0$, $x_1 = \Delta x$, $x_2 = 2\Delta x$, $\cdots$,

$x_n = n\Delta x = \dfrac{\pi}{2}$. Since sin x is increasing on $\left[0, \dfrac{\pi}{2}\right]$, the upper sum, U, is the sum of the areas of the

circumscribed rectangles of areas $f(x_1)\Delta x = (\sin \Delta x)\Delta x$, $f(x_2)\, \Delta x = (\sin 2\Delta x)\Delta x$, $\cdots$, $f(x_n)\Delta x = (\sin n\Delta x)\Delta x$.

Then $U = (\sin \Delta x + \sin 2\Delta x + \cdots + \sin n\Delta x)\Delta x = \left[\dfrac{\cos \dfrac{\Delta x}{2} - \cos\left(n + \dfrac{1}{2}\right)\Delta x}{2 \sin \dfrac{\Delta x}{2}}\right]\Delta x =$

$\left[\dfrac{\cos \dfrac{\pi}{4n} - \cos\left(n + \dfrac{1}{2}\right)\dfrac{\pi}{2n}}{2 \sin \dfrac{\pi}{4n}}\right]\left(\dfrac{\pi}{2n}\right) = \dfrac{\pi\left(\cos \dfrac{\pi}{4n} - \cos\left(\dfrac{\pi}{2} + \dfrac{\pi}{4n}\right)\right)}{4n \sin \dfrac{\pi}{4n}} = \dfrac{\cos \dfrac{\pi}{4n} - \cos\left(\dfrac{\pi}{2} + \dfrac{\pi}{4n}\right)}{\dfrac{\sin \dfrac{\pi}{4n}}{\dfrac{\pi}{4n}}}$

b) Then the area is $\displaystyle\int_0^{\pi/2} \sin x \, dx = \lim_{n \to \infty} \left(\dfrac{\cos \dfrac{\pi}{4n} - \cos\left(\dfrac{\pi}{2} + \dfrac{\pi}{4n}\right)}{\dfrac{\sin \dfrac{\pi}{4n}}{\dfrac{\pi}{4n}}}\right) = \dfrac{1 - \cos \dfrac{\pi}{2}}{1} = 1$.

SECTION 4.5   THE FUNDAMENTAL THEOREM OF CALCULUS

1. $\displaystyle\int_1^2 (2x + 5) \, dx = \left[x^2 + 5x\right]_1^2 =$

   $[4 + 10] - [1 + 5] = 8$

3. $\displaystyle\int_0^3 (4 - x^3) \, dx = \left[4x - \dfrac{x^4}{4}\right]_0^3 =$

   $\left(4(3) - \dfrac{3^4}{4}\right) - (0) = -\dfrac{33}{4}$

5. $\displaystyle\int_0^1 (x^2 + \sqrt{x}) \, dx = \left[\dfrac{x^3}{3} + \dfrac{2}{3}x^{3/2}\right]_0^1 =$

   $\left(\dfrac{1}{3} + \dfrac{2}{3}\right) - (0) = 1$

7. $\displaystyle\int_1^{32} x^{-6/5} \, dx = \left[-5x^{-1/5}\right]_1^{32} =$

   $\left(-\dfrac{5}{2}\right) - (-5) = \dfrac{5}{2}$

9. $\displaystyle\int_{0}^{\pi} \sin x \, dx = [-\cos x]_{0}^{\pi} =$

$(-\cos \pi) - (-\cos 0) = -(-1) - (-1) = 2$

11. $\displaystyle\int_{0}^{\pi/3} 2\sec^2 x \, dx = [2\tan x]_{0}^{\pi/3} =$

$(2\tan (\pi/3)) - (2\tan 0) = 2\sqrt{3} - 0 = 2\sqrt{3}$

13. $\displaystyle\int_{\pi/4}^{3\pi/4} \csc \theta \cot \theta \, d\theta = \left[-\csc \theta\right]_{\pi/4}^{3\pi/4} = (-\csc (3\pi/4)) - (-\csc (\pi/4)) = -\sqrt{2} - (-\sqrt{2}) = 0$

15. $\displaystyle\int_{\pi/2}^{0} 4\sin^2 t \, dt = \int_{\pi/2}^{0} 4\left(\frac{1-\cos 2t}{2}\right) dt = \int_{\pi/2}^{0} (2 - \cos 2t) \, dt = \left[2t - \frac{\sin 2t}{2}\right]_{\pi/2}^{0} = \left(2(0) - \frac{\sin 2(0)}{2}\right) -$

$\left(2(\pi/2) - \frac{\sin 2(\pi/2)}{2}\right) = -\pi$

17. $\displaystyle\int_{-\pi/2}^{\pi/2} (8y^2 + \sin y) \, dy = \left[\frac{8y^3}{3} - \cos y\right]_{-\pi/2}^{\pi/2} = \left(\frac{8(\pi/2)^3}{3} - \cos \frac{\pi}{2}\right) - \left(\frac{8(-\pi/2)^3}{3} - \cos \left(-\frac{\pi}{2}\right)\right) = \frac{2\pi^3}{3}$

19. $\displaystyle\int_{-1}^{1} (r + 1)^2 \, dr = \int_{-1}^{1} (r^2 + 2r + 1) \, dr = \left[\frac{r^3}{3} + r^2 + r\right]_{-1}^{1} = \left(\frac{1}{3} + 1 + 1\right) - \left(\frac{-1}{3} + 1 - 1\right) = \frac{8}{3}$

21. $\displaystyle\int_{-1}^{0} \left(\frac{u^7}{2} - u^{15}\right) du = \left[\frac{u^8}{16} - \frac{u^{16}}{16}\right]_{-1}^{0} = (0) - \left(\frac{(-1)^8}{16} - \frac{(-1)^{16}}{16}\right) = 0$

23. $\displaystyle\int_{1}^{\sqrt{2}} \frac{s^2 + \sqrt{s}}{s^2} \, ds = \int_{1}^{\sqrt{2}} \left(1 + s^{-3/2}\right) ds = \left[s - 2s^{-1/2}\right]_{1}^{\sqrt{2}} = \left(\sqrt{2} - 2\left(\sqrt{2}\right)^{-1/2}\right) - \left(1 - 2(1)^{-1/2}\right) = \sqrt{2} - \sqrt[4]{8} + 1$

25. $\displaystyle\int_{-4}^{4} |x| \, dx = \int_{-4}^{0} -x \, dx + \int_{0}^{4} x \, dx = \left[-\frac{x^2}{2}\right]_{-4}^{0} + \left[\frac{x^2}{2}\right]_{0}^{4} = (0) - \left(-\frac{(-4)^2}{2}\right) + \left(\frac{4^2}{2} - (0)\right) = 16$

27.

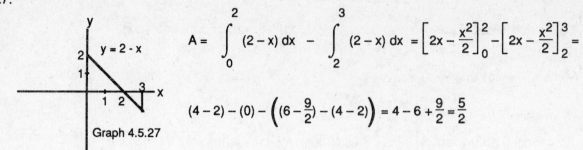

$$A = \int_0^2 (2-x)\,dx \; - \; \int_2^3 (2-x)\,dx = \left[2x - \frac{x^2}{2}\right]_0^2 - \left[2x - \frac{x^2}{2}\right]_2^3 =$$

$$(4-2) - (0) - \left(\left(6 - \frac{9}{2}\right) - (4-2)\right) = 4 - 6 + \frac{9}{2} = \frac{5}{2}$$

Graph 4.5.27

29.

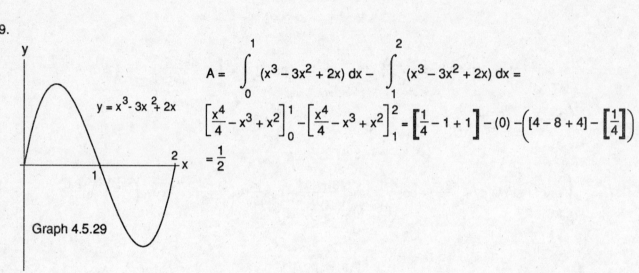

$$A = \int_0^1 (x^3 - 3x^2 + 2x)\,dx - \int_1^2 (x^3 - 3x^2 + 2x)\,dx =$$

$$\left[\frac{x^4}{4} - x^3 + x^2\right]_0^1 - \left[\frac{x^4}{4} - x^3 + x^2\right]_1^2 = \left[\frac{1}{4} - 1 + 1\right] - (0) - \left([4 - 8 + 4] - \left[\frac{1}{4}\right]\right)$$

$$= \frac{1}{2}$$

Graph 4.5.29

31. $A = \int_0^1 x^2\,dx + \int_1^2 (2-x)\,dx = \left[\frac{x^3}{3}\right]_0^1 + \left[2x - \frac{x^2}{2}\right]_1^2 = \left(\frac{1}{3}\right) - (0) + (4-2) - \left(2 - \frac{1}{2}\right) = \frac{5}{6}$

33. $A = \int_0^\pi (2 - (1 + \cos x))\,dx = \int_0^\pi (1 - \cos x)\,dx = [x - \sin x]_0^\pi = (\pi - \sin \pi) - (0) = \pi$

35. $F(x) = \frac{hx^2}{2b} \Rightarrow F(b) - F(0) = \frac{h(b)^2}{2b} - \frac{h(0)^2}{2b} = \frac{1}{2}bh$

37. $\frac{dy}{dx} = \sqrt{1 + x^2}$

39. $\frac{dy}{dx} = \sin\left(\sqrt{x}\right)^2 \left(\frac{d}{dx}\left(\sqrt{x}\right)\right) = \frac{1}{2}x^{-1/2} \sin x$

41. $\frac{dy}{dx} = \frac{d}{dx} \int_x^4 \sqrt{t^4 + 7}\,dt = -\sqrt{x^4 + 7}$

43. $\dfrac{dy}{dx} = \dfrac{d}{dx} \displaystyle\int_{\sin x}^{0} \dfrac{dt}{\sqrt{1-t^2}} = \dfrac{-1}{\sqrt{1-\sin^2 x}}(\cos x) = \dfrac{-1}{\sqrt{\cos^2 x}}(\cos x) = \dfrac{-\cos x}{|\cos x|} = \dfrac{-\cos x}{\cos x} = -1$ since $|x| < \dfrac{\pi}{2}$.

45. d, since $y' = \dfrac{1}{x}$ and $y(\pi) = \displaystyle\int_{\pi}^{\pi} \dfrac{1}{t}\,dt - 3 = -3$ 47. b, since $y' = \sec x$ and $y(0) = \displaystyle\int_{0}^{0} \sec t\,dt + 4 = 4$

49. $\dfrac{dy}{dx} = \sec x \Rightarrow y = \displaystyle\int_{2}^{x} \sec t\,dt + C.$ $y = 3$ when $x = 2 \Rightarrow 3 = \displaystyle\int_{2}^{2} \sec t\,dt + C \Rightarrow 3 = C.$

$\therefore y = \displaystyle\int_{2}^{x} \sec t\,dt + 3.$

51.

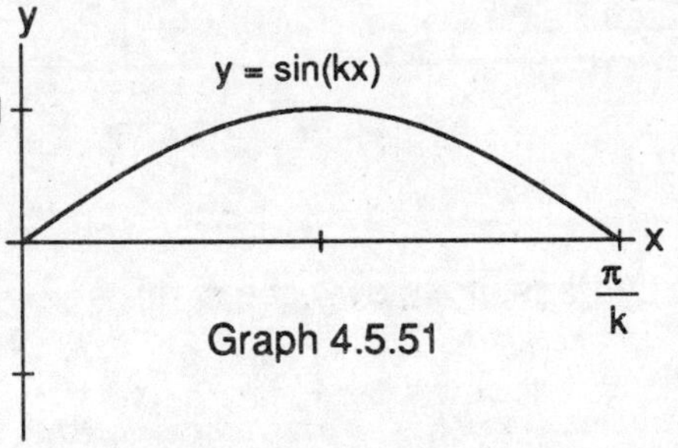

Graph 4.5.51

$A = \displaystyle\int_{0}^{\pi/k} \sin kx\,dx = \left[-\dfrac{1}{k}\cos kx \right]_{0}^{\pi/k} = \left[-\dfrac{1}{k}\cos\left(k\left(\dfrac{\pi}{k}\right)\right) \right] -$

$\left[-\dfrac{1}{k}\cos\left(k(0)\right) \right] = \left(-\dfrac{1}{k}\cos \pi \right) - \left(-\dfrac{1}{k}\cos 0 \right) = \dfrac{2}{k}$

53. a) Average value $= \dfrac{1}{(1/60) - 0} \displaystyle\int_{0}^{1/60} V_{max} \sin 120\pi t\,dt = 60\left[\dfrac{-V_{max}\cos 120\pi t}{120\pi} \right]_{0}^{1/60} =$

$60\left[\dfrac{-V_{max}\cos 2\pi}{120\pi} - \left(\dfrac{-V_{max}\cos 0}{120\pi} \right) \right] = 0$

b) $V_{rms} = \dfrac{V_{max}}{\sqrt{2}} \Rightarrow V_{max} = \sqrt{2}\,V_{rms} \Rightarrow V_{max} = \sqrt{2}(240 \text{ volts rms}) = 360 \text{ volts (rounded)}.$

c) $V_{ar}^2 = \dfrac{1}{(1/60) - 0} \displaystyle\int_{0}^{1/60} \left(V_{max}\right)^2 \sin^2 120\pi t\,dt = 60\left(V_{max}\right)^2 \displaystyle\int_{0}^{1/60} \dfrac{1 + \cos 240\pi t}{2}\,dt =$

$30\left(V_{max}\right)^2 \displaystyle\int_{0}^{1/60} (1 + \cos 240\pi t)\,dt = 30\left(V_{max}\right)^2 \left[t + \dfrac{\sin 240\pi t}{240\pi} \right]_{0}^{1/60} = 30\left(V_{max}\right)^2 \left[\dfrac{1}{60} + \dfrac{\sin 4\pi}{240\pi} - (0) \right] =$

$\dfrac{\left(V_{max}\right)^2}{2}$

55. $\dfrac{d}{dx}\displaystyle\int_{1}^{x} f(t)\ dt = f(x) \Rightarrow f(x) = \dfrac{d}{dx}\left(x^2 - 2x + 1\right) = 2x - 2$

57. $\displaystyle\lim_{x \to 0}\frac{1}{x^3}\int_{0}^{x}\frac{t^2}{t^4 + 1}\ dt = \lim_{x \to 0}\frac{\displaystyle\int_{0}^{x}\frac{t^2}{t^4 + 1}\ dt}{x^3} = \lim_{x \to 0}\frac{\dfrac{d}{dx}\left(\displaystyle\int_{0}^{x}\frac{t^2}{t^4 + 1}\ dt\right)}{\dfrac{d}{dx}\left(x^3\right)} = \lim_{x \to 0}\frac{\dfrac{x^2}{x^4 + 1}}{3x^2} = \lim_{x \to 0}\frac{1}{3(x^4 + 1)} = \frac{1}{3}$

59. $g(x) = \displaystyle\int_{0}^{x} f(t)\ dt \Rightarrow g'(x) = f(x) \Rightarrow g'(1) = f(1) = 0.$ Since $f'(x) > 0$ for all x, $g''(x) > 0$ for all x. Since $f'(x)$

   exists for all x, f is continuous for all $x \Rightarrow g'$ is continuous for all $x \Rightarrow g$ is continuous for all x.

   a) True, since g is continuous for all x   b) True, see above.

   c) True, since $g'(1) = 0$ d) False, $g''(x) > 0$

   e) True, $g''(x) > 0$ for all x   f) False, see Part e

   g) True, since $g'(1) = 0$

61. $F(x) = \displaystyle\int_{0}^{x} t \cos \pi t\ dt = \frac{x}{\pi}\sin \pi x + \frac{1}{\pi^2}\cos \pi x$ (using Integration by Parts: $u = t, dv = \cos \pi t\ dt$)

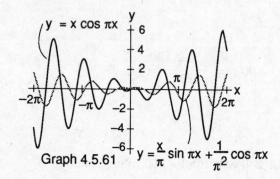

Graph 4.5.61 $y = \dfrac{x}{\pi}\sin \pi x + \dfrac{1}{\pi^2}\cos \pi x$

63. $F(x) = \displaystyle\int_{0}^{x} (t^3 - 4t^2 + 3t)\ dt = \frac{x^4}{4} - \frac{4}{3}x^3 + \frac{3}{2}x^2$ 65. The limit is $3x^2$

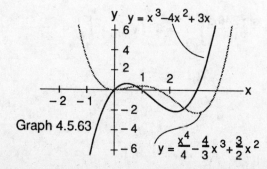

Graph 4.5.63

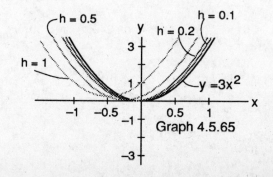

Graph 4.5.65

SECTION 4.6 INTEGRATION BY SUBSTITUTION

1. $\int x\sin(2x^2)\,dx = \int \frac{1}{4}\sin u\,du = -\frac{1}{4}\cos u + C = -\frac{1}{4}\cos 2x^2 + C$

$u = 2x^2 \Rightarrow du = 4x\,dx \Rightarrow \frac{1}{4}du = x\,dx$

3. $\int \left(1-\cos\frac{t}{2}\right)^2\sin\frac{t}{2}\,dt = \int 2u^2\,du = = \frac{2}{3}u^3 + C = \frac{2}{3}\left(1-\cos\frac{t}{2}\right)^3 + C$

$u = 1-\cos\frac{t}{2} \Rightarrow du = \frac{1}{2}\sin\frac{t}{2}\,dt \Rightarrow \frac{1}{2}du = dx$

5. $\int \frac{9r^2\,dr}{\sqrt{1-r^3}} = \int -3u^{-1/2}\,du = -3(2)u^{1/2} + C = -6\left(1-r^3\right)^{1/2} + C$

$u = 1-r^3 \Rightarrow du = -3r^2\,dr \Rightarrow -3\,du = 9r^2\,dr$

7. a) $\int \csc^2 2\theta\cot 2\theta\,d\theta = -\int \frac{1}{2}u\,du$ b) $\int \csc^2 2\theta\cot 2\theta\,d\theta = \int -\frac{1}{2}u\,du = -\frac{1}{2}\left(\frac{u^2}{2}\right) + C$

$u = \cot 2\theta$ $= -\frac{1}{2}\left(\frac{u^2}{2}\right) + C$ $u = \csc 2\theta$ $= -\frac{u^2}{4} + C$

$du = -2\csc^2 2\theta\,d\theta$ $= -\frac{u^2}{4} + C$ $du = -2\csc 2\theta\cot 2\theta\,d\theta$ $= -\frac{\csc^2 2\theta}{4} + C$

$-\frac{1}{2}du = \csc^2 2\theta\,d\theta$ $= -\frac{\cot^2 2\theta}{4} + C$ $-\frac{1}{2}du = \csc 2\theta\cot 2\theta\,d\theta$

9. $\int_0^{\pi/6} \frac{\sin 2x}{\cos^2 2x}\,dx = \frac{1}{2}\int_1^{1/2} -u^{-2}\,du = \frac{1}{2}\left[\frac{1}{u}\right]_1^{1/2} = \frac{1}{2}\left[\frac{1}{\frac{1}{2}}\right] - \frac{1}{2}\left[\frac{1}{1}\right] = \frac{1}{2}$

$u = \cos 2x \Rightarrow du = -2\sin 2x\,dx \Rightarrow -\frac{1}{2}du = \sin 2x\,dx$ and $x = 0 \Rightarrow u = 1, x = \frac{\pi}{6} \Rightarrow u = \frac{1}{2}$

11. $\int_{-\pi/2}^{\pi/2} \frac{\cos t}{(2+\sin t)^2}\,dt = \int_1^3 u^{-2}\,du = \left[-\frac{1}{u}\right]_1^3 = \left[-\frac{1}{3}\right] - \left[-\frac{1}{1}\right] = \frac{2}{3}$

$u = 2+\sin t \Rightarrow du = \cos t\,dt$ and $t = -\frac{\pi}{2} \Rightarrow u = 1, t = \frac{\pi}{2} \Rightarrow u = 3$

13. $\displaystyle\int \frac{dx}{(1-x)^2} = \int -u^{-2}\, du = \frac{1}{u} + C$

$u = 1 - x$ $\qquad\qquad = \dfrac{1}{1-x} + C$

$du = -dx$

$-du = dx$

15. $\displaystyle\int \sec^2(x+2)\, dx = \int \sec^2 u\, du = \tan u + C$

$u = x + 2$ $\qquad\qquad\qquad = \tan(x+2) + C$

$du = dx$

17. $\displaystyle\int 8r(r^2 - 1)^{1/3}\, dr = \int 4u^{1/3}\, du$

$u = r^2 - 1$ $\qquad\qquad = 4\left(\dfrac{3}{4}\right)u^{4/3} + C$

$du = 2r\, dr$ $\qquad\qquad = 3u^{4/3} + C$

$4\, du = 8r\, dr$ $\qquad\qquad = 3(r^2 - 1)^{4/3} + C$

19. $\displaystyle\int \sec\left[\theta + \frac{\pi}{2}\right]\tan\left[\theta + \frac{\pi}{2}\right] d\theta =$

$\qquad\displaystyle\int \sec u \tan u\, du = \sec u + C$

$\qquad = \sec\left[\theta + \dfrac{\pi}{2}\right] + C$

$u = \theta + \dfrac{\pi}{2} \Rightarrow du = d\theta$

21. $\displaystyle\int \frac{6x^3}{\sqrt[4]{1+x^4}}\, dx = \int \frac{3}{2}\, u^{-1/4}\, du = \frac{3}{2}\left(\frac{4}{3}\right)u^{3/4} + C = 2u^{3/4} + C = 2(1+x^4)^{3/4} + C$

$u = 1 + x^4 \Rightarrow du = 4x^3\, dx \Rightarrow \dfrac{3}{2}\, du = 6x^3\, dx$

23. a) $\displaystyle\int_0^3 \sqrt{y+1}\, dy = \int_1^4 u^{1/2}\, du = \left[\frac{2}{3}u^{3/2}\right]_1^4$

$u = y + 1$ $\qquad = \left[\dfrac{2}{3}(4)^{3/2}\right] - \left[\dfrac{2}{3}(1)^{3/2}\right]$

$du = dy$ $\qquad = \left(\dfrac{2}{3}(8)\right) - \left(\dfrac{2}{3}(1)\right) = \dfrac{14}{3}$

$x = 0 \Rightarrow u = 1,\ x = 3 \Rightarrow u = 4$

b) $\displaystyle\int_{-1}^1 \sqrt{y+1}\, dy = \int_0^1 u^{1/2}\, du = \left[\frac{2}{3}u^{3/2}\right]_0^1$

$\qquad = \left[\left(\dfrac{2}{3}(1)^{3/2}\right) - (0)\right] = \dfrac{2}{3}$

Use same substitution for u as in part a)

$x = -1 \Rightarrow u = 0,\ x = 0 \Rightarrow u = 1$

25. a) $\displaystyle\int_0^{\pi/4} \tan x \sec^2 x\, dx = \int_0^1 u\, du = \left[\frac{u^2}{2}\right]_0^1$

$u = \tan x$ $\qquad = \dfrac{1^2}{2} - 0 = \dfrac{1}{2}$

$du = \sec^2 x\, dx$

$x = 0 \Rightarrow u = 0,\ x = \dfrac{\pi}{4} \Rightarrow u = 1$

b) $\displaystyle\int_{-\pi/4}^0 \tan x \sec^2 x\, dx = \int_{-1}^0 u\, du = \left[\frac{u^2}{2}\right]_{-1}^0$

Use same substitution as in part a) $\qquad = (0) - \left(\dfrac{1}{2}\right)$

$x = -\dfrac{\pi}{4} \Rightarrow u = -1,\ x = 0 \Rightarrow u = 0$ $\qquad = -\dfrac{1}{2}$

27. a) $\displaystyle\int_0^1 \frac{x^3}{\sqrt{x^4+9}}\,dx = \int_9^{10} \frac{1}{4}u^{-1/2}\,du$

b) $\displaystyle\int_{-1}^0 \frac{x^3}{\sqrt{x^4+9}}\,dx = \int_{10}^9 \frac{1}{4}u^{-1/2}\,du =$

$u = x^4 + 9$ $\qquad = \left[\dfrac{1}{4}(2)u^{1/2}\right]_9^{10}$

Use same substitution as in part a)

$du = 4x^3\,dx$ $\qquad = \left(\dfrac{1}{2}(10)^{1/2}\right) - \left(\dfrac{1}{2}(9)^{1/2}\right)$

$x = -1 \Rightarrow u = 10,\ x = 0 \Rightarrow u = 9$

$\dfrac{1}{4}\,du = x^3\,dx$ $\qquad = \dfrac{\sqrt{10}-3}{2}$

$= -\displaystyle\int_9^{10} \frac{1}{4}u^{-1/2}\,du = \dfrac{3-\sqrt{10}}{2}$

$x = 0 \Rightarrow u = 9,\ x = 1 \Rightarrow u = 10$

29. a) $\displaystyle\int_0^{\sqrt{7}} x\left(x^2+1\right)^{1/3}\,dx = \int_1^8 \frac{1}{2}u^{1/3}\,du$

b) $\displaystyle\int_{-\sqrt{7}}^0 x(x^2+1)^{1/3}\,dx = \int_8^1 \frac{1}{2}u^{1/3}\,du$

$u = x^2 + 1$ $\qquad = \left[\dfrac{1}{2}\left(\dfrac{3}{4}\right)u^{4/3}\right]_1^8$

Use same substitution as in part a)

$du = 2x\,dx$ $\qquad = \left(\dfrac{3}{8}(8)^{4/3}\right) - \left(\dfrac{3}{8}(1)^{4/3}\right)$

$x = -\sqrt{7} \Rightarrow u = 8,\ x = 0 \Rightarrow u = 1$

$\dfrac{1}{2}\,du = x\,dx$ $\qquad = \dfrac{45}{8}$

$= -\displaystyle\int_1^8 \frac{1}{2}u^{1/3}\,du = -\dfrac{45}{8}$

$x = 0 \Rightarrow u = 1,\ x = \sqrt{7} \Rightarrow u = 8$

31. a) $\displaystyle\int_0^{\pi/6} (1-\cos 3x)\sin 3x\,dx = \int_0^1 \frac{1}{3}u\,du$

b) $\displaystyle\int_{\pi/6}^{\pi/3} (1-\cos 3x)\sin 3x\,dx = \int_1^2 \frac{1}{3}u\,du$

$u = 1 - \cos 3x$ $\qquad = \left[\dfrac{1}{3}\left(\dfrac{u^2}{2}\right)\right]_0^1$

Use same substitution as in part a)

$du = 3\sin 3x\,dx$ $\qquad = \left[\dfrac{1}{6}(1)^2\right] - \left[\dfrac{1}{6}(0)^2\right]$

$x = \dfrac{\pi}{6} \Rightarrow u = 1,\ x = \dfrac{\pi}{3} \Rightarrow u = 2$

$\dfrac{1}{3}\,du = \sin 3x\,dx$ $\qquad = \dfrac{1}{6}$

$= \left[\dfrac{1}{3}\left(\dfrac{u^2}{2}\right)\right]_1^2 = \left(\dfrac{1}{6}(2)^2\right) - \left(\dfrac{1}{6}(1)^2\right) = \dfrac{1}{2}$

$x = 0 \Rightarrow u = 0$

$x = \dfrac{\pi}{6} \Rightarrow u = 1$

33. a) $\displaystyle\int_0^{2\pi} \frac{\cos x}{\sqrt{2+\sin x}}\,dx = \int_2^2 u^{-1/2}\,du = 0$

b) $\displaystyle\int_{-\pi}^{\pi} \frac{\cos x}{\sqrt{2+\sin x}}\,dx = \int_2^2 u^{-1/2}\,du = 0$

$u = 2 + \sin x$

Use same substitution as in part a)

$du = \cos x\,dx$

$x = -\pi \Rightarrow u = 2$

$x = 0 \Rightarrow u = 2,\ x = 2\pi \Rightarrow u = 2$

$x = \pi \Rightarrow u = 2$

35. $\displaystyle\int_0^1 \sqrt{t^5 + 2t}\;(5t^4 + 2)\;dt = \int_0^3 u^{1/2}\;du = \left[\frac{2}{3}u^{3/2}\right]_0^3 = \left(\frac{2}{3}(3)^{3/2}\right) - \left(\frac{2}{3}(0)^{3/2}\right) = 2\sqrt{3}$

   $u = t^5 + 2t \Rightarrow du = (5t^4 + 2)dt$

   $t = 0 \Rightarrow u = 0,\; t = 1 \Rightarrow u = 3$

37. $\displaystyle\int_0^{\pi/2} \cos^3 2x \sin 2x\;dx = \int_1^{-1} -\frac{1}{2}u^3\;du = \left[-\frac{1}{2}\left(\frac{u^4}{4}\right)\right]_1^{-1} = \left(-\frac{(-1)^4}{8}\right) - \left(-\frac{1^4}{8}\right) = 0$

   $u = \cos 2x \Rightarrow du = -2\sin 2x\;dx \Rightarrow -\frac{1}{2}\;du = \sin 2x\;dx$

   $x = 0 \Rightarrow u = 1,\; x = \dfrac{\pi}{2} \Rightarrow u = -1$

39. $\displaystyle\int_0^{\pi} \frac{8\sin t}{\sqrt{5 - 4\cos t}}\;dt = \int_1^9 2u^{-1/2}\;du = \left[2(2u^{1/2})\right]_1^9 = \left(4(9)^{1/2}\right) - \left(4(1)^{1/2}\right) = 8$

   $u = 5 - 4\cos t \Rightarrow du = 4\sin t\;dt \Rightarrow 2du = 8\sin t\;dt$

   $t = 0 \Rightarrow u = 1,\; t = \pi \Rightarrow u = 9$

41. $\displaystyle\int_0^1 15x^2\sqrt{5x^3 + 4}\;dx = \int_4^9 u^{1/2}\;du = \left[\frac{2}{3}u^{3/2}\right]_4^9 = \left(\frac{2}{3}(9)^{3/2}\right) - \left(\frac{2}{3}(4)^{3/2}\right) = \frac{38}{3}$

   $u = 5x^3 + 4 \Rightarrow du = 15x^2 dx$

   $x = 0 \Rightarrow u = 4,\; x = 1 \Rightarrow u = 9$

43. $A = -\displaystyle\int_{-2}^0 x\sqrt{4 - x^2}\;dx + \int_0^2 x\sqrt{4 - x^2}\;dx = -\int_0^4 -\frac{1}{2}u^{1/2}\;du + \int_4^0 -\frac{1}{2}u^{1/2}\;du = 2\int_0^4 \frac{1}{2}u^{1/2}\;du$

   $u = 4 - x^2$ $\qquad\qquad\qquad\qquad = \displaystyle\int_0^4 u^{1/2}\;du = \left[\frac{2}{3}u^{3/2}\right]_0^4 = \left(\frac{2}{3}(4)^{3/2}\right) - \left(\frac{2}{3}(0)^{3/2}\right) = \frac{16}{3}$

   $du = -2x\;dx \qquad\qquad x = -2 \Rightarrow u = 0$

   $-\dfrac{1}{2}\;du = dx \qquad\quad x = 0 \Rightarrow u = 4 \qquad x = 2 \Rightarrow u = 0$

45. Since $\sin^2 x = 1 - \cos^2 x$, $\sin^2 x + C_1 = 1 - \cos^2 x + C_1 \Rightarrow 1 + C_1 = C_2$, the constant in part b)

   Since $\sin^2 x = \dfrac{1}{2} - \dfrac{1}{2}\cos 2x$, $\sin^2 x + C_1 = \dfrac{1}{2} - \dfrac{\cos 2x}{2} + C_1 \Rightarrow \dfrac{1}{2} + C_1 = C_3$, the constant in part c)

   Since $1 + C_1 = C_2$ and $\dfrac{1}{2} + C_1 = C_3$, $C_2 - 1 = C_3 - \dfrac{1}{2} \Rightarrow C_2 = C_3 + \dfrac{1}{2}$

   $\therefore$ the integrals are the same since they differ only by a constant.

47. $s = \int 24t(3t^2 - 1)^3\, dt = \int 4u^3\, du = \left[\frac{4u^4}{4}\right] + C = u^4 + C = (3t^2 - 1)^4 + C$

$u = 3t^2 - 1 \Rightarrow du = 6t\, dt \Rightarrow 4\, du = 24t\, dt$

$s = (3t^2 - 1)^4 + C$ and $s(0) = 0 \Rightarrow 0 = (3(0)^2 - 1)^4 + C \Rightarrow 0 = 1 + C \Rightarrow C = -1 \quad \therefore s = (3t^2 - 1)^4 - 1$

49. $s = \int 6\sin(t + \pi)\, dt = \int 6\sin u\, du = -6\cos u + C = -6\cos(t + \pi) + C$

$u = t + \pi \Rightarrow du = dt$

$s = -6\cos(t + \pi) + C$ and $s(0) = 0 \Rightarrow 0 = -6\cos(0 + \pi) + C \Rightarrow 0 = -6\cos \pi + C \Rightarrow 0 = 6 + C \Rightarrow$

$C = -6 \quad \therefore s = -6\cos(t + \pi) - 6$

51. $\displaystyle\int_1^3 \frac{\sin 2x}{x}\, dx = \int_2^6 \frac{\sin u}{u/2}\left(\frac{1}{2}\, du\right) = \int_2^6 \frac{\sin u}{u}\, du = [F(u)]_2^6 = F(6) - F(2)$

$u = 2x \Rightarrow x = \dfrac{u}{2}$ and $du = 2\, dx.$ $\quad x = 1 \Rightarrow u = 2,\ x = 3 \Rightarrow u = 6$

53. a) f odd $\Rightarrow f(-x) = -f(x).$ Then $\displaystyle\int_{-1}^0 f(x)\, dx = \int_1^0 f(-u)\,(-du) = \int_1^0 -f(u)\,(-du) = \int_1^0 f(u)\, du =$

$\displaystyle -\int_0^1 f(u)\, du = -3.$ $\qquad$ Let $u = -x \Rightarrow du = -dx;\ x = -1 \Rightarrow u = 1,\ x = 0 \Rightarrow u = 0$

b) f even $\Rightarrow f(-x) = f(x).$ Then $\displaystyle\int_{-1}^0 f(x)\, dx = \int_1^0 f(-u)\,(-du) = -\int_1^0 f(u)\, du = \int_0^1 f(u)\, du = 3$

Let $u = -x \Rightarrow du = -dx;\ x = -1 \Rightarrow u = 1,\ x = 0 \Rightarrow u = 0$

55. a) $\displaystyle\int_0^{\pi/4} \frac{18\tan^2 x\, \sec^2 x}{(2 + \tan^3 x)^2}\, dx = \int_0^1 \frac{18u^2}{(2 + u^3)^2}\, du = \int_0^1 \frac{6}{(2 + v)^2}\, dv = \int_2^3 \frac{6}{w^2}\, dw = \int_2^3 6w^{-2}\, dw$

$u = \tan x$	$v = u^3$	$w = 2 + v$	$= \left[-6w^{-1}\right]_2^3 = \left[-\dfrac{6}{w}\right]_2^3$
$du = \sec^2 x\, dx$	$dv = 3u^2\, du$	$dw = dv$	$= \left(-\dfrac{6}{3}\right) - \left(-\dfrac{6}{2}\right) = 1$
$x = 0 \Rightarrow u = 0$	$6\, dv = 18u^2\, du$	$v = 0 \Rightarrow w = 2$	
$x = \dfrac{\pi}{4} \Rightarrow u = 1$	$u = 0 \Rightarrow v = 0$	$v = 1 \Rightarrow w = 3$	
	$u = 1 \Rightarrow v = 1$		

55. (Continued from previous page)

b) $\displaystyle\int_0^{\pi/4} \frac{18\tan^2 x \sec^2 x}{(2+\tan^3 x)^2}\, dx = \int_0^1 \frac{6}{(2+u)^2}\, du = \int_2^3 \frac{6}{v^2}\, dv = \int_2^3 6v^{-2}\, dv = \left[-\frac{6}{v}\right]_2^3 = 1$

$u = \tan^3 x$ $\qquad\qquad v = 2 + u$

$du = 3\tan^2 x \sec^2 x\, dx$ $\qquad dv = du$

$6\, du = 18\tan^2 x \sec^2 x\, dx$ $\qquad u = 0 \Rightarrow v = 2,\ u = 1 \Rightarrow v = 3$

$x = 0 \Rightarrow u = 0,\ x = \dfrac{\pi}{4} \Rightarrow u = 1$

c) $\displaystyle\int_0^{\pi/4} \frac{18\tan^2 x \sec^2 x}{(2+\tan^3 x)^2}\, dx = \int_2^3 \frac{6}{u^2}\, du = \int_2^3 6u^{-2}\, du = \left[-\frac{6}{u}\right]_2^3 = 1$

$u = 2 + \tan^3 x$ $\qquad\qquad x = 0 \Rightarrow u = 2$

$du = 3\tan^2 x \sec^2 x\, dx$ $\qquad x = \dfrac{\pi}{4} \Rightarrow u = 3$

$6\, du = 18\tan^2 x \sec^2 x\, dx$

57. $\displaystyle\int \frac{1}{x^2} \sin\frac{1}{x} \cos\frac{1}{x}\, dx = \int u(-du) = -\int u\, du = -\frac{u^2}{2} + C = -\frac{\sin^2(1/x)}{2} + C$

$u = \sin\dfrac{1}{x} \Rightarrow du = -\dfrac{1}{x^2}\cos\dfrac{1}{x}\, dx \Rightarrow -du = \dfrac{1}{x^2}\cos\dfrac{1}{x}\, dx$

59. $\displaystyle\int \frac{\cos\sqrt{\theta}}{\sqrt{\theta}\,\sin^2\sqrt{\theta}}\, d\theta = \int \frac{1}{u^2}(2\, du) = 2\int \frac{1}{u^2}\, du = -\frac{2}{u} + C = -\frac{2}{\sin\sqrt{\theta}} + C$

$u = \sin\sqrt{\theta} \Rightarrow du = \dfrac{1}{2}\left(\dfrac{\cos\sqrt{\theta}}{\sqrt{\theta}}\right) d\theta \Rightarrow 2\, du = \dfrac{\cos\sqrt{\theta}}{\sqrt{\theta}}\, d\theta$

61. $\displaystyle\int_{a-c}^{b-c} f(x+c)\, dx = \int_a^b f(u)\, du = \int_a^b f(x)\, dx$

Let $u = x + c \Rightarrow du = dx;\ x = a - c \Rightarrow u = a,\ x = b - c \Rightarrow u = b$

SECTION 4.7 NUMERICAL INTEGRATION

1. a) For $n = 4$, $h = \dfrac{b-a}{n} = \dfrac{2-0}{4} = \dfrac{1}{2} \Rightarrow \dfrac{h}{2} = \dfrac{1}{4}$

x_i	x_i	$f(x_i)$	m	$mf(x_i)$
x_0	0	0	1	0
x_1	$\dfrac{1}{2}$	$\dfrac{1}{2}$	2	1
x_2	1	1	2	2
x_3	$\dfrac{3}{2}$	$\dfrac{3}{2}$	2	3
x_4	2	2	1	2

$\sum mf(x_i) = 8 \Rightarrow T = \dfrac{1}{4}(8) = 2$

b) $h = \dfrac{1}{2} \Rightarrow \dfrac{h}{3} = \dfrac{1}{6}$

x_i	x_i	$f(x_i)$	m	$mf(x_i)$
x_0	0	0	1	0
x_1	$\dfrac{1}{2}$	$\dfrac{1}{2}$	4	2
x_2	1	1	2	2
x_3	$\dfrac{3}{2}$	$\dfrac{3}{2}$	4	6
x_4	2	2	1	2

$\sum mf(x_i) = 12 \Rightarrow S = \dfrac{1}{6}(12) = 2$

c) $\displaystyle\int_0^2 x\, dx = \left[\dfrac{x^2}{2}\right]_0^2 = \dfrac{4}{2} - (0) = 2$

3. a) $n = 4 \Rightarrow h = \dfrac{b-a}{n} = \dfrac{2-0}{4} = \dfrac{1}{2} \Rightarrow \dfrac{h}{2} = \dfrac{1}{4}$

x_i	x_i	$f(x_i)$	m	$mf(x_i)$
x_0	0	0	1	0
x_1	$\dfrac{1}{2}$	$\dfrac{1}{8}$	2	$\dfrac{1}{4}$
x_2	1	1	2	2
x_3	$\dfrac{3}{2}$	$\dfrac{27}{8}$	2	$\dfrac{27}{4}$
x_4	2	8	1	8

$\sum mf(x_i) = 17 \Rightarrow T = \dfrac{1}{4}(17) = \dfrac{17}{4}$

b) $h = \dfrac{1}{2} \Rightarrow \dfrac{h}{3} = \dfrac{1}{6}$

x_i	x_i	$f(x_i)$	m	$mf(x_i)$
x_0	0	0	1	0
x_1	$\dfrac{1}{2}$	$\dfrac{1}{8}$	4	$\dfrac{1}{2}$
x_2	1	1	2	2
x_3	$\dfrac{3}{2}$	$\dfrac{27}{8}$	4	$\dfrac{27}{2}$
x_4	2	8	1	8

$\sum mf(x_i) = 24 \Rightarrow S = \dfrac{1}{6}(24) = 4$

c) $\displaystyle\int_0^2 x^3\, dx = \left[\dfrac{x^4}{4}\right]_0^2 = \dfrac{16}{4} - (0) = 4$

5. a) $n = 4 \Rightarrow h = \dfrac{b-a}{n} = \dfrac{4-0}{4} = 1 \Rightarrow \dfrac{h}{2} = \dfrac{1}{2}$

x_i	x_i	$f(x_i)$	m	$mf(x_i)$
x_0	0	0	1	0
x_1	1	1	2	2
x_2	2	$\sqrt{2}$	2	$2\sqrt{2}$
x_3	3	$\sqrt{3}$	2	$2\sqrt{3}$
x_4	4	2	1	2

$\sum mf(x_i) = 4 + 2\sqrt{2} + 2\sqrt{3} \Rightarrow$

$T = \dfrac{1}{2}(4 + 2\sqrt{2} + 2\sqrt{3}) = 2 + \sqrt{2} + \sqrt{3}$

≈ 5.1463

b) $h = 1 \Rightarrow \dfrac{h}{3} = \dfrac{1}{3}$

x_i	x_i	$f(x_i)$	m	$mf(x_i)$
x_0	0	0	1	0
x_1	1	1	4	4
x_2	2	$\sqrt{2}$	2	$2\sqrt{2}$
x_3	3	$\sqrt{3}$	4	$4\sqrt{3}$
x_4	4	2	1	2

$\sum mf(x_i) = 6 + 2\sqrt{2} + 4\sqrt{3} \Rightarrow$

$S = \dfrac{1}{3}(6 + 2\sqrt{2} + 4\sqrt{3}) = 2 + \dfrac{2}{3}\sqrt{2} + \dfrac{4}{3}\sqrt{3}$

≈ 5.252

c) $\displaystyle\int_0^4 \sqrt{x}\, dx = \left[\dfrac{2}{3}x^{3/2}\right]_0^4 = \dfrac{2}{3}(4)^{3/2} - 0 = \dfrac{16}{3}$

7. $f(x) = \dfrac{1}{x} \Rightarrow f'(x) = -\dfrac{1}{x^2} \Rightarrow f''(x) = \dfrac{2}{x^3}$ which is decreasing on $[1,2]$. $\therefore |f''(x)| \le 2$ on $[1,2]$ since $f''(1) = 2$

is the maximum value of f'' on the interval. $\therefore M = 2$. $n = 10 \Rightarrow h = \dfrac{b-a}{n} = \dfrac{2-1}{10} = \dfrac{1}{10}$

Then $|E_T| \le \dfrac{2-1}{12}\left(\dfrac{1}{10}\right)^2 (2) = \dfrac{1}{600} \approx 0.001667$

9. a) $|E_T| \le 10^{-4} \Rightarrow \dfrac{b-a}{12}(h^2)M \le 10^{-4}$. $h = \dfrac{b-a}{n} = \dfrac{2-0}{n} = \dfrac{2}{n}$. $f(x) = x \Rightarrow f'(x) = 1 \Rightarrow f''(x) = 0 \Rightarrow M = 0$

For $\dfrac{2}{12}\left(\dfrac{2}{n}\right)^2 (0) \le 10^{-4}$, let $n = 1$

b) $|E_S| \le 10^{-4} \Rightarrow \dfrac{b-a}{180}(h)^4 M \le 10^{-4}$. $h = \dfrac{2}{n}$; $f^{(4)}(x) = 0 \Rightarrow M = 0$. For $\dfrac{2}{180}\left(\dfrac{2}{n}\right)^4 (0) \le 10^{-4}$, let $n = 2$

(Remember n must be even for Simpson's Rule)

11. a) $|E_T| \le 10^{-4} \Rightarrow \dfrac{b-a}{12}(h^2)M \le 10^{-4}$. $h = \dfrac{b-a}{n} = \dfrac{2-0}{n} = \dfrac{2}{n}$; $f(x) = x^3 \Rightarrow f'(x) = 3x^2 \Rightarrow f''(x) = 6x$ which

is increasing on $[0,2] \Rightarrow |6x| \le 12$ on $[0,2] \Rightarrow M = 12$. For $|E_T| \le 10^{-4}$, $\dfrac{2-0}{12}\left(\dfrac{2}{n}\right)^2 (12) \le 10^{-4} \Rightarrow$

$\dfrac{8}{n^2} \le 10^{-4} \Rightarrow \dfrac{n^2}{8} \ge 10000 \Rightarrow n^2 \ge 80000 \Rightarrow n \ge \sqrt{80000} \approx 282.8$ $\therefore$ let $n = 283$

b) $|E_S| \le 10^{-4} \Rightarrow \dfrac{b-a}{180}(h)^4 M \le 10^{-4}$. $h = \dfrac{2}{n}$; $f'''(x) = 6 \Rightarrow f^{(4)}(x) = 0 \Rightarrow M = 0$. For

$\dfrac{2}{180}\left(\dfrac{2}{n}\right)^2 (0) \le 10^{-4}$, let $n = 2$ (n must be even)

13. a) $|E_T| \leq 10^{-4} \Rightarrow \frac{b-a}{12}(h^2)M \leq 10^{-4}$. $h = \frac{b-a}{n} = \frac{4-1}{n} = \frac{3}{n}$. $f(x) = \sqrt{x} = x^{1/2} \Rightarrow f'(x) = \frac{1}{2}x^{-1/2} \Rightarrow f''(x)$

$= -\frac{1}{4}x^{-3/2}$ whose absolute value is decreasing on $[1,4] \Rightarrow \left|f''(x)\right| \leq \frac{1}{4}$ on $[1,4] \Rightarrow M = \frac{1}{4}$. For

$|E_T| \leq 10^{-4}$, $\frac{4-1}{12}\left(\frac{3}{n}\right)^2\left(\frac{1}{4}\right) \leq 10^{-4} \Rightarrow \left(\frac{9}{16}\right)\left(\frac{1}{n^2}\right) \leq 10^{-4} \Rightarrow n^2 \geq \frac{9}{16}(10^4) \Rightarrow n \geq \frac{3}{4}(10^2) \Rightarrow n \geq 75$

$\therefore$ let $n = 75$

b) $|E_S| \leq 10^{-4} \Rightarrow \frac{b-a}{180}(h)^4 M \leq 10^{-4}$. $h = \frac{3}{n}$. $f'''(x) = \frac{3}{8}x^{-5/2} \Rightarrow f^{(4)}(x) = -\frac{15}{16}x^{-7/2}$ whose absolute

value is decreasing on $[1,4] \Rightarrow \left|f^{(4)}(x)\right| \leq \frac{15}{16}$ on $[1,4] \Rightarrow M = \frac{15}{16}$. For $|E_S| \leq 10^{-4}$,

$\frac{4-1}{180}\left(\frac{3}{n}\right)^4\left(\frac{15}{16}\right) \leq 10^{-4} \Rightarrow \left(\frac{81}{64}\right)\left(\frac{1}{n^4}\right) \leq 10^{-4} \Rightarrow n^4 \geq \frac{81}{64}(10^4) \Rightarrow n \geq 10.6$ $\therefore$ let $n = 12$

(n must be even)

15. Using Simpson's Rule, $h = 200 \Rightarrow \frac{h}{3} = \frac{200}{3}$

x_i	x_i	$f(x_i)$	m	$mf(x_i)$
x_0	0	0	1	0
x_1	200	520	4	2080
x_2	400	800	2	1600
x_3	600	1000	4	4000
x_4	800	1140	2	2280
x_5	1000	1160	4	4640
x_6	1200	1110	2	2220
x_7	1400	860	4	3440
x_8	1600	0	1	0

$\sum mf(x_i) = 20260 \Rightarrow$ Area $\approx \frac{200}{3}(20260)$

$= 1\ 350\ 666.667$ ft^2. Since the average

depth $= 20$ ft, Volume ≈ 20(Area) $\approx$

$27\ 013\ 333.33$ ft^3

Number of fish $=$ Volume$/1000 = 27013$ (to the

nearest fish)

Maximum to be caught $= 75\%$ of $27013 =$

20260

Number of licenses $= \frac{20260}{20} = 1013$

17. Using Simpson's Rule, $h = \frac{b-a}{n} = \frac{24-0}{6} = \frac{24}{6} = 4$

x_i	x_i	y_i	m	my_i
x_0	0	0	1	0
x_1	4	18.75	4	75
x_2	8	24	2	48
x_3	12	26	4	104
x_4	16	24	2	48
x_5	20	18.75	4	75
x_6	24	0	1	0

$\sum my_i = 350 \Rightarrow$

$S = \frac{4}{3}(350) = \frac{1400}{3} \approx 466.7$ in^2

19. For a No. 22 flashbulb, $A \approx \frac{5}{2}[4.2 + 2(3.0) + 2(1.7) + 2(0.7) + 2(0.35) + 2(0.2) + 0]$ where $\frac{b-a}{n} = \frac{70-20}{10} = 5$
$\Rightarrow A \approx \frac{5}{2}(16.1) = 40.25$ lumens.

For a No. 31 flashbulb, $A \approx \frac{5}{2}[1.0 + 2(1.2) + 2(1.0) + 2(0.9) + 2(1.0) + 2(1.1) + 2(1.3) + 2(1.4) + 2(1.3) +$
$2(1.0) + 0.8]$ where $\frac{b-a}{n} = \frac{70-20}{10} = 5 \Rightarrow A \approx \frac{5}{2}(22.2) = 55.5$ lumens. We are assuming two things: first,

the flash goes off at the same instant the button is pushed but the shutter opens 20 ms later; second, for the

Number 22 flashbulb, the bulb produces no light after 50 ms but the shutter remains open for another 20 ms

and, therefore, from 50 ms to 70 ms, the bulb produces 0 lumens/ms.

21. $h = \frac{b-a}{n} = \frac{2-0}{2} = 1$

x_n	x_n	y_n	m	my_n
x_0	0	0	1	0
x_1	1	1	4	4
x_2	2	8	1	8

$\sum_{2} my_n = 12 \Rightarrow A = \frac{1}{3}(12) = 4.$

$\int_0^2 x^3 \, dx = \left[\frac{x^4}{4}\right]_0^2 = 4$

Exercises 23–26 were done using the Calculus Toolkit Integral Evaluator with n = 50

23. S = 3.13791233   24. S = 1.08942941   25. S = 1.37074081   26. 0.828116331

SECTION 4.M MISCELLANEOUS EXERCISES

1. $y = \int \frac{x^2+1}{x^2} \, dx = \int \left(1 + \frac{1}{x^2}\right) dx = \int (1 + x^{-2}) \, dx = x - x^{-1} + C = x - \frac{1}{x} + C.$ $y = -1$ when $x = 1 \Rightarrow$
$1 - \frac{1}{1} + C = -1 \Rightarrow C = -1.$ $\therefore y = x - \frac{1}{x} - 1$

3. $s' = \int \frac{4t}{(1+t^2)^2} \, dt = \int u^{-2} (2 \, du) = 2 \int u^{-2} \, du = 2(-u^{-1}) + C = -\frac{2}{u} + C = -\frac{2}{1+t^2} + C.$
$\left(\text{Let } u = 1 + t^2 \Rightarrow du = 2t \, dt \Rightarrow 2 \, du = 4t \, dt\right)$ $s = 0$ when $t = 0 \Rightarrow -\frac{2}{1+0^2} + C = 0 \Rightarrow C = 2$
$\therefore s = -\frac{2}{1+t^2} + 2$

5. $y' = \int \left(15\sqrt{x} + \frac{3}{\sqrt{x}}\right) dx = \int (15x^{1/2} + 3x^{-1/2}) \, dx = 10x^{3/2} + 6x^{1/2} + C.$ $y' = 8$ when $x = 1 \Rightarrow 10(1)^{3/2} +$
$6(1)^{1/2} + C = 8 \Rightarrow C = -8.$ $\therefore y' = 10x^{3/2} + 6x^{1/2} - 8.$ Then $y = \int (10x^{3/2} + 6x^{1/2} - 8) \, dx =$
$4x^{5/2} + 4x^{3/2} - 8x + C.$ $y = 0$ when $x = 1 \Rightarrow 4(1)^{5/2} + 4(1)^{3/2} - 8(1) + C = 0 \Rightarrow C = 0.$
$\therefore y = 4x^{5/2} + 4x^{3/2} - 8x$

7. $\displaystyle\int_0^1 \sqrt{1+x^4}\ dx = F(1) - F(0)$

9. $\dfrac{d^2y}{dx^2} = 0 \Rightarrow \dfrac{dy}{dx} = C.\ \ \dfrac{dy}{dx} = 1$ when $x = 0 \Rightarrow \dfrac{dy}{dx} = 1.\ \ \dfrac{dy}{dx} = 1 \Rightarrow y = x + C.\ \ y = 0$ when $x = 0 \Rightarrow 0 = 0 + C \Rightarrow$

$C = 0.\ \ \therefore\ y = x$

11. Acceleration due to gravity is 32 ft/sec^2 downward (-32 ft/sec^2) $\Rightarrow v = \displaystyle\int -32\ dt = -32t + v_0$ where v_0 is the

initial velocity $\Rightarrow v = -32t + 32 \Rightarrow s = \displaystyle\int (-32t + 32)\ dt = -16t^2 + 32t + C.$ If the release point is $s = 0$, then

$C = 0 \Rightarrow s = -16t^2 + 32t.$ When will $s = 17$ft? $17 = -16t^2 + 32t \Rightarrow 16t^2 - 32t + 17 = 0.$ The discriminant of

this quadratic equation is -64 which says $\cdots$ DUCK!!!

13. Step 1: $v(t) = \displaystyle\int \dfrac{d^2s}{dt^2}\ dt = \int -k\ dt = -kt + C.\ \ v(0) = 44 \Rightarrow 44 = -k(0) + C \Rightarrow C = 44.\ \ \therefore\ v(t) = -kt + 44$

$s(t) = \displaystyle\int \dfrac{ds}{dt}\ dt = \int (-kt + 44)dt = \dfrac{-kt^2}{2} + 44t + C.\ \ s(0) = 0 \Rightarrow 0 = \dfrac{-k(0)^2}{2} + 44(0) + C \Rightarrow C = 0$

$\therefore\ s(t) = -\dfrac{kt^2}{2} + 44t$

Step 2: $-kt^* + 44 = 0 \Rightarrow t^* = \dfrac{44}{k}$

Step 3: $\dfrac{-k\left(\dfrac{44}{k}\right)^2}{2} + 44\left(\dfrac{44}{k}\right) = 45 \Rightarrow -k\left(\dfrac{44^2}{k^2}\right) + 2\left(\dfrac{44^2}{k}\right) = 90 \Rightarrow -\left(\dfrac{44^2}{k}\right) + 2\left(\dfrac{44^2}{k}\right) = 90 \Rightarrow \dfrac{44^2}{k} = 90 \Rightarrow$

$90k = 44^2 \Rightarrow k = \dfrac{44^2}{90} \approx 21.5$ ft/sec^2

15. a) Before the chute opens, $a = -32$ ft/sec^2. Since the helicopter is hovering, $v_0 = 0$ ft/sec. $\therefore\ v =$

$\displaystyle\int -32\ dt = -32t + v_0 = -32t.$ Let $s_0 = 6400$ ft $\Rightarrow s = \displaystyle\int -32t\ dt = -16t^2 + s_0 = -16t^2 + 6400.$ At $t = 4$

sec, $s = -16(4)^2 + 6400 = 6144$ ft when the chute opens.

b) For B, $s_0 = 7000$ ft, $v_0 = 0$, $a = -32$ ft/sec$^2 \Rightarrow v = \displaystyle\int -32\ dt = -32t + v_0 = -32t \Rightarrow s = \displaystyle\int -32t\ dt =$

$-16t^2 + s_0 = -16t^2 + 7000.$ At $t = 13$ sec, $s = -16(13)^2 + 7000 = 4296$ ft.

c) After the chutes open, $v = -16$ ft/sec $\Rightarrow s = \displaystyle\int -16\ dt = -16t + s_0.$ For A, $s_0 = 6144$ ft; for B, $s_0 = 4296$

ft. $\therefore\$ for A, $s = -16t + 6144$; for B, $s = -16t + 4296.$ When they hit the ground, $s = 0 \Rightarrow$ for A , $0 = -16t +$

$6144 \Rightarrow t = \dfrac{6144}{16} = 384$ seconds and for B, $0 = -16t + 4296 \Rightarrow t = \dfrac{4296}{16} = 268.5$ seconds to hit the ground

after the chutes open. B's chute opens 54 seconds after A's opens $\Rightarrow$ B hits the ground first.

17. $\displaystyle\int_1^2 \frac{1}{x}\, dx$

19. a) True, $\displaystyle\int_5^2 f(x)\, dx = -\int_2^5 f(x)\, dx$　　b) True, $\displaystyle\int_{-2}^5 (f(x) + g(x))\, dx = \int_{-2}^5 f(x)\, dx +$

$\displaystyle\int_{-2}^5 g(x)\, dx = \int_{-2}^2 f(x)\, dx + \int_2^5 f(x)\, dx + \int_{-2}^5 g(x)\, dx$

c) False, $\displaystyle\int_{-2}^5 f(x)\, dx > \int_{-2}^5 g(x)\, dx \Rightarrow f(x) \lneqq g(x)$

21. $\displaystyle\int_2^3 \left(t - \frac{2}{t}\right)\left(t + \frac{2}{t}\right) dt = \int_2^3 \left(t^2 - \frac{4}{t^2}\right) dt = \int_2^3 (t^2 - 4t^{-2}) dt = \left[\frac{t^3}{3} + 4t^{-1}\right]_2^3 = \left[\frac{t^3}{3} + \frac{4}{t}\right]_2^3 =$

$\left[e^u\right]_1^2 = e^2 - e^1 = e^2 - e$ 　　　　$\left(\frac{3^3}{3} + \frac{4}{3}\right) - \left(\frac{2^3}{3} + \frac{4}{2}\right) = \left(9 + \frac{4}{3}\right) - \left(\frac{8}{3} + 2\right) = \frac{17}{3}$

$u = \sec x \Rightarrow du = \sec x \tan x\, dx$

$x = 0 \Rightarrow u = 1,\ x = \frac{\pi}{3} \Rightarrow u = 2$

23. $\displaystyle\int_{-1}^0 \frac{12\, dx}{(2 - 3x)^2} = \int_5^2 12\left(\frac{1}{u^2}\right)\left(-\frac{1}{3} du\right) = -4 \int_5^2 u^{-2}\, du = -4[-u^{-1}]_5^2 = \frac{4}{2} - \frac{4}{5} = \frac{6}{5}$

$u = 2 - 3x \Rightarrow du = -3\, dx \Rightarrow -\frac{1}{3} du = dx.\ x = -1 \Rightarrow u = 5;\ x = 0 \Rightarrow u = 2$

25. $\displaystyle\int_0^1 \frac{dr}{\sqrt[3]{(7 - 5r)^2}} = \int_0^1 (7 - 5r)^{-2/3}\, dr = \int_7^2 u^{-2/3}\left(-\frac{1}{5} du\right) = -\frac{1}{5}\left[3u^{1/3}\right]_7^2 = -\frac{3}{5}\sqrt[3]{2} + \frac{3}{5}\sqrt[3]{7}$

$u = 7 - 5r \Rightarrow du = -5\, dr \Rightarrow -\frac{1}{5} du = dr.\ r = 0 \Rightarrow u = 7;\ r = 1 \Rightarrow u = 2$

27. $\displaystyle\int_1^{25} \frac{dv}{\sqrt{v}\sqrt{2\sqrt{v} - 1}} = \int_1^9 \frac{1}{\sqrt{u}}\, du = \int_1^9 u^{-1/2}\, du = [2u^{1/2}]_1^9 = 2\left(9^{1/2}\right) - 2\left(1^{1/2}\right) = 4$

$u = 2\sqrt{v} - 1 \Rightarrow du = 2\left(\frac{1}{2} v^{-1/2}\right) dv = \frac{1}{\sqrt{v}}\, dv.\ v = 1 \Rightarrow u = 2\sqrt{1} - 1 = 1;\ v = 25 \Rightarrow u = 2\sqrt{25} - 1 = 9$

29. $\displaystyle\int_{\pi}^{3\pi} \cot^2 \frac{x}{6}\, dx = \int_{\pi/6}^{\pi/2} 6\cot^2 u\, du = \int_{\pi/6}^{\pi/2} (\csc^2 u - 1)\, du = \left[6(-\cot u - u)\right]_{\pi/6}^{\pi/2} = 6\left(-\cot \frac{\pi}{2} - \frac{\pi}{2}\right) -$

$6\left(-\cot \frac{\pi}{6} - \frac{\pi}{6}\right) = 6\sqrt{3} - 2\pi.$

$u = \dfrac{x}{6} \Rightarrow du = \dfrac{1}{6}\, dx \Rightarrow 6\, du = dx \quad x = \pi \Rightarrow u = \dfrac{\pi}{6},\, x = 3\pi \Rightarrow u = \dfrac{\pi}{2}$

31. $\displaystyle\int_{0}^{\pi/2} 5(\sin t)^{3/2} \cos t\, dt = \int_{0}^{1} 5u^{3/2}\, du = \left[5\left(\frac{2}{5}\right)u^{5/2}\right]_{0}^{1} = \left[2u^{5/2}\right]_{0}^{1} = 2(1)^{5/2} - 2(0)^{5/2} = 2$

$u = \sin t \Rightarrow du = \cos t\, dt \quad t = 0 \Rightarrow u = 0,\, t = \dfrac{\pi}{2} \Rightarrow u = 1$

33. $\displaystyle\int_{-\pi/2}^{\pi/2} 15 \sin^4 3x \cos 3x\, dx = \int_{1}^{-1} u^4 (5\, du) = 5\left[\frac{u^5}{5}\right]_{1}^{-1} = (-1)^5 - 1^5 = -2$

$u = \sin 3x \Rightarrow du = 3 \cos 3x\, dx \Rightarrow 5\, du = 15 \cos 3x\, dx. \quad x = -\dfrac{\pi}{2} \Rightarrow u = \sin\left(-\dfrac{3\pi}{2}\right) = 1;\, x = \dfrac{\pi}{2} \Rightarrow$

$u = \sin\left(\dfrac{3\pi}{2}\right) = -1$

35. $\displaystyle\int_{0}^{1} \pi z^2 \sec^2\left[\frac{\pi z^3}{3}\right] dz = \int_{0}^{\pi/3} \sec^2 u\, du = [\tan u]_{0}^{\pi/3} = \tan \frac{\pi}{3} - \tan 0 = \sqrt{3}$

$u = \dfrac{\pi z^3}{3} \Rightarrow du = \pi z^2\, dz. \quad z = 0 \Rightarrow u = 0;\, z = 1 \Rightarrow u = \dfrac{\pi}{3}$

37. $\displaystyle\int_{0}^{\pi/3} \frac{\tan \theta}{\sqrt{2 \sec \theta}}\, d\theta = \int_{0}^{\pi/3} \frac{\sec \theta \tan \theta}{\sec \theta \sqrt{2 \sec \theta}}\, d\theta = \int_{0}^{\pi/3} \frac{\sec \theta \tan \theta}{\sqrt{2}(\sec \theta)^{3/2}}\, d\theta = \int_{1}^{2} \frac{1}{\sqrt{2}\, u^{3/2}}\, du = \frac{1}{\sqrt{2}} \int_{1}^{2} u^{-3/2}\, du =$

$\dfrac{1}{\sqrt{2}}\left[\dfrac{u^{-1/2}}{-1/2}\right]_{1}^{2} = \left[-\dfrac{2}{\sqrt{2u}}\right]_{1}^{2} = -\dfrac{2}{\sqrt{2(2)}} - \left(-\dfrac{2}{\sqrt{2(1)}}\right) = -1 + \sqrt{2}$

$u = \sec \theta \Rightarrow du = \sec \theta \tan \theta\, d\theta. \quad \theta = 0 \Rightarrow u = \sec 0 = 1;\, \theta = \dfrac{\pi}{3} \Rightarrow u = \sec \dfrac{\pi}{3} = 2$

39. $y_{av} = \dfrac{1}{365 - 0} \displaystyle\int_{0}^{365} \left[37 \sin\left(\frac{2\pi}{365}(x - 101)\right) + 25\right] dx = \dfrac{1}{365}\left[-37\left(\frac{365}{2\pi} \cos\left(\frac{2\pi}{365}(x - 101)\right) + 25x\right)\right]_{0}^{365}$

$= \dfrac{1}{365}\left[\left(-37\left(\frac{365}{2\pi}\right) \cos\left[\frac{2\pi}{365}(365 - 101)\right] + 25(365)\right) - \left(-37\left(\frac{365}{2\pi}\right) \cos\left[\frac{2\pi}{365}(0 - 101)\right] + 25(0)\right)\right] =$

$-\dfrac{37}{2\pi} \cos \dfrac{2\pi}{365}(264) + 25 + \dfrac{37}{2\pi} \cos \dfrac{2\pi}{365}(-101) = -\dfrac{37}{2\pi}\left(\cos \dfrac{2\pi}{365}(264) - \cos \dfrac{2\pi}{365}(-101)\right) + 25 = 25°F$

41. $\int_1^b f(x)\, dx = \sqrt{b^2 + 1} - \sqrt{2} \Rightarrow \dfrac{d}{db} \int_1^b f(x)\, dx = f(b) = \dfrac{1}{2}\left(b^2 + 1\right)^{-1/2}(2b) = \dfrac{b}{\sqrt{b^2 + 1}} \Rightarrow f(x) = \dfrac{x}{\sqrt{x^2 + 1}}$

43. If $x = \displaystyle\int_0^y \dfrac{1}{\sqrt{1 + 4t^2}}\, dt$, then $\dfrac{d}{dx}(x) = \dfrac{d}{dx}\left[\displaystyle\int_0^y \dfrac{1}{\sqrt{1 + 4t^2}}\, dt\right] = \dfrac{d}{dy}\left[\displaystyle\int_0^y \dfrac{1}{\sqrt{1 + 4t^2}}\, dt\right]\left(\dfrac{dy}{dx}\right) \Rightarrow$

$1 = \dfrac{1}{\sqrt{1 + 4y^2}}\left(\dfrac{dy}{dx}\right) \Rightarrow \dfrac{dy}{dx} = \sqrt{1 + 4y^2}$. Then $\dfrac{d^2y}{dx^2} = \dfrac{d}{dx}\left(\sqrt{1 + 4y^2}\right) = \dfrac{d}{dy}\left(\sqrt{1 + 4y^2}\right)\left(\dfrac{dy}{dx}\right) =$

$\dfrac{1}{2}(1 + 4y^2)^{-1/2}\left(8y\,\dfrac{dy}{dx}\right) = \dfrac{4y(dy/dx)}{\sqrt{1 + 4y^2}} = \dfrac{4y\left(\sqrt{1 + 4y^2}\right)}{\sqrt{1 + 4y^2}} = 4y$. Thus $\dfrac{d^2y}{dx^2} = 4y$, and the constant is 4.

45. a) If $\displaystyle\int_0^{x^2} f(t)\, dt = x\cos \pi x$, then $\dfrac{d}{dx}\displaystyle\int_0^{x^2} f(t)\, dt = \cos \pi x - \pi x \sin \pi x \Rightarrow f(x^2)(2x) = \cos \pi x - \pi x \sin \pi x \Rightarrow$

$f(x^2) = \dfrac{\cos \pi x - \pi x \sin \pi x}{2x}$. $x = 2 \Rightarrow f(4) = \dfrac{\cos 2\pi - 2\pi \sin 2\pi}{4} = \dfrac{1}{4}$

b) $\displaystyle\int_0^{f(x)} t^2\, dt = \left[\dfrac{t^3}{3}\right]_0^{f(x)} = \dfrac{1}{3}(f(x))^3 \Rightarrow \dfrac{1}{3}(f(x))^3 = x\cos \pi x \Rightarrow (f(x))^3 = 3x\cos \pi x \Rightarrow f(x) = \sqrt[3]{3x\cos \pi x}$

$\therefore f(4) = \sqrt[3]{3(4)\cos 4\pi} = \sqrt[3]{12}$

47. $|E_s| \le \dfrac{3-1}{180}(h)^4 M$ where $h = \dfrac{3-1}{n} = \dfrac{2}{n}$. $f(x) = \dfrac{1}{x} = x^{-1} \Rightarrow f'(x) = -x^{-2} \Rightarrow f''(x) = 2x^{-3} \Rightarrow f'''(x) = -6x^{-4}$

$\Rightarrow f^{(4)}(x) = 24x^{-5}$ which is decreasing on $[1,3] \Rightarrow$ maximum of $f^{(4)}(x)$ on $[1,3]$ is $f^{(4)}(1) = 24 \Rightarrow M = 24$

For $|E_s| \le 0.0001$, $\dfrac{3-1}{180}\left(\dfrac{2}{n}\right)^4 (24) \le 0.0001 \Rightarrow \dfrac{768}{180}\left(\dfrac{1}{n^4}\right) \le 0.0001 \Rightarrow \dfrac{1}{n^4} \le (0.0001)\dfrac{180}{768} \Rightarrow$

$n^4 \ge 10000\left(\dfrac{768}{180}\right) \Rightarrow n \ge 14.37 \Rightarrow n \ge 16$ (n must be even) $\Rightarrow h \le \dfrac{2}{16} \Rightarrow h \le \dfrac{1}{8}$

49. $h = \dfrac{b-a}{n} = \dfrac{\pi - 0}{6} = \dfrac{\pi}{6} \Rightarrow \dfrac{h}{2} = \dfrac{\pi}{12}$

x_i	x_i	$f(x_i)$	m	$mf(x_i)$
x_0	0	0	1	0
x_1	$\pi/6$	1/2	2	1
x_2	$\pi/3$	3/2	2	3
x_3	$\pi/2$	2	2	4
x_4	$2\pi/3$	3/2	2	3
x_5	$5\pi/6$	1/2	2	1
x_6	π	0	1	0

$\displaystyle\sum_{i=0}^{6} mf(x_i) = 12$

$T = \dfrac{\pi}{12}(12) = \pi$

49. (Continued)

x_i	x_i	$f(x_i)$	m	$mf(x_i)$
x_0	0	0	1	0
x_1	$\pi/6$	1/2	4	2
x_2	$\pi/3$	3/2	2	3
x_3	$\pi/2$	2	4	8
x_4	$2\pi/3$	3/2	2	3
x_5	$5\pi/6$	1/2	4	2
x_6	π	0	1	0

$$\sum_{i=0}^{6} mf(x_i) = 18, \ \frac{h}{3} = \frac{\pi}{18}$$

$$\therefore S = \frac{\pi}{18}(18) = \pi$$

51. Cost is given at $2.10/ft^2

x_i	x_i	$f(x_i)$	m	$mf(x_i)$
x_0	0	0	1	0
x_1	15	36	2	72
x_2	30	54	2	108
x_3	45	51	2	102
x_4	60	49.5	2	99
x_5	75	54	2	108
x_6	90	64.4	2	128.8
x_7	105	67.5	2	135
x_8	120	42	1	42

$$\sum_{i=0}^{8} mf(x_i) = 794.8 \quad h = 15 \Rightarrow \frac{h}{2} = \frac{15}{2}$$

$$\therefore T = \frac{15}{2}(794.8) = 5961 \ ft^2$$

Total cost = $2.10(Area) = $2.10(5961) =

$12,518.10 (approximately)

The job cannot be done for $11,000.

53. $f(x) = \displaystyle\int_{1/x}^{x} \frac{1}{t}\,dt \Rightarrow f'(x) = \frac{1}{x}\left(\frac{dx}{dx}\right) - \frac{1}{1/x}\left(\frac{d}{dx}\left(\frac{1}{x}\right)\right) = \frac{1}{x} - x\left(-\frac{1}{x^2}\right) = \frac{1}{x} + \frac{1}{x} = \frac{2}{x}$

55. $g(y) = \displaystyle\int_{\sqrt{y}}^{2\sqrt{y}} \sin t^2\,dt \Rightarrow g'(y) = \sin\left(2\sqrt{y}\right)^2 \left(\frac{d}{dy}\left(2\sqrt{y}\right)\right) - \sin\left(\sqrt{y}\right)^2 \left(\frac{d}{dy}\left(\sqrt{y}\right)\right) = \frac{\sin 4y}{\sqrt{y}} - \frac{\sin y}{2\sqrt{y}}$

57.

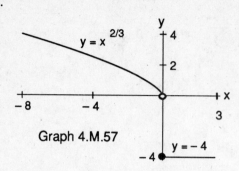

Graph 4.M.57

$$\int_{-8}^{3} f(x)\,dx = \int_{-8}^{0} x^{2/3}\,dx + \int_{0}^{3} -4\,dx = \left[\frac{3}{5} x^{5/3}\right]_{-8}^{0} + \left[-4x\right]_{0}^{3} =$$

$$\left(0 - \frac{3}{5}(-8)^{5/3}\right) + (-4(3) - 0) = \frac{96}{5} - 12 = \frac{36}{5}$$

59.

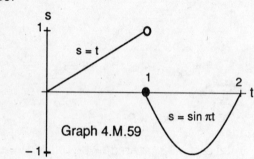

Graph 4.M.59

$$\int_{0}^{2} f(t)\,dt = \int_{0}^{1} t\,dt + \int_{1}^{2} \sin \pi t\,dt = \left[\frac{t^2}{2}\right]_{0}^{1} + \left[-\frac{1}{\pi}\cos \pi t\right]_{1}^{2} =$$

$$\left(\frac{1}{2} - 0\right) + \left(-\frac{1}{\pi}\cos 2\pi - \left(-\frac{1}{\pi}\cos \pi\right)\right) = \frac{1}{2} - \frac{2}{\pi}$$

61.

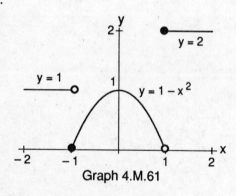

Graph 4.M.61

$$\int_{-2}^{2} f(x)\,dx = \int_{-2}^{-1} dx + \int_{-1}^{1} (1-x^2)\,dx + \int_{1}^{2} 2\,dx = [x]_{-2}^{-1} +$$

$$\left[x - \frac{x^3}{3}\right]_{-1}^{1} + [2x]_{1}^{2} = -1 - (-2)\left(1 - \frac{1^3}{3}\right) - \left(-1 - \frac{(-1)^3}{3}\right) +$$

$$2(2) - 2(1) = 1 + \frac{2}{3} - \left(-\frac{2}{3}\right) + 4 - 2 = \frac{13}{3}$$

63. Ave. value $= \dfrac{1}{b-a} \displaystyle\int_{a}^{b} f(x)\,dx = \dfrac{1}{2-0} \int_{0}^{2} f(x)\,dx = \dfrac{1}{2}\left[\int_{0}^{1} x\,dx + \int_{1}^{2} (x-1)\,dx\right] = \dfrac{1}{2}\left[\dfrac{x^2}{2}\right]_{0}^{1} + \dfrac{1}{2}\left[\dfrac{x^2}{2} - x\right]_{1}^{2} =$

$$\frac{1}{2}\left[\frac{1^2}{2} - 0 + \left(\frac{2^2}{2} - 2\right) - \left(\frac{1^2}{2} - 1\right)\right] = \frac{1}{2}$$

65. Let $f(x) = x^5$ on $[0,1]$. Partition $[0,1]$ into n regular subintervals $\Rightarrow \Delta x = \dfrac{1-0}{n} = \dfrac{1}{n}$. Then $\dfrac{1}{n}, \dfrac{2}{n}, \cdots, \dfrac{n}{n}$ are the

right hand end points of the subintervals. Since f is increasing on $[0,1]$, $U = \displaystyle\sum_{j=1}^{\infty} \left(\frac{j}{n}\right)^5 \frac{1}{n}$ is the upper sum

65. (Continued)

$$\text{for } f(x) = x^5 \text{ on } [0,1] \Rightarrow \lim_{n \to \infty} \sum_{j=1}^{\infty} \left(\frac{i}{n}\right)^5 \frac{1}{n} = \lim_{n \to \infty} \frac{1}{n}\left[\left(\frac{1}{n}\right)^5 + \left(\frac{2}{n}\right)^5 + \cdots + \left(\frac{n}{n}\right)^5\right] =$$

$$\lim_{n \to \infty}\left[\frac{1^5 + 2^5 + \cdots + n^5}{n^6}\right] = \int_0^1 x^5 \, dx = \left[\frac{x^6}{6}\right]_0^1 = \frac{1}{6}$$

67. Let $y = f(x)$ on $[0,1]$. Partition $[0,1]$ into n regular subintervals $\Rightarrow \Delta x = \frac{1-0}{n} = \frac{1}{n}$. Then $\frac{1}{n}, \frac{2}{n}, \cdots, \frac{n}{n}$ are the

right hand end points of the subintervals. Since f is continuous on $[0,1]$, $\sum_{j=1}^{\infty} f\left(\frac{i}{n}\right)\frac{1}{n}$ is a Riemann sum

of $y = f(x)$ on $[0,1] \Rightarrow \lim_{n \to \infty} \sum_{j=1}^{\infty} f\left(\frac{i}{n}\right)\frac{1}{n} = \lim_{n \to \infty} \frac{1}{n}\left[f\left(\frac{1}{n}\right) + f\left(\frac{2}{n}\right) + \cdots + f\left(\frac{n}{n}\right)\right] = \int_0^1 f(x) \, dx$

69. a) Let the polygon be inscribed in a circle of radius r. If we draw a radius from the center of the circle
        (and the polygon) to each vertex of the polygon, we have n isosceles triangles formed (the equal sides
        are equal to r, the radius of the circle) and a vertex angle of θ_n where $\theta_n = \frac{2\pi}{n}$. The area of each

    triangle is $A_n = \frac{1}{2} r^2 \sin\theta_n \Rightarrow$ the area of the polygon is $A = nA_n = \frac{nr^2}{2}\sin\theta_n = \frac{nr^2}{2}\sin\frac{2\pi}{n}$.

    b) $\lim_{n \to \infty} A = \lim_{n \to \infty} \frac{nr^2}{2}\sin\frac{2\pi}{n} = \lim_{n \to \infty} \frac{n\pi r^2}{2\pi}\sin\frac{2\pi}{n} = \lim_{n \to \infty} (\pi r^2)\frac{\sin(2\pi/n)}{2\pi/n} = \lim_{2\pi/n \to 0} (\pi r^2)\frac{\sin(2\pi/n)}{2\pi/n} = \pi r^2$

CHAPTER 5

APPLICATIONS OF DEFINITE INTEGRALS

5.1 APPLICATIONS OF DEFINITE INTEGRALS

1. $A = \int_0^\pi 1 - \cos^2 x \, dx = \int_0^\pi \sin^2 x \, dx = \frac{1}{2}\int_0^\pi 1 - \cos 2x \, dx = \frac{1}{2}\left[x - \frac{\sin 2x}{2}\right]_0^\pi = \frac{\pi}{2}$

3. $A = \int_0^1 1 - 2x^{1/2} + x \, dx = \left[x - \frac{4x^{3/2}}{3} + \frac{x^2}{2}\right]_0^1 = \frac{1}{6}$

5. $A = \int_0^1 x^{1/3} - x^{1/2} \, dx = \left[\frac{3x^{4/3}}{4} - \frac{2x^{3/2}}{3}\right]_0^1 = \frac{1}{12}$

7. $A = 2\int_0^2 2 - (x^2 - 2) \, dx = 2\int_0^2 4 - x^2 \, dx = 2\left[4x - \frac{x^3}{3}\right]_0^2 = \frac{32}{3}$

9. $A = \int_0^2 8x - x^4 \, dx = \left[4x^2 - \frac{x^5}{5}\right]_0^2 = \frac{48}{5}$

11. $A = \int_0^2 4x - 2x^2 \, dx = \left[2x^2 - \frac{2x^3}{3}\right]_0^2 = \frac{8}{3}$

13. $A = 2\int_0^1 (4 - 4x^2) - (x^4 - 1) \, dx = 2\left[-\frac{x^5}{5} - \frac{4x^3}{3} + 5x\right]_0^1 = \frac{104}{15}$

15. $A = \int_{-1}^0 \frac{x+6}{5} - \sqrt{-x} \, dx + \int_0^4 \frac{x+6}{5} - \sqrt{x} \, dx + \int_4^9 \sqrt{x} - \left(\frac{x}{5} + \frac{6}{5}\right) dx = \int_{-1}^0 \frac{x}{5} + \frac{6}{5} - (-x)^{1/2} \, dx +$

    $\int_0^4 \frac{x}{5} + \frac{6}{5} - (x)^{1/2} \, dx + \int_4^9 x^{1/2} - \left(\frac{x}{5} + \frac{6}{5}\right) dx = \left[\frac{x^2}{10} + \frac{6x}{5} + \frac{2}{3}(-x)^{3/2}\right]_{-1}^0 + \left[\frac{x^2}{10} + \frac{6x}{5} + \frac{2}{3}(x)^{3/2}\right]_0^4 +$

    $\left[\frac{2}{3}x^{3/2} - \frac{x^2}{10} - \frac{6x}{5}\right]_4^9 = \frac{34}{15}$

17. $A = 2\int_0^1 (x^4 - 4x^2 + 4) - x^2 \, dx + 2\int_1^2 x^2 - (x^4 - 4x^2 + 4) \, dx = 2\left[\frac{x^5}{5} - \frac{5x^3}{3} + 4x\right]_0^1 +$

    $2\left[-\frac{x^5}{5} + \frac{5x^3}{3} - 4x\right]_1^2 = 8$

19. $A = \int_0^3 2y^2 \, dy = \left[\frac{2y^3}{3}\right]_0^3 = 18$

21. $A = \int_{-4}^5 \left(\frac{y+16}{4}\right) - \left(\frac{y^2-4}{4}\right) dy = \frac{1}{4}\int_{-4}^5 -y^2 + y + 20 \, dy = \frac{1}{4}\left[-\frac{y^3}{3} + \frac{y^2}{2} + 20y\right]_{-4}^5 = \frac{243}{8}$

23. $A = 2\int_0^1 \left(2 - 3y^2\right) - \left(-y^2\right) dy = 4\left[y - \frac{y^3}{3}\right]_0^1 = \frac{8}{3}$

25. $A = 2\int_0^1 |y| \sqrt{1 - y^2} - \left(y^2 - 1\right) dy = \int_0^1 \left(1 - y^2\right)^{1/2} 2y - 2y^2 + 2\ dy =$

$\left[-\frac{2}{3}\left(1 - y^2\right)^{3/2} - \frac{2y^3}{3} + 2y \right]_0^1 = 2$

27. $A = \int_0^\pi (2\sin x - \sin 2x)\ dx = \left[-2\cos x + \frac{\cos 2x}{2} \right]_0^\pi = 4$

29. $A = \int_{-\pi/4}^{\pi/4} \sec^2 x - \tan^2 x\ dx = \int_{-\pi/4}^{\pi/4} 1\ dx = \frac{\pi}{2}$, since $\sec^2 x = 1 + \tan^2 x$

31. $A = 2\int_0^1 \left(1 - y^2\right) - \cos\left(\frac{\pi y}{2}\right) dy = 2\left[y - \frac{y^3}{3} - \frac{2}{\pi}\sin\left(\frac{\pi y}{2}\right) \right]_0^1 = \frac{4\pi - 12}{3\pi}$

33. $A = \int_0^{\pi/2} 3(\sin y)\sqrt{\cos y}\ dy = \left[-2\cos^{3/2} y \right]_0^{\pi/2} = 2$

35. $A = \int_1^2 2 - (x - 2)^2 - x\ dx = \left[-2x + \frac{3x^2}{2} - \frac{x^3}{3} \right]_1^2 = \frac{1}{6}$

37. $A = \int_{-2}^1 \left(4 - y^2\right) - (y + 2)\ dy = \left[-\frac{y^3}{3} - \frac{y^2}{2} + 2y \right]_{-2}^1 = \frac{9}{2}$

39. $A = \int_0^1 x\ dx + \int_1^2 x^{-2}\ dx = \left[\frac{x^2}{2} \right]_0^1 + \left[\frac{-1}{x} \right]_1^2 = 1$

41. $A = \int_0^{\pi/4} \cos x - \sin x\ dx = \left[\sin x + \cos x \right]_0^{\pi/4} = \sqrt{2} - 1$

43. a) $\int_0^c \sqrt{y}\ dy = \frac{1}{2}\int_0^4 \sqrt{y}\ dy \Rightarrow \left[\frac{2y^{3/2}}{3} \right]_0^c = \frac{1}{2}\left[\frac{2y^{3/2}}{3} \right]_0^4 \Rightarrow \frac{2}{3}c^{3/2} = \frac{1}{2}\left[\frac{16}{3} \right] \Rightarrow c^{3/2} = 4 \Rightarrow c = 4^{2/3}$

    b) $\int_0^{\sqrt{c}} c - x^2\ dx = \frac{1}{2}\int_0^2 4 - x^2\ dx \Rightarrow \left[cx - \frac{x^3}{3} \right]_0^{\sqrt{c}} = \frac{1}{2}\left[4x - \frac{x^3}{3} \right]_0^2 \Rightarrow \frac{2c^{3/2}}{3} = \frac{8}{3} \Rightarrow c = 4^{2/3}$

45. $A = \int_a^b 2f(x)\ dx - \int_a^b f(x)\ dx = 2\int_a^b f(x)\ dx - \int_a^b f(x)\ dx = \int_a^b f(x)\ dx = 4$

5.2 VOLUMES: SLICING, DISKS, AND WASHERS

1. $V = \pi \displaystyle\int_0^2 (x^2)^2 \, dx = \left[\dfrac{\pi x^5}{5}\right]_0^2 = \dfrac{32\pi}{5}$

3. $V = \pi \displaystyle\int_0^1 (x - x^2)^2 \, dx = \pi\left[\dfrac{x^3}{3} - \dfrac{x^4}{2} + \dfrac{x^5}{5}\right]_0^1 = \dfrac{\pi}{30}$

5. $V = \pi \displaystyle\int_0^{\pi/2} (\sqrt{\cos x})^2 \, dx = [\pi \sin x]_0^{\pi/2} = \pi$

7. $V = \pi \displaystyle\int_0^4 4 - y \, dy = \pi\left[4y - \dfrac{y^2}{2}\right]_0^4 = 8\pi$

9. $V = 2\pi \displaystyle\int_0^1 (1 - y^2)^2 \, dy = 2\pi\left[y - \dfrac{2y^3}{3} + \dfrac{y^5}{5}\right]_0^1 = \dfrac{16\pi}{15}$

11. $V = \pi \displaystyle\int_0^{\pi/2} 2 \sin 2y \, dy = \pi[-\cos 2y]_0^{\pi/2} = 2\pi$

13. $V = \pi \displaystyle\int_0^1 1^2 - x^2 \, dx = \pi\left[x - \dfrac{x^3}{3}\right]_0^1 = \dfrac{2\pi}{3}$

15. $V = \pi \displaystyle\int_0^2 4^2 - (x^2)^2 \, dx = \pi\left[16x - \dfrac{x^5}{5}\right]_0^2 = \dfrac{128\pi}{5}$

17. $V = \pi \displaystyle\int_{-1}^2 (x + 3)^2 - (x^2 + 1)^2 \, dx = \pi\left[-\dfrac{x^5}{5} - \dfrac{x^3}{3} + 3x^2 + 8x\right]_{-1}^2 = \dfrac{117\pi}{5}$

19. $V = 2\pi \displaystyle\int_0^{\pi/4} (\sqrt{2})^2 - \sec^2 x \, dx = 2\pi[2x - \tan x]_0^{\pi/4} = \pi(\pi - 2)$

21. $V = \pi \displaystyle\int_0^1 (y + 1)^2 - 1^2 \, dy = \pi\left[\dfrac{y^3}{3} + y^2\right]_0^1 = \dfrac{4\pi}{3}$

23. $V = \pi \displaystyle\int_0^4 2^2 - (\sqrt{y})^2 \, dy = \pi\left[4y - \dfrac{y^2}{2}\right]_0^4 = 8\pi$

25. $V = 2\pi \displaystyle\int_0^5 (\sqrt{25 - y^2})^2 \, dy = 2\pi\left[25y - \dfrac{y^3}{3}\right]_0^5 = \dfrac{500\pi}{3}$

27. $V = 2\pi \displaystyle\int_0^{\pi/2} 1^2 - \cos x \, dx = 2\pi[x - \sin x]_0^{\pi/2} = \pi^2 - 2\pi$

29. a) $V = \pi \displaystyle\int_0^4 2^2 - (\sqrt{x})^2 \, dx = \pi\left[4x - \dfrac{x^2}{2}\right]_0^4 = 8\pi$

    b) $V = \pi \displaystyle\int_0^2 (y^2)^2 \, dy = \left[\dfrac{\pi y^5}{5}\right]_0^2 = \dfrac{32\pi}{5}$

    c) $V = \pi \displaystyle\int_0^4 (2 - \sqrt{x})^2 \, dx = \pi\left[4x - \dfrac{8}{3}x^{3/2} + \dfrac{x^2}{2}\right]_0^4 = \dfrac{8\pi}{3}$

    d) $V = \pi \displaystyle\int_0^2 4^2 - (4 - y^2)^2 \, dy = \pi\left[\dfrac{8y^3}{3} - \dfrac{y^5}{5}\right]_0^2 = \dfrac{224\pi}{15}$

31.   a) $V = 2\pi \int_0^1 (1 - x^2)^2 \, dx = 2\pi \left[x - \frac{2}{3}x^3 + \frac{1}{5}x^5 \right]_0^1 = \frac{16\pi}{15}$

      b) $V = 2\pi \int_0^1 (2 - x^2)^2 - 1^2 \, dx = 2\pi \left[3x - \frac{4x^3}{3} + \frac{x^5}{5} \right]_0^1 = \frac{56\pi}{15}$

      c) $V = 2\pi \int_0^1 2^2 - (1 + x^2)^2 \, dx = 2\pi \left[3x - \frac{2x^3}{3} - \frac{x^5}{5} \right]_0^1 = \frac{64\pi}{15}$

33. $V = \pi \int_0^\pi (c - \sin x)^2 \, dx = \pi \int_0^\pi c^2 - 2c \sin x + \sin^2 x \, dx = \pi \int_0^\pi c^2 - 2c \sin x + \frac{1 - \cos 2x}{2} \, dx =$

$\pi \left[c^2 x + 2c \cos x + \frac{x}{2} - \frac{\sin 2x}{4} \right]_0^\pi = \pi \left(c^2 \pi - 4c + \frac{\pi}{2} \right)$. If $V(c) = \pi \left(c^2 \pi - 4c + \frac{\pi}{2} \right)$, then $V'(c) =$

$\pi(2c\pi - 4) \Rightarrow$ that $c = \frac{2}{\pi}$ is the critical point. $V''(c) = 2\pi^2 > 0 \Rightarrow$ that $c = \frac{2}{\pi}$ minimizes the volume.

35. $V = \pi \int_{-16}^{-7} (\sqrt{256 - y^2})^2 \, dy = \pi \left[256y - \frac{y^3}{3} \right]_{-16}^{-7} = 1053\pi \ cm^3$

37. $V = \pi \int_{-a}^a \left(b + \sqrt{a^2 - y^2} \right)^2 \, dy - \pi \int_{-a}^a \left(b - \sqrt{a^2 - y^2} \right)^2 \, dy = 4b\pi \int_{-a}^a \sqrt{a^2 - y^2} \, dy = 2a^2 b\pi^2$

39.   a) $V = 2\pi \int_0^a \left(\sqrt{a^2 - x^2} \right)^2 \, dx = 2\pi \left[a^2 x - \frac{x^3}{3} \right]_0^a = \frac{4a^3 \pi}{3}$

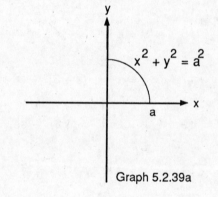

Graph 5.2.39a

      b) $V = \pi \int_0^h \left(\frac{rx}{h} \right)^2 \, dx = \frac{\pi r^2}{h^2} \int_0^h x^2 \, dx = \frac{\pi r^2}{h^2} \left[\frac{x^3}{3} \right]_0^h = \frac{\pi r^2 h}{3}$

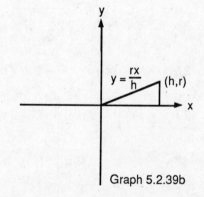

Graph 5.2.39b

41.   For each $0 \leq a \leq 12$, one half of the vertical distance between $y = x$ and $y = \frac{x}{2}$ is $\frac{a}{4}$. This distance is also the radius of the corresponding circular cross section. A right circular cone of height 12 and radius 3 is formed by rotating $y = \frac{1}{4}x$ about the x-axis when $0 \leq x \leq 12$. Since these solids have corresponding cross sections of the same size, by Cavalieri's theorem, they have the same volume.

43.   Area of the cross section is 2x. $V = \int_0^4 2x\, dx = \left[x^2\right]_0^4 = 16$

45.   Area of the cross section is $4(1 - x^2)$. $V = 2\int_0^1 4(1 - x^2)\, dx = 8\left[x - \frac{x^3}{3}\right]_0^1 = \frac{16}{3}$

47.   Area of the cross section is $\left(\frac{\sec x - \tan x}{2}\right)^2 \pi$. $V = \int_{-\pi/3}^{\pi/3} \left(\frac{\sec x - \tan x}{2}\right)^2 \pi\, dx =$

$\frac{1}{4}\int_{-\pi/3}^{\pi/3} 2\sec^2 x - \tan x \sec x - 1\, dx = \frac{1}{4}[2\tan x - \sec x - x]_{-\pi/3}^{\pi/3} = \frac{24\sqrt{3} - \pi}{24}$

49.   Area of the cross section is $\frac{1}{2}(2\sqrt{1 - y^2})(2\sqrt{1 - y^2}) = 2(1 - y^2)$. $V = 2\int_0^1 2(1 - y^2)\, dy =$

$4\left[y - \frac{y^3}{3}\right]_0^1 = \frac{8}{3}$

51. $\int_0^h s^2\, dx = s^2 h$. The volume does not depend on the number of revolutions.

5.3 CYLINDRICAL SHELLS – AN ALTERNATIVE TO WASHERS

1. $V = 2\pi \int_0^2 \left(x - \left(-\frac{x}{2}\right)\right)(x)\, dx = \left[\pi x^3\right]_0^2 = 8\pi$

3. $V = 2\pi \int_0^1 (x^2 + 1)x\, dx = 2\pi\left[\frac{x^4}{4} + \frac{x^2}{2}\right]_0^1 = \frac{3\pi}{2}$

5. $V = 2\pi \int_{1/2}^2 \left(\frac{1}{x}\right)(x)\, dx = [2\pi x]_{1/2}^2 = 3\pi$

7. $V = 2\pi \int_0^1 (2y)(y)\, dy = 4\pi\left[\frac{y^3}{3}\right]_0^1 = \frac{4\pi}{3}$

9. $V = 2\pi \int_0^2 \left((y + 2) - y^2\right)y\, dy = 2\pi\left[\frac{y^3}{3} + y^2 - \frac{y^4}{4}\right]_0^2 = \frac{16\pi}{3}$

11. $V = 2\pi \int_0^2 (2y - y^2)y\, dy = 2\pi\left[\frac{2y^3}{3} - \frac{y^4}{4}\right]_0^2 = \frac{8\pi}{3}$

13. $V = 2\pi \int_0^{\sqrt{3}} x\sqrt{x^2 + 1}\, dx = \pi \int_0^{\sqrt{3}} (x^2 + 1)^{1/2}(2x)\, dx = \frac{2\pi}{3}\left[(x^2 + 1)^{3/2}\right]_0^{\sqrt{3}} = \frac{14\pi}{3}$

15.   a) $V = 2\pi \int_0^1 12(y^2 - y^3)y \, dy = 24\pi \left[\dfrac{y^4}{4} - \dfrac{y^5}{5} \right]_0^1 = \dfrac{6\pi}{5}$

     b) $V = 2\pi \int_1^0 (1 - y)(12)\left(y^2 - y^3\right) dy = 24\pi \int_0^1 y^2 - 2y^3 + y^4 \, dy = 24\pi \left[\dfrac{y^3}{3} - \dfrac{y^4}{2} + \dfrac{y^5}{5} \right]_0^1 = \dfrac{4\pi}{5}$

     c) $V = 2\pi \int_0^1 \left(\dfrac{8}{5} - y \right)(12)\left(y^2 - y^3\right) dy = 24\pi \int_0^1 \dfrac{8}{5} y^2 - \dfrac{13}{5} y^3 + y^4 \, dy =$

     $24\pi \left[\dfrac{8x^3}{15} - \dfrac{14y^4}{20} + \dfrac{y^5}{5} \right]_0^1 = 2\pi$

     d) $V = 2\pi \int_0^1 \left(y + \dfrac{2}{5} \right)(12)\left(y^2 - y^3\right) dy = 24\pi \int_0^1 -y^4 + \dfrac{3}{5} y^3 + \dfrac{2}{5} y^2 \, dy =$

     $24\pi \left[-\dfrac{y^5}{5} + \dfrac{3y^4}{20} + \dfrac{2y^3}{15} \right]_0^1 = 2\pi$

17.   a) $V = 2\pi \int_1^2 y(y - 1) \, dy = 2\pi \left[\dfrac{y^3}{3} - \dfrac{y^2}{2} \right]_1^2 = \dfrac{5\pi}{3}$ b) $V = 2\pi \int_1^2 x(2 - x) \, dx = 2\pi \left[x^2 - \dfrac{x^3}{3} \right]_1^2 = \dfrac{4\pi}{3}$

     c) $V = 2\pi \int_1^2 \left(\dfrac{10}{3} - x \right)(2 - x) \, dx = 2\pi \left[\dfrac{20x}{3} - \dfrac{8x^2}{3} + \dfrac{x^3}{3} \right]_1^2 = 2\pi$

     d) $V = 2\pi \int_1^2 (y - 1)^2 \, dy = 2\pi \left[\dfrac{(y - 1)^3}{3} \right]_1^2 = \dfrac{2\pi}{9}$

19.   a) $V = 2\pi \int_0^1 y\left(1 - (y - y^3)\right) dy = 2\pi \left[\dfrac{y^2}{2} - \dfrac{y^3}{3} + \dfrac{y^5}{5} \right]_0^1 = \dfrac{11\pi}{15}$

     b) $V = \pi - \pi \int_0^1 (y - y^3)^2 \, dy = \pi - \pi \left[\dfrac{y^3}{3} - \dfrac{2y^5}{5} + \dfrac{y^7}{7} \right]_0^1 = \pi - \dfrac{8\pi}{105} = \dfrac{97\pi}{105}$

     c) $V = \pi \int_0^1 \left[1 - (y - y^3) \right]^2 \, dy = \pi \left[y - y^2 + \dfrac{y^4}{2} + \dfrac{y^3}{3} - \dfrac{2y^5}{5} + \dfrac{y^7}{7} \right]_0^1 = \dfrac{121\pi}{210}$

     d) $V = 2\pi \int_0^1 (1 - y)(1 - y + y^3) \, dy = 2\pi \left[y - y^2 + \dfrac{y^3}{3} + \dfrac{y^4}{4} - \dfrac{y^5}{5} \right]_0^1 = \dfrac{23\pi}{30}$

21.   a) $V = \pi \int_0^2 (4x)^2 - (x^3)^2 \, dx = \pi \left[\dfrac{16x^3}{3} - \dfrac{x^7}{7} \right]_0^2 = \dfrac{512\pi}{21}$

     b) $V = \pi \int_0^2 (8 - x^3)^2 - (8 - 4x)^2 \, dx = \pi \int_0^2 64x - 16x^2 - 16x^3 + x^6 \, dx =$

     $\pi \left[32x^2 - \dfrac{16x^3}{3} - 4x^4 + \dfrac{x^7}{7} \right]_0^2 = \dfrac{832\pi}{21}$

23.   a) $V = 2\pi \int_0^1 \left((2x - x^2) - x\right)(x) \, dx = 2\pi \left[\dfrac{x^3}{3} - \dfrac{x^4}{4} \right]_0^1 = \dfrac{\pi}{6}$

     b) $V = 2\pi \int_0^1 (1 - x)\left((2x - x^2) - x\right) dx = 2\pi \left[\dfrac{x^2}{2} - \dfrac{2x^3}{3} + \dfrac{x^4}{4} \right]_0^1 = \dfrac{\pi}{6}$

25.   By the shell method we have $2\pi b^3 = 2\pi \int_0^b x \, f(x) \, dx \Rightarrow x^3 = \int_0^x t \, f(t) \, dt$ where $x > 0$. By the

     Fundamental Theorem of Calculus we have $3x^2 = x \, f(x) \Rightarrow f(x) = 3x$.

5.4 LENGTHS OF CURVES IN THE PLANE

1. $y = x^2 \Rightarrow 1 + (y')^2 = 1 + 4x^2 \Rightarrow L = \int_{-1}^{2} \sqrt{1 + 4x^2}\, dx$

3. $y^2 + 2y = 2x + 1 \Rightarrow y' = \dfrac{1}{y+1} \Rightarrow 1 + \left(\dfrac{dx}{dy}\right)^2 = 1 + (y+1)^2 \Rightarrow L = \int_{-1}^{3} \sqrt{1 + (y+1)^2}\, dy$

5. $x = \sin y \Rightarrow \dfrac{dx}{dy} = \cos y \Rightarrow 1 + \left(\dfrac{dx}{dy}\right)^2 = 1 + \cos^2 y \Rightarrow L = \int_{0}^{\pi} \sqrt{1 + \cos^2 y}\, dy$

7. $y = \int_{0}^{x} \tan t\, dt \Rightarrow y' = \tan x \Rightarrow 1 + (y')^2 = 1 + \tan^2 x = \sec^2 x \Rightarrow L = \int_{0}^{\pi/6} \sec x\, dx$

9. $L = \int_{0}^{3} \sqrt{1 + \left(\dfrac{dy}{dx}\right)^2}\, dx = \int_{0}^{3} \sqrt{1 + x^2(x^2+2)}\, dx = \int_{0}^{3} \sqrt{(x^2+1)^2}\, dx = \dfrac{1}{3}\left[x^3 + x\right]_{0}^{3} = 12$

11. $L = \int_{0}^{3} \sqrt{1 + \left(\dfrac{dx}{dy}\right)^2}\, dy = \int_{0}^{3} \sqrt{1 + y}\, dy = \left[\dfrac{2(1+y)^{3/2}}{3}\right]_{0}^{3} = \dfrac{14}{3}$

13. $y = \dfrac{x^3}{3} + \dfrac{1}{4x},\ 1 \le x \le 3 \Rightarrow \dfrac{dy}{dx} = y^2 - \dfrac{1}{4x^2} \Rightarrow 1 + \left(\dfrac{dy}{dx}\right)^2 = x^4 + \dfrac{1}{2} + \dfrac{1}{16x^4} = \left(x^2 + \dfrac{1}{4x^2}\right)^2$

$L = \int_{1}^{3} \sqrt{1 + \left(\dfrac{dy}{dx}\right)^2}\, dx = \int_{1}^{3}\left(x^2 + \dfrac{1}{4x^2}\right) dx = \left[\dfrac{x^3}{3} - \dfrac{1}{4x}\right]_{1}^{3} = \dfrac{53}{6}$

15. $x = \dfrac{y^4}{4} + \dfrac{1}{8y^2} \Rightarrow \dfrac{dx}{dy} = y^3 - \dfrac{1}{4y^3} \Rightarrow 1 + \left(\dfrac{dx}{dy}\right)^2 = 1 + y^6 - \dfrac{1}{2} + \dfrac{1}{16y^6} = y^6 + \dfrac{1}{2} + \dfrac{1}{16y^6} = \left(y^3 + \dfrac{1}{4y^3}\right)^2$

$L = \int_{1}^{2} \sqrt{1 + \left(\dfrac{dx}{dy}\right)^2}\, dy = \int_{1}^{2}\left|y^3 + \dfrac{1}{4y^3}\right| dy = \int_{1}^{2} y^3 + \dfrac{1}{4y^3}\, dy = \left[\dfrac{y^4}{4} - \dfrac{1}{8y^2}\right]_{1}^{2} = \dfrac{123}{32}$

17. $y = \dfrac{3}{4}x^{4/3} - \dfrac{3}{8}x^{2/3} \Rightarrow y' = x^{1/3} - \dfrac{1}{4}x^{-1/3} \Rightarrow 1 + (y')^2 = \left(x^{1/3} + \dfrac{1}{4}x^{-1/3}\right)^2 \Rightarrow$

$L = \int_{1}^{8} x^{1/3} + \dfrac{1}{4}x^{-1/3}\, dx = \left[\dfrac{3}{4}x^{4/3} + \dfrac{3}{8}x^{2/3}\right]_{1}^{8} = \dfrac{99}{8}$

19. $x = \int_{0}^{y} \sqrt{\sec^4 t - 1}\, dt \Rightarrow 1 = \sqrt{\sec^4 y - 1}\ y' \Rightarrow \dfrac{dx}{dy} = \sqrt{\sec^4 y - 1} \Rightarrow 1 + \left(\dfrac{dx}{dy}\right)^2 = \sec^4 y \Rightarrow$

$L = \int_{-\pi/4}^{\pi/4} \sec^2 y\, dy = \left[\tan y\right]_{-\pi/4}^{\pi/4} = 2$

21. $x^{2/3} + y^{2/3} = 1 \Rightarrow \dfrac{dy}{dx} = -\dfrac{y^{1/3}}{x^{1/3}} = -\dfrac{(1 - x^{2/3})^{1/2}}{x^{1/3}} \Rightarrow 1 + \left(\dfrac{dy}{dx}\right)^2 = \dfrac{1}{x^{2/3}}.\ L = 8\int_{\sqrt{2}/4}^{1} \sqrt{1 + \left(\dfrac{dy}{dx}\right)^2}\, dx =$

$8\int_{\sqrt{2}/4}^{1} x^{-1/3}\, dx = \left[12x^{2/3}\right]_{\sqrt{2}/4}^{1} = 6$

23. $y = \int_0^x \sqrt{\cos 2t}\, dt \Rightarrow \dfrac{dy}{dt} = \sqrt{\cos 2t} \Rightarrow 1 + \left(\dfrac{dy}{dt}\right)^2 = 1 + \cos 2t. \quad L = \int_0^{\pi/4} \sqrt{1 + \left(\dfrac{dy}{dt}\right)^2}\, dt =$

$\int_0^{\pi/4} \sqrt{2 \cos^2 t}\, dt = \sqrt{2} \int_0^{\pi/4} \cos t\, dt = \sqrt{2}\left[\sin t\right]_0^{\pi/4} = 1$

25. $y = \dfrac{b}{a} x \Rightarrow 1 + (y')^2 = \dfrac{a^2 + b^2}{a^2} \Rightarrow L = \int_0^a \sqrt{\dfrac{a^2 + b^2}{a^2}}\, dx = \dfrac{\sqrt{a^2 + b^2}}{a} (a - 0) = \sqrt{a^2 + b^2}$

27. Use the Calculus Tool Kit. $L = \int_{-25}^{25} \sqrt{1 + \dfrac{\pi^2}{4} \sin^2\left(\dfrac{\pi x}{50}\right)}\, dx$ where $\dfrac{\pi^2}{4} \approx 2.4674011$ and

$\dfrac{\pi x}{50} \approx 0.062831853. \quad L \approx 73.1847737 \approx 73.18. \quad \text{Cost} = (73.18)(300)(1.75) = \$38419.50.$

5.5 AREAS OF SURFACES OF REVOLUTION

1. $y = \tan x \Rightarrow 1 + (y')^2 = 1 + \sec^4 x \Rightarrow S = 2\pi \int_{-\pi/4}^{\pi/4} \tan x \sqrt{1 + \sec^4 x}\, dx$

3. $xy = 1 \Rightarrow x = \dfrac{1}{y} \Rightarrow 1 + \left(\dfrac{dx}{dy}\right)^2 = 1 + y^{-4} \Rightarrow S = 2\pi \int_1^2 \dfrac{1}{y} \sqrt{1 + y^{-4}}\, dy$

5. $x^{1/2} + y^{1/2} = 3 \Rightarrow y' = -\dfrac{3 - x^{1/2}}{x^{1/2}} = 1 - 3x^{-1/2} \Rightarrow 1 + (y')^2 = 1 + \left(1 - 3x^{-1/2}\right)^2 \Rightarrow$

$S = 2\pi \int_1^4 \left(3 - \sqrt{x}\right)^2 \sqrt{1 + \left(1 - 3x^{-1/2}\right)^2}\, dx$

7. $x = \int_0^y \tan t\, dt \Rightarrow \dfrac{dx}{dy} = \tan y \Rightarrow 1 + \left(\dfrac{dx}{dy}\right)^2 = 1 + \tan^2 y = \sec^2 y \Rightarrow$

$S = 2\pi \int_0^{\pi/3} \left(\int_0^y \tan t\, dt\right) \sec y\, dy$

9. $S = 2\pi \int_0^4 \dfrac{x}{2} \sqrt{1 + \left(\dfrac{1}{2}\right)^2}\, dx = \dfrac{\pi\sqrt{5}}{2} \int_0^4 x\, dx = \dfrac{\pi\sqrt{5}}{2} \left[\dfrac{x^2}{2}\right]_0^4 = 4\pi\sqrt{5}$

11. $S = 2\pi \int_1^3 \dfrac{1}{2}(x + 1) \sqrt{1 + \left(\dfrac{1}{2}\right)^2}\, dx = \dfrac{\pi\sqrt{5}}{2} \int_1^3 (x + 1)\, dx = \dfrac{\pi\sqrt{5}}{2} \left[\dfrac{x^2}{2} + x\right]_1^3 = 3\pi\sqrt{5}$

13. $S = 2\pi \int_0^2 \dfrac{x^3}{9} \sqrt{1 + \left(\dfrac{x^2}{3}\right)^2}\, dx = \dfrac{\pi}{54} \int_0^2 (9 + x^4)^{1/2}(4x^3)\, dx = \dfrac{\pi}{81} \left[(9 + x^4)^{3/2}\right]_0^2 = \dfrac{98\pi}{81}$

15. $S = 2\pi \int_0^2 \sqrt{2x - x^2} \sqrt{1 + \left(\dfrac{1 - x}{\sqrt{2x - x^2}}\right)^2}\, dx = 2\pi \int_0^2 \sqrt{2x - x^2 + 1 - 2x + x^2}\, dx =$

$2\pi \int_0^2 dx = \left[2\pi x\right]_0^2 = 4\pi$

17. $S = 2\pi \int_0^1 \frac{y^3}{3} \sqrt{1 + (y^2)^2}\, dy = \frac{\pi}{9}\left[(1 + y^4)^{3/2}\right]_0^1 = \frac{\pi(\sqrt{8} - 1)}{9}$

19. $S = 2\pi \int_0^{15/4} 2\sqrt{4 - y}\left(\frac{\sqrt{5 - y}}{\sqrt{4 - y}}\right) dy = 4\pi \int_0^{15/4} (5 - y)^{1/2}\, dy = -\frac{8\pi}{3}\left[(5 - y)^{3/2}\right]_0^{15/4} = \frac{35\pi\sqrt{5}}{3}$

21. $S = 2\pi \int_1^2 y\, ds = 2\pi \int_1^2 y\sqrt{dx^2 + dy^2} = 2\pi \int_1^2 y\sqrt{(y^3 - y^{-3}/4)^2 + 1}\, dy = 2\pi \int_1^2 y^4 + \frac{1}{4} y^{-2}\, dy =$

$2\pi\left[\frac{y^5}{5} - \frac{y^{-1}}{4}\right]_1^2 = \frac{253\pi}{20}$

23. $S = 4\pi \int_0^a \sqrt{a^2 - x^2}\sqrt{1 + x^2(a^2 - x^2)^{-1}}\, dx = 4\pi \int_0^a \sqrt{a^2 - x^2}\sqrt{\frac{a^2 - x^2 + x^2}{a^2 - x^2}}\, dx =$

$4\pi \int_0^a a\, dx = 4\pi a^2$

25. a) $S = 2\pi \int_{-\pi/2}^{\pi/2} (\cos x)\sqrt{1 + \sin^2 x}\, dx$

   b) $S \approx 14.4236$

27. $x^2 + y^2 = 256 \Rightarrow ds = \frac{16}{y}\, dx$. Assume that the interior and exterior surfaces are approximately equal.

   Let S represent one of these surfaces. $S = 2\pi \int_{-3\sqrt{23}}^{3\sqrt{23}} y\, ds = 2\pi \int_{-3\sqrt{23}}^{3\sqrt{23}} y \frac{16}{y}\, dy = 192\sqrt{23}\pi$ cm$^2 \approx$

   2892.78 cm^2. The volume of one color is $\left(\frac{1}{10}\text{ cm}\right)(2892.78\text{ cm}^2) = 289.278$ cm$^3 = .289278$ lt/wok.

   Order 1446.39 liters of each color.

29. $y = x \Rightarrow 1 + (y')^2 = 2 \Rightarrow S = 2\pi \int_{-1}^0 - x\sqrt{2}\, dx + 2\pi \int_0^2 x\sqrt{2}\, dx = -2\sqrt{2}\pi\left[\frac{x^2}{2}\right]_{-1}^0 +$

$2\sqrt{2}\pi\left[\frac{x^2}{2}\right]_0^2 = 5\sqrt{2}\pi$

31. $y = \sin x \Rightarrow 1 + (y')^2 = 1 + \cos^2 x \Rightarrow S = 2\pi \int_0^\pi (\sin x)\sqrt{1 + \cos^2 x}\, dx \approx 14.4$

33. $y = x + \sin 2x \Rightarrow y' = 1 + 2\cos 2x \Rightarrow 1 + (y')^2 = 1 + (1 + 2\cos 2x)^2 \Rightarrow$

$S = 4\pi \int_0^{2\pi/3} (x + \sin 2x)\sqrt{1 + (1 + 2\cos 2x)^2}\, dx \approx 54.9$, since $(x + \sin 2x)$ is odd.

5.6 MOMENTS AND CENTERS OF MASS

1. $\overline{x} = \dfrac{50x + (5)(40)}{90} = 0 \Rightarrow x = -4$. The child is 4 ft from the fulcrum.

3. $M_o = \displaystyle\int_0^2 4x\,dx = \left[2x^2\right]_0^2 = 8$, $M = \displaystyle\int_0^2 4\,dx = [4x]_0^2 = 8 \Rightarrow \overline{x} = \dfrac{8}{8} = 1$

5. $M_o = \displaystyle\int_0^4 x\left(1 + \dfrac{x}{4}\right)^2 dx = \int_0^4 x + \dfrac{x^2}{2} + \dfrac{x^3}{16}\,dx = \left[\dfrac{x^2}{2} + \dfrac{x^3}{6} + \dfrac{x^4}{64}\right]_0^4 = \dfrac{68}{3}$, $M = \displaystyle\int_0^4 \left(1 + \dfrac{x}{4}\right)^2 dx =$

$\displaystyle\int_0^4 1 + \dfrac{x}{2} + \dfrac{x^2}{16}\,dx = \left[x + \dfrac{x^2}{4} + \dfrac{x^3}{48}\right]_0^4 = \dfrac{28}{3} \Rightarrow \overline{x} = \dfrac{17}{7}$

7. $M_x = \delta\displaystyle\int_{-2}^2 \left(\dfrac{4 + x^2}{2}\right)(4 - x^2)\,dx = \delta\int_0^2 16 - x^4\,dx = \delta\left[16x - \dfrac{x^5}{5}\right]_0^2 = \dfrac{128\delta}{5}$, $M =$

$2\delta\displaystyle\int_0^2 4 - x^2\,dx = 2\delta\left[4x - \dfrac{x^3}{3}\right]_0^2 = \dfrac{32\delta}{3} \Rightarrow \overline{y} = \dfrac{12}{5}$, and by symmetry $\overline{x} = 0$

9. $M_x = \delta\displaystyle\int_0^2 \left(\dfrac{x - x^2 - x}{2}\right)(x - x^2 + x)\,dx = \dfrac{\delta}{2}\int_0^2 x^4 - 2x^3\,dx = \dfrac{\delta}{2}\left[\dfrac{x^5}{5} - \dfrac{x^4}{2}\right]_0^2 = -\dfrac{4\delta}{5}$, $M_y =$

$\delta\displaystyle\int_0^2 x(2x - x^2)\,dx = \delta\left[\dfrac{2x^3}{3} - \dfrac{x^4}{4}\right]_0^2 = \dfrac{4\delta}{3}$, $M = \delta\displaystyle\int_0^2 (x - x^2) + x\,dx = \delta\left[x^2 - \dfrac{x^3}{3}\right]_0^2 = \dfrac{4\delta}{3} \Rightarrow$

$\overline{x} = 1$ and $\overline{y} = -\dfrac{3}{5}$

11. $M_x = \delta\displaystyle\int_0^1 y(y - y^3)\,dy = \delta\left[\dfrac{y^3}{3} - \dfrac{y^5}{5}\right]_0^1 = \dfrac{2\delta}{15}$, $M_y = \delta\displaystyle\int_0^1 \dfrac{x}{2}(y - y^3)\,dy = \delta\int_0^1 \dfrac{(y - y^3)^2}{2}\,dy =$

$\dfrac{\delta}{2}\left[\dfrac{y^3}{3} - \dfrac{2y^5}{5} + \dfrac{y^7}{7}\right]_0^1 = \dfrac{4\delta}{105}$, $M = \delta\displaystyle\int_0^1 y - y^3\,dy = \delta\left[\dfrac{y^2}{2} - \dfrac{y^4}{4}\right]_0^1 = \dfrac{\delta}{4} \Rightarrow \overline{x} = \dfrac{16}{105}$ and $\overline{y} = \dfrac{8}{15}$

13. $M_x = \delta\displaystyle\int_{-\pi/2}^{\pi/2} \dfrac{\cos^2 x}{2}\,dx = \dfrac{\delta}{4}\int_{-\pi/2}^{\pi/2} 1 + \cos x\,dx = \dfrac{\delta}{4}\left[x + \dfrac{\sin 2x}{2}\right]_{-\pi/2}^{\pi/2} = \dfrac{\pi\delta}{4}$, $M = \delta\displaystyle\int_{-\pi/2}^{\pi/2} \cos x\,dx =$

$\delta[\sin x]_{-\pi/2}^{\pi/2} = 2\delta \Rightarrow \overline{y} = \dfrac{\pi}{8}$ and by symmetry $\overline{x} = 0$

15. $M_x = \delta\displaystyle\int_0^2 \dfrac{(2x - x^2)^2 - (2x^2 - 4x)^2}{2}\,dx = \dfrac{\delta}{2}\int_0^2 -3x^4 + 12x^3 - 12x^2\,dx =$

$-\dfrac{3\delta}{2}\left[\dfrac{x^5}{5} - x^4 + \dfrac{4x^3}{3}\right]_0^2 = -\dfrac{8\delta}{5}$, $M_y = \delta\displaystyle\int_0^2 x\left[(2x - x^2) - (2x^2 - 4x)\right]dx = -\delta\int_0^2 3x^3 - 6x^2\,dx =$

$-\delta\left[\dfrac{3x^4}{4} - 2x^3\right]_0^2 = 4\delta$, $M = \delta\displaystyle\int_0^2 (2x - x^2) - (2x^2 - 4x)\,dx = \delta\left[3x^2 - x^3\right]_0^2 = 4\delta \Rightarrow$

$\overline{x} = 1$ and $\overline{y} = -\dfrac{2}{5}$

17. $M_y = \delta \int_0^3 x(3 - \sqrt{9-x^2})\, dx = \delta \int_0^3 3x\, dx + \frac{\delta}{2}\int_0^3 \sqrt{9-x^2}(-2x)\, dx =$

$\delta\left[\frac{3x^2}{2}\right]_0^3 + \frac{\delta}{3}\left[(9-x^2)^{3/2}\right]_0^3 = \frac{9\delta}{2}$, $M = \delta\int_0^3 3 - \sqrt{9-x^2}\, dx = \delta\int_0^3 3\, dx - \delta\int_0^3 \sqrt{9-x^2}\, dx$ (interpret

integral as area of a quarter circle) $= \delta[3x]_0^3 - \delta\left(\frac{9\pi}{4}\right) = \frac{\delta(36-9\pi)}{4} \Rightarrow$

$\overline{x} = \frac{2}{4-\pi}$ and by symmetry $\overline{y} = \overline{x}$

19. $y = x^{1/2} \Rightarrow dy = \frac{1}{2}x^{-1/2} \Rightarrow ds = \sqrt{(dx)^2 + (dy)^2} = \sqrt{1 + \frac{1}{4x}}\, dx$; $M_x = \delta\int_0^2 \sqrt{x}\sqrt{1 + \frac{1}{4x}}\, dx =$

$\delta\int_0^2 \sqrt{x + \frac{1}{4}}\, dx = \frac{2\delta}{3}\left[\left(x + \frac{1}{4}\right)^{3/2}\right]_0^2 = \frac{13\delta}{6}$

21. From example 6 we have $M_x = \int_0^\pi (a^2 \sin\theta)(k\sin\theta)\, d\theta = a^2 k\int_0^\pi \sin^2\theta\, d\theta =$

$\frac{a^2 k}{2}\int_0^\pi 1 - \cos 2\theta\, d\theta = \frac{a^2 k}{2}\left[\theta - \frac{\sin 2\theta}{2}\right]_0^\pi = \frac{a^2 k\pi}{2}$, $M_y = \int_0^\pi (a^2\cos\theta)(k\sin\theta)\, d\theta =$

$a^2 k\int_0^\pi \sin\theta\cos\theta\, d\theta = \frac{a^2 k}{2}\left[\sin^2\theta\right]_0^\pi = 0$ and $M = \int_0^\pi ak\sin\theta\, d\theta = ak\left[-\cos\theta\right]_0^\pi = 2ak.$

$\therefore\ \overline{x} = \frac{M_y}{M} = 0$ and $\overline{y} = \frac{M_y}{M} = \frac{a\pi}{4}$

23. $\overline{y} = \frac{1}{3}(3) = 1$ and by symmetry $\overline{x} = 0$

25. By symmetry $\overline{x} = \overline{y}$. The centroid is located at the intersection of medians: $y = x$ and $y = -\frac{x}{2} + \frac{a}{2}$.

$\therefore\ \overline{x} = \overline{y} = \frac{a}{3}$

27. Consider the curve as an infinite number of line segments joined together. From the derivation of arc length we have that the length of a particular segment is $ds = \sqrt{(dx)^2 + (dy)^2}$. This implies

$M_x = \int \delta y\, ds$, $M_y = \int \delta x\, ds$ and $M = \int \delta\, ds$. If δ is constant, then $\overline{x} = \frac{M_y}{M} = \frac{\int x\, ds}{\int ds} = \frac{\int x\, ds}{\text{length}}$ and

$\overline{y} = \frac{M_x}{M} = \frac{\int y\, ds}{\int ds} = \frac{\int y\, ds}{\text{length}}$.

29. A generalization of example 6 yields $\overline{y} = \dfrac{\int_{\pi/2-\alpha}^{\pi/2+\alpha} a^2\sin\theta\, d\theta}{\int_{\pi/2-\alpha}^{\pi/2+\alpha} a\, d\theta}$. $a^2\int_{\pi/2-\alpha}^{\pi/2+\alpha} \sin\theta\, d\theta =$

$a^2\left[-\cos\theta\right]_{\pi/2-\alpha}^{\pi/2+\alpha} = 2a^2\sin\alpha$ and $\int_{\pi/2-\alpha}^{\pi/2+\alpha} a\, d\theta = a\left[\left(\frac{\pi}{2}+\alpha\right) - \left(\frac{\pi}{2}-\alpha\right)\right] = 2\alpha a.$

$\therefore\ \overline{y} = \frac{2a^2\sin\alpha}{2\alpha a} = \frac{a\sin\alpha}{\alpha} = \frac{ac}{s}.$

5.7 WORK

1. $W = \int_0^{20} 40 - 2x \, dx = \left[40x - x^2\right]_0^{20} = 400$ ft · lb.

3. $W = \int_0^{50} (0.74)(50 - x) \, dx = (0.74)\left[50x - \dfrac{x^2}{2}\right]_0^{50} = 925$ N · m

5. $W = \int_0^{180} 4.5(180 - x) \, dx = 4.5\left[180x - \dfrac{x^2}{2}\right]_0^{180} = 72900$ ft · lb.

7. 2 N $= k(2$ cm$) \Rightarrow k = 1$ N/cm. 4 N $= (1$ N/cm$)(x) \Rightarrow x = 4$ cm, the distance a 4 N force will stretch the rubber band. The work needed to stretch the rubber band 4 cm is $W = \int_0^4 x \, dx = \left[\dfrac{x^2}{2}\right]_0^4 = 8$ N · m.

9. a) $21714 = k \cdot 3 \Rightarrow k = 7238$ lb/in

   b) $W = \int_0^{1/2} 7238x \, dx = \left[7238 \dfrac{x^2}{2}\right]_0^{1/2} = 904.75$ in · lb.; $W = \int_{1/2}^1 7238x \, dx =$

   $\left[7238 \dfrac{x^2}{2}\right]_{1/2}^1 = 2714.25$ in · lb.

11. a) $W = 62.5 \int_0^{20} (20 - y)(10)(12) \, dy = 120(62.5) \int_0^{20} 20 - y \, dy =$

   $7500\left[20y - \dfrac{y^2}{2}\right]_0^{20} = 1500000$ ft · lb.

   b) $\dfrac{1500000}{250}$ sec $= 6000$ sec $= 100$ min $= 1$ hr and 40 min.

   c) $W = 7500 \int_{10}^{20} 20 - y \, dy = 7500\left[20y - \dfrac{y^2}{2}\right]_{10}^{20} = 375000$ ft · lb, the work required to drop the

   water level 10 ft. The time needed to lower the water 10 feet is $\dfrac{37500 \text{ ft} \cdot \text{lb}}{\dfrac{60 \text{ sec}}{1 \text{ min}} \dfrac{250 \text{ ft} \cdot \text{lb}}{1 \text{ sec}}} = 25$ min.

13. $W = 62.5 \int_0^{10} 25\pi y \, dy = 1562.5\pi\left[\dfrac{y^2}{2}\right]_0^{10} \approx 245436.93$ ft · lb.

15. $W = \omega \int_0^{30} (30 - y)\pi 10^2 \, dy = 100\pi\omega\left[30y - \dfrac{y^2}{2}\right]_0^{30} = 45000\pi\omega \approx 7238229.48$ ft · lb.

17. $W = \omega \int_0^{16} \pi(\sqrt{y})^2(16 - y) \, dy = \omega\pi \int_0^{16} 16y - y^2 \, dy = \omega\pi\left[8y^2 - \dfrac{y^3}{3}\right]_0^{16} = \dfrac{2048\omega\pi}{3}$ N · m $\approx$

137258.28 N · m

19. $W = \omega \int_0^{10} \left(\sqrt{100-y^2}\right)^2 \pi(12-y)\,dy = \omega\pi \int_0^{10} 1200 - 100y - 12y^2 + y^3\,dy =$

$\omega\pi \left[1200y - 50y^2 - 4y^3 + \dfrac{y^4}{4} \right]_0^{10} = 5500\omega\pi \approx 967610.54 \text{ ft} \cdot \text{lb.}$

The cost is $(967610.54)(0.005) \approx \$4838.05$.

21. $90 \text{ MPH} = \dfrac{90 \text{ mi}}{1 \text{ hr}} \cdot \dfrac{1 \text{ hr}}{60 \text{ min}} \cdot \dfrac{1 \text{ min}}{60 \text{ sec}} \cdot \dfrac{5280 \text{ ft}}{1 \text{ mi}} = 132 \text{ ft/sec}; \quad m = \dfrac{0.3125 \text{ lb}}{32 \text{ ft/sec}^2} = \dfrac{0.3125}{32} \text{ slugs};$

$W = \left(\dfrac{1}{2}\right)\left(\dfrac{0.3125 \text{ lb}}{32 \text{ ft/sec}^2}\right)(132 \text{ ft/sec})^2 = 85.1 \text{ ft} \cdot \text{lb.}$

23. weight $= 2 \text{ oz} = \dfrac{1}{8} \text{ lb.}, \ m = \dfrac{1/8}{32} \text{ slugs} = \dfrac{1}{256} \text{ slugs}, \ 124 \text{ MPH} \approx 181.87 \text{ ft/sec};$

$W = \left(\dfrac{1}{2}\right)\left(\dfrac{1}{256} \text{ slugs}\right)(181.87 \text{ ft/sec})^2 \approx 64.6 \text{ ft} \cdot \text{lb.}$

25. weight $= 6.5 \text{ oz} = \dfrac{6.5}{16} \text{ lb.}, \ m = \dfrac{6.5}{16(32)} \text{ slugs}; \ W = \left(\dfrac{1}{2}\right)\left(\dfrac{6.5}{16(32)} \text{ slugs}\right)(132 \text{ ft/sec})^2 \approx 110.6 \text{ ft} \cdot \text{lb.}$

27. Let $a = 6370000$ and $b = 35780000$. $\int_a^b \dfrac{1000MG}{r^2}\,dr = 1000MG \int_a^b r^{-2}\,dr = -1000MG\left[\dfrac{1}{r}\right]_a^b =$

$1000MG\left(\dfrac{b-a}{ab}\right) = (1000)(5.975)(10^{24})(6.6720)(10^{-11})\left(\dfrac{29410000}{(35780000)(6370000)}\right) =$

$0.000005144 \times 10^{16} = 5.144 \times 10^{10} \text{ N} \cdot \text{m}$

5.8 FLUID PRESSURES AND FLUID FORCES

1. $F = \omega \int_0^3 (7-y)(2y)\,dy = 2\omega\left[\dfrac{7y^2}{2} - \dfrac{y^3}{3}\right]_0^3 = 45\omega = 2812.5 \text{ lb.}$

3. a) $F = 2\omega \int_0^3 \left(\dfrac{2y}{3}\right)(3-y)\,dy = \dfrac{4\omega}{3}\left[\dfrac{3y^2}{2} - \dfrac{y^3}{3}\right]_0^3 = 6\omega = 375 \text{ lb. against each end. No, the}$

length is of no concern.

b) Let x be the amount we must lower the water. From part a) we have

$2\omega \int_0^{3-x} \left(\dfrac{2y}{3}\right)(3-x-y)\,dy = \dfrac{375}{2} \text{ lb} \Rightarrow \dfrac{4\omega}{3}\left[\dfrac{3y^2}{2} - \dfrac{xy^2}{2} - \dfrac{y^3}{3}\right]_0^{3-x} = \dfrac{375}{2} \text{ lb.} \Rightarrow$

$9(3-x)^2 - 3x(3-x)^2 - 2(3-x)^3 = \dfrac{9(375)}{4w} \Rightarrow 9(3-x)^2 - 3x(3-x)^2 - (3-x) - \dfrac{9(375)}{4(62.5)} = 0 \Rightarrow$

$x = 0.618898$ ft, by Newton's method $\Rightarrow$ lower the water about 7.43 inches.

5. $F = 2\omega \int_{-1}^0 (3-y)(-y)\,dy = 2\omega\left[-\dfrac{3y^2}{2} + \dfrac{y^3}{3}\right]_{-1}^0 = \dfrac{11\omega}{3} \approx 229.17 \text{ lb.}$

7.   a)   The force on one side is $F = \omega \int_0^{11/6} (11/6 - y)(4) \, dy = 4\omega \left[\frac{11y}{6} - \frac{y^2}{2}\right]_0^{11/6} = \frac{121\omega}{18} \approx 420.14$ lb.

The force on the sides is 840.28 lb. The force on one end is $F = \omega \int_0^{11/6} \left(\frac{11}{6} - y\right) 2 \, dy =$

$2\omega \left[\frac{11y}{6} - \frac{y^2}{2}\right]_0^{11/6} = \frac{121\omega}{36} \approx 210.07$ lb. The force on the ends is 420.14 lb.

b)   The volume of the tank is 16 ft^3 and the volume of the water is $\frac{44}{3}$ ft^3. When the tank is on its

end the water is $\frac{11}{3}$ ft deep. $F = \omega \int_0^{11/3} (2)\left(\frac{11}{3} - y\right) dy = 2\omega \left[\frac{11y}{3} - \frac{y^2}{2}\right]_0^{11/3} =$

$\frac{121\omega}{9}$ lb. ≈ 840.28 lb.

9. $F = \frac{\omega}{1728} \int_0^{7.75} (7.75 - y)(3.75) \, dy = \frac{3.75\omega}{1728} \left[7.75y - \frac{y^2}{2}\right]_0^{7.75} = \frac{(3.75)(7.75^2)\omega}{2(1728)} \approx 4.2$ lb.

11.   a) $F = 2\omega \int_0^1 (2 - y)\sqrt{y} \, dy = 2\omega \int_0^1 2y^{1/2} - y^{3/2} \, dy = 2\omega \left[\frac{4y^{3/2}}{3} - \frac{2y^{5/2}}{5}\right]_0^1 = \frac{28\omega}{15} \approx 93.33$ lb.

b)   Let h represent the depth of the water. $F = 2\omega \int_0^1 (h - y)\sqrt{y} \, dy = 160 \Rightarrow$

$\omega \int_0^1 hy^{1/2} - y^{3/2} \, dy = 80 \Rightarrow \omega \left[\frac{2hy^{3/2}}{3} - \frac{2y^{5/2}}{5}\right]_0^1 = 80 \Rightarrow \omega \left(\frac{2h}{3} - \frac{2}{5}\right) = 80 \Rightarrow h = \frac{(8)(15)}{\omega} + \frac{3}{5} =$

$\frac{(8)(15)}{50} + \frac{3}{5} = 3$ ft.

13.   a) $F = 2\omega \int_0^1 (6 - y)y \, dy = 2\omega \int_0^1 (6y - y^2) \, dy = 2\omega \left[3y^2 - \frac{y^3}{3}\right]_0^1 = \frac{16\omega}{3} = 333\frac{1}{3}$ lb.

b)   Let h represent the depth of the water. $520 = 2\omega \int_0^1 (h - y)y \, dy = 2\omega \left[\frac{hy^2}{2} - \frac{y^3}{3}\right]_0^1 =$

$\frac{2\omega(3h - 2)}{6} \Rightarrow 3h - 2 = \frac{(6)(260)}{\omega} \Rightarrow h = \frac{520}{\omega} + \frac{2}{3} \approx 9$ ft.

15.   The force on the moveable side when the water reaches the top of the tank is $F =$

$2\omega \int_{-2}^0 \left(4 - y^2\right)^{1/2}(-y) \, dy = \omega \int_{-2}^0 \left(4 - y^2\right)^{1/2}(-2y) \, dy = \frac{2\omega}{3} \left[\left(4 - y^2\right)^{3/2}\right]_{-2}^0 = \frac{16\omega}{3}$. The

force compressing the spring when the tank is full is $\frac{16\omega}{3} \Rightarrow \frac{16\omega}{3} = 100x \Rightarrow$ the distance the

moveable side moves is $\frac{16\omega}{300} = \frac{10}{3}$ ft. $\therefore$ the tank will overflow.

5.9 The Basic Pattern. Other Modeling Applications

129

5.9 THE BASIC PATTERN. OTHER MODELING APPLICATIONS

1. a)

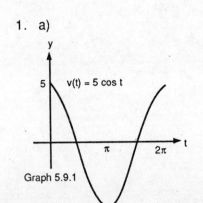

Graph 5.9.1

b)
$$\int_0^{2\pi} |5\cos t|\, dt = \int_0^{\pi/2} 5\cos t\, dt - \int_{\pi/2}^{3\pi/2} 5\cos t\, dt +$$

$$\int_{3\pi/2}^{2\pi} 5\cos t\, dt = 20 \text{ m}$$

c)
$$\int_0^{2\pi} 5\cos t\, dt = \left[5\sin t\right]_0^{2\pi} = 0 \text{ m}$$

3. a)

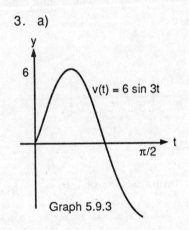

Graph 5.9.3

b)
$$\int_0^{\pi/2} |6\sin 3t|\, dt = \int_0^{\pi/3} 6\sin 3t\, dt - \int_{\pi/3}^{\pi/2} 6\sin 3t\, dt = 6 \text{ m}$$

c)
$$\int_0^{\pi/2} 6\sin 3t\, dt = 2 \text{ m}$$

5. a)

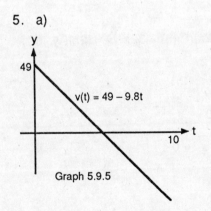

Graph 5.9.5

b)
$$\int_0^{10} |49 - 9.8t|\, dt = \int_0^5 49 - 9.8t\, dt -$$

$$\int_5^{10} 49 - 9.8t\, dt = 245 \text{ m}$$

c)
$$\int_0^{10} 49 - 9.8t\, dt = 0 \text{ m}$$

7. a)

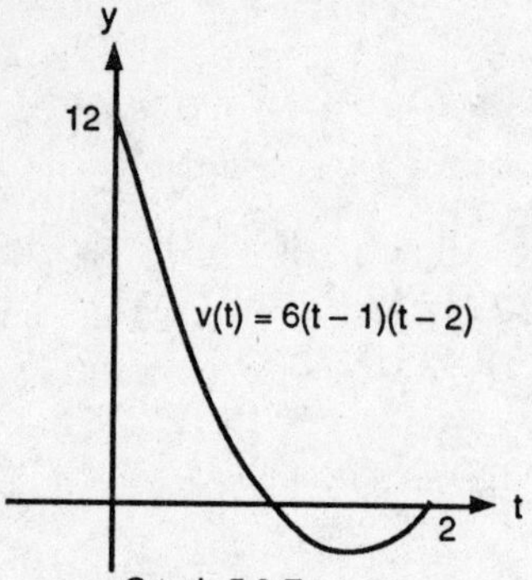

Graph 5.9.7

b) $\quad 6\int_0^2 |t^2 - 3t + 2|\, dt = 6\int_0^1 t^2 - 3t + 2\, dt -$

$\quad\quad 6\int_1^2 t^2 - 3t + 2\, dt = 6\text{ m}$

c) $\quad 6\int_0^2 t^2 - 3t + 2\, dt = 4\text{ m}$

9. a) $v(0) = \dfrac{dx}{dt}\Big|_{t=0} = t^2 - 6t + 8\Big|_{t=0} = 8 > 0 \Rightarrow$ a movement to the right

b) $t^2 - 6t + 8 = (t - 4)(t - 2) \Rightarrow$ the sign pattern

$\begin{array}{ccc} +\!+\!+\!+ & -\!-\!-\!- & +\;+\!+\!+ \\ \hline 0 \quad\quad 2 \quad\quad 4 \end{array}\; v$ which indicates

the particle is moving to the left when $2 < t < 4$

c) $s(3) = \dfrac{1}{3}t^3 - 3t^2 + 8t\Big|_{t=3} = 6\text{ m}$

d) distance $= \displaystyle\int_0^3 |t^2 - 6t + 8|\, dt = \int_0^2 t^2 - 6t + 8\, dt - \int_2^3 t^2 - 6t + 8\, dt = \left[\dfrac{t^3}{3} - 3t^2 + 8t\right]_0^2 -$

$\left[\dfrac{t^3}{3} - 3t^2 + 8t\right]_2^3 = \dfrac{22}{3}\text{ m}$

11. Out of 450 squares, about 136 are not shrimp-colored. Consequently, around 70% of the granular material are shrimp-colored.

13. $S = \displaystyle\int_0^{\sqrt{3}} 2\pi\dfrac{x}{\sqrt{3}}\, dx = \dfrac{2\pi}{\sqrt{3}}\left[\dfrac{x^2}{2}\right]_0^{\sqrt{3}} = \sqrt{3}\pi$

15. The centroid of the square is located at (2,2). $V = (2\pi)(\overline{y})(A) = (2\pi)(2)(8) = 32\pi$, $S = (2\pi)(\overline{y})(L) = (2\pi)(20)(4\sqrt{8}) = 32\sqrt{2}\pi$.

17. The centroid is located at (2,0), $V = (2\pi)(\overline{y})(A) = (2\pi)(2)(\pi) = 4\pi^2$

19. $S = 2\pi\overline{y}L \Rightarrow 4\pi a^2 = (2\pi\overline{y})(\pi a) \Rightarrow \overline{y} = \dfrac{2a}{\pi}$. By symmetry $\overline{x} = 0$

21. $V = 2\pi\overline{y}A \Rightarrow \dfrac{4}{3}\pi ab^2 = (2\pi\overline{y})\left(\dfrac{\pi ab}{2}\right) \Rightarrow \overline{y} = \dfrac{4b}{3\pi}$. By symmetry $\overline{x} = 0$

5.9 The Basic Pattern. Other Modeling Applications

131

23. $V = 2\pi \overline{y} A = (2\pi)$(area of the region)(distance from the centroid, $(0, 4a/3\pi)$, to the line $y = x - a$).

We must find the distance from $(0, 4a/3\pi)$ to $y = x - a$. The line containing the centroid and

perpendicular to $y = x - a$ has a slope of -1 and contains the point $(0, 4a/3\pi)$. This line is

$y = -x + \dfrac{4a}{3\pi}$. The intersection of $y = x - a$ and $y = -x + \dfrac{4a}{3\pi}$ is $\left(\dfrac{4a + 3a\pi}{6\pi}, \dfrac{4a - 3a\pi}{6\pi}\right)$. The

distance from the centroid to the line $y = x - a$ is $\sqrt{\left(\dfrac{4a + 3a\pi}{6\pi}\right)^2 + \left(\dfrac{4a}{3\pi} - \dfrac{4a}{6\pi} + \dfrac{3a\pi}{6\pi}\right)^2} =$

$\dfrac{\sqrt{2}(4a + 3a\pi)}{6\pi}$. $\therefore V = \left(\dfrac{\pi a^2}{2}\right)\left(\dfrac{\sqrt{2}(4a + 3a\pi)}{6\pi}\right)(2\pi) = \dfrac{\sqrt{2}\pi a^3(4 + 3\pi)}{6}$.

25. From example 4 and Pappus's Theorem for Volumes we have the moment about the x–axis is

$M_x = \overline{y} M = \left(\dfrac{4a}{3\pi}\right)\left(\dfrac{\pi a^2}{2}\right) = \dfrac{2a^3}{3}$.

5.M MISCELLANEOUS EXERCISES

1. $A = \displaystyle\int_{-2}^{1} (3 - x^2) - (x + 1)\, dx = \left[-\dfrac{x^3}{3} - \dfrac{x^2}{2} + 2x\right]_{-2}^{1} = \dfrac{9}{2}$

3. $A = \displaystyle\int_{0}^{\pi/4} (x - \sin x)\, dx = \left[\cos x + \dfrac{x^2}{2}\right]_{0}^{\pi/4} = \dfrac{-32 + 16\sqrt{2} + \pi^2}{32} \approx 0.0155$

5. $A = \displaystyle\int_{1}^{2} \sqrt{y} - (2 - y)\, dy = \left[\dfrac{2y^{3/2}}{3} - 2y + \dfrac{y^2}{2}\right]_{1}^{2} = \dfrac{8\sqrt{2} - 7}{6}$

7. $f(x) = x^3 - 3x^2 = x^2(x - 3) \Rightarrow f'(x) = 3x^2 - 6x = 3x(x - 2)$ and

    $\begin{array}{ccc} {\scriptstyle +++} & {\scriptstyle ----} & {\scriptstyle +++} \\ \hline & | & | \\ & 0 & 2 \end{array}\ f'(x)$

    implies at $x = 0$, $f(0) = 0$ a maximum and at $x = 2$, $f(2) =$

    -4 a minimum. $A = -\displaystyle\int_{0}^{3} x^3 - 3x^2\, dx = -\left[\dfrac{x^4}{4} - x^3\right]_{0}^{3} = \dfrac{27}{4}$

9. The area above the x–axis is $A_1 = \displaystyle\int_{0}^{1} y^{2/3} - y\, dy = \left[\dfrac{3y^{5/3}}{5} - \dfrac{y^2}{2}\right]_{0}^{1} = \dfrac{1}{10}$, the area below the

    x–axis is $A_2 = \displaystyle\int_{-1}^{0} y^{2/3} - y\, dy = \left[\dfrac{3y^{5/3}}{5} - \dfrac{y^2}{2}\right]_{-1}^{0} = \dfrac{11}{10} \Rightarrow$ the total area is $A_1 + A_2 = \dfrac{6}{5}$

11. $V = \dfrac{\pi}{4}\displaystyle\int_{0}^{1} x - 2x^{5/2} + x^4\, dx = \dfrac{\pi}{4}\left[\dfrac{x^2}{2} - \dfrac{4}{7}x^{7/2} + \dfrac{x^5}{5}\right]_{0}^{1} = \dfrac{9\pi}{280} \approx 0.10097$

13. $V = \displaystyle\int_{\pi/4}^{5\pi/4} \pi(\sin x - \cos x)^2\, dx = \pi\displaystyle\int_{\pi/4}^{5\pi/4} 1 - 2\sin x \cos x\, dx = \pi\left[x - \sin^2 x\right]_{\pi/4}^{5\pi/4} = \pi^2$

15. $V = \pi \int_0^4 \left(\sqrt{x} - \frac{x^2}{8}\right)^2 dx = \pi \left[\frac{x^2}{2} - \frac{x^{7/2}}{14} + \frac{x^5}{320}\right]_0^4 = \frac{72\pi}{35}$

17. a) $V = 2\pi \int_0^1 \left(3x^4\right)^2 dx = \left[2\pi x^9\right]_0^1 = 2\pi$

b) $V = 2\pi \int_0^1 x\left(3x^4\right) dx = \pi \left[x^6\right]_0^1 = \pi$

c) $V = 2\pi \int_{-1}^1 (1 - x) \, 3x^4 \, dx = 2\pi \left[\frac{3x^5}{5} - \frac{x^6}{2}\right]_{-1}^1 = \frac{12\pi}{5}$

d) $V = 2\pi \int_0^1 3^2 - \left(3 - 3x^4\right)^2 dx = 18\pi \int_0^1 1 - 1 + 2x^4 - x^8 \, dx = 18\pi \left[\frac{2x^5}{5} - \frac{x^9}{9}\right]_0^1 = \frac{26\pi}{5}$

19. a) $V = \pi \int_1^5 (\sqrt{x} - 1)^2 dx = \pi \left[\frac{x^2}{2} - x\right]_1^5 = 8\pi$

b) $V = 2\left(50\pi - \pi \int_0^2 (y^2 + 1)^2 dy\right) = 100\pi - 2\pi \int_0^2 y^4 + 2y^2 + 1 \, dy =$

$100\pi - 2\pi \left[\frac{y^5}{5} + \frac{2y^3}{3} + y\right]_0^2 = \frac{1088\pi}{15} \approx 227.87$

c) $V = 2\pi \int_0^2 \left(5 - (y^2 + 1)\right)^2 dy = 2\pi \left[16y - \frac{8y^3}{3} + \frac{y^5}{5}\right]_0^2 = \frac{512\pi}{15} \approx 107.233$

21. $V = \pi \int_0^{\pi/3} \tan^2 x \, dx = \pi \int_0^{\pi/3} \sec^2 x - 1 \, dx = \pi [\tan x - x]_0^{\pi/3} = \frac{\pi(3\sqrt{3} - \pi)}{3}$

23. a) $V = \pi \int_0^2 \left(x^2 - 2x\right)^2 dx = \pi \left[\frac{x^5}{5} - x^4 + \frac{4x^3}{3}\right]_0^2 = \frac{16\pi}{15}$

b) $V = 2\pi - \pi \int_0^2 \left(1 + (x^2 - 2x)\right)^2 dx = 2\pi - \pi \left[\frac{x^5}{5} - x^4 + 2x^3 - 2x^2 + x\right]_0^2 = 2\pi - \frac{2\pi}{5} = \frac{8\pi}{5}$

c) $V = 2\pi \int_0^2 (2 - x)(-1)\left(x^2 - 2x\right) dx = 2\pi \left[\frac{x^4}{4} - \frac{4x^3}{3} + 2x^2\right]_0^2 = \frac{8\pi}{3}$

d) $V = \pi \int_0^2 \left(2 - \left(x^2 - 2x\right)\right)^2 dx - 8\pi = \pi \left[\frac{x^5}{5} - x^4 + 4x^2 + 4x\right]_0^2 - 8\pi = \frac{32\pi}{5}$

25. $V = \frac{32\pi}{3} - 2\pi \int_0^1 \left(\sqrt{4 - x^2}\right)^2 - \left(\sqrt{3}\right)^2 dx = \frac{32\pi}{3} - 2\pi \left[x - \frac{x^3}{3}\right]_0^1 = \frac{28\pi}{3}$

27. $V = \pi \int_a^b f(x)^2 dx = b^2 - ab \Rightarrow \pi \int_a^x f(t)^2 dt = x^2 - ax$ for all $x > a \Rightarrow \pi f(x)^2 = 2x - a \Rightarrow f(x) = \sqrt{\frac{2x - a}{\pi}}$

29. $y = x^{1/2} - \dfrac{x^{3/2}}{3} \Rightarrow y' = \dfrac{1}{2}x^{-1/2} - \dfrac{1}{2}x^{1/2} \Rightarrow 1 + (y')^2 = \left(\dfrac{x^{-1/2} + x^{1/2}}{2}\right)^2.\ L = \displaystyle\int_0^3 \sqrt{1 + (y')^2}\, dx =$

$\dfrac{1}{2}\displaystyle\int_0^3 x^{-1/2} + x^{1/2}\, dx = \dfrac{1}{2}\left[2x^{1/2} + \dfrac{2}{3}x^{3/2}\right]_0^3 = 2\sqrt{3}$

31. $y = \dfrac{5}{12}x^{6/5} - \dfrac{5}{8}x^{4/5} \Rightarrow y' = \dfrac{x^{1/5}}{2} - \dfrac{x^{-1/5}}{2} \Rightarrow 1 + (y')^2 = \left(\dfrac{1}{2}x^{1/5} + \dfrac{1}{2}x^{-1/5}\right)^2;\ L = \displaystyle\int_1^{32} \sqrt{1 + (y')^2}\, dx =$

$\dfrac{1}{2}\displaystyle\int_1^{32} x^{1/5} + x^{-1/5}\, dx = \dfrac{1}{2}\left[\dfrac{5}{6}x^{6/5} + \dfrac{5}{4}x^{4/5}\right]_1^{32} = \dfrac{285}{8}$

33. $y = \sqrt{2x+1} \Rightarrow 1 + (y')^2 = \dfrac{2x+2}{2x+1}.\ S = \displaystyle\int_0^{12} 2\pi\sqrt{2x+1}\ \dfrac{\sqrt{2x+2}}{\sqrt{2x+1}}\, dx = \pi\displaystyle\int_0^{12}(2x+2)^{1/2}(2)\, dx =$

$\dfrac{(2\pi)}{3}\left[(2x+2)^{3/2}\right]_0^{12} = \dfrac{2^{5/2}\pi(13^{3/2} - 1)}{3} \approx 271.739$

35. $x = \dfrac{y^{3/2}}{3} - y^{1/2} \Rightarrow \dfrac{dx}{dy} = \dfrac{y^{1/2}}{2} - \dfrac{y^{-1/2}}{2} \Rightarrow 1 + \left(\dfrac{dx}{dy}\right)^2 = \left(\dfrac{1}{2}y^{1/2} + \dfrac{1}{2}y^{-1/2}\right)^2 \Rightarrow$

$A = 2\pi\displaystyle\int_4^9\left(\dfrac{1}{3}y^{3/2} - y^{1/2}\right)\left(\dfrac{1}{2}y^{1/2} + \dfrac{1}{2}y^{-1/2}\right)dy = \pi\displaystyle\int_4^9 \dfrac{1}{3}y^2 - \dfrac{2}{3}y - 1\ dy = \pi\left[\dfrac{y^3}{9} - \dfrac{y^2}{3} - y\right]_4^9 = \dfrac{425\pi}{9}$

37. $y = 2\sqrt{x} \Rightarrow ds = \sqrt{\dfrac{1}{x} + 1}\, dx \Rightarrow A = \displaystyle\int_0^3 2\sqrt{x}\sqrt{\dfrac{1}{x} + 1}\, dx = \dfrac{4}{3}\left[(1+x)^{3/2}\right]_0^3 = \dfrac{28}{3}$

39. $M_x = \displaystyle\int_{-2}^2 \dfrac{(x^2)^2}{2}\, dx = \left[\dfrac{x^5}{10}\right]_{-2}^2 = \dfrac{32}{5},\ M = 2\displaystyle\int_0^2 x^2\, dx = \left[\dfrac{2x^3}{3}\right]_0^2 = \dfrac{16}{3}.\ \overline{y} = \dfrac{M_x}{M} = \dfrac{6}{5}$, and by

symmetry $\overline{x} = 0$

41. $M_x = 3\displaystyle\int_0^4\left(\dfrac{\sqrt{x} + x/2}{2}\right)(\sqrt{x} - x/2)\, dx = \dfrac{3}{2}\left[\dfrac{x^2}{2} - \dfrac{x^3}{12}\right]_0^4 = 4,\ M_y = 3\displaystyle\int_0^4 x(\sqrt{x} - x/2)\, dx = 3\left[\dfrac{2x^{5/2}}{5} - \dfrac{x^6}{6}\right]_0^4 =$

$\dfrac{32}{5},\ M = 3\displaystyle\int_0^2(2y - y^2)\, dy = 3\left[y^2 - \dfrac{y^3}{3}\right]_0^2 = 4.\ \overline{x} = \dfrac{M_y}{M} = \dfrac{8}{5}$ and $\overline{y} = \dfrac{M_x}{M} = 1$

43. $M_x = \displaystyle\int_{-1}^1 \dfrac{(1 - x^n)^2}{2}\, dx = \dfrac{1}{2}\displaystyle\int_{-1}^1 1 - 2x^n + x^{2n}\, dx = \dfrac{1}{2}\left[x - \dfrac{2x^{n+1}}{n+1} + \dfrac{x^{2n+1}}{2n+1}\right]_{-1}^1 = \dfrac{2n^2}{(n+1)(2n+1)}$,

$M = \displaystyle\int_{-1}^1\left(1 - x^n\right)dx = \left[x - \dfrac{x^{n+1}}{n+1}\right]_{-1}^1 = \dfrac{2n}{n+1} \Rightarrow \overline{y} = \dfrac{M_x}{M} = \dfrac{2n^2}{(n+1)(2n+1)}\cdot\dfrac{n+1}{2n} = \dfrac{n}{2n+1}$,

$\overline{x} = 0$, by symmetry; $\displaystyle\lim_{n\to\infty} \overline{y} = \lim_{n\to\infty}\dfrac{n}{2n+1} = \dfrac{1}{2} \Rightarrow$ the limiting position is $\left(0, \dfrac{1}{2}\right)$

45. Weight of the equipment is (10 kg)(9.8 m/sec^2) = 98 newtons. The work needed to pull up the
     equipment is (98 N)(40 m) = 3920 N · m. The work needed to pull up the rope is W =
     $\int_0^{40} (0.8)(40-x)\,dx = 0.8\left[40x - \dfrac{x^2}{2}\right]_0^{40} = 640$ N · m. The total amount of work is

     3920 N · m + 640 N · m = 4560 N · m.

47. The work it takes to stretch the spring 1 ft is W = $\int_0^1 20x\,dx = \left[10x^2\right]_0^1 = 10$ ft · lb. The work

     needed to stretch the spring an additional foot is W = $\int_1^2 20x\,dx = \left[10x^2\right]_1^2 = 30$ ft · lb.

49. W = $\int_a^b \dfrac{k}{x^2}\,dx = \left[-\dfrac{k}{x}\right]_a^b = \dfrac{k(b-a)}{ab}$

51. W = $57\int_{-4}^0 (10-y)(20)\left(2\sqrt{16-y^2}\right)dy = 2280\int_{-4}^0 10\sqrt{16-y^2}\,dy +$

     $1140\int_{-4}^0 \left(16-y^2\right)^{1/2}(-2y)\,dy = 22800($ area of a quarter circle having a radius of 4 $) +$

     $\dfrac{2}{3}(1140)\left[\left(16-y^2\right)^{3/2}\right]_{-4}^0 = 22800(4\pi) + 48640 = 335153.25$ ft · lb

53. F = ma $\Rightarrow$ ma = $t^2 \Rightarrow$ m$\dfrac{d^2x}{dt^2} = t^2 \Rightarrow \dfrac{dx}{dt} = \dfrac{t^3}{3m} + c_1$, but $\dfrac{dx}{dt} = 0$ at t = 0 $\Rightarrow \dfrac{dx}{dt} = \dfrac{t^3}{3m} \Rightarrow x = \dfrac{t^4}{12m} + c_2$, but

     x = 0 at t = 0 $\Rightarrow x = \dfrac{t^4}{12m}$. $\therefore$ W = $\int_0^h F(t)\,dx = \int_0^h t^2\,dx = \int_0^h \sqrt{12mx}\,dx = \dfrac{2}{3}\left[\dfrac{(12mx)^{3/2}}{12m}\right]_0^h =$

     $\dfrac{4h\sqrt{3mh}}{3}$.

55. F = $2\omega\int_0^2 (2-y)(2y)\,dy = 4\omega\left[y^2 - \dfrac{y^3}{3}\right]_0^2 = \dfrac{16\omega}{3} \approx 333.33$ lb

57. F = $2\omega\int_0^4 (9-y)\left(\dfrac{\sqrt{y}}{2}\right)dy = \omega\left[6y^{3/2} - \dfrac{2}{5}y^{5/2}\right]_0^4 = \dfrac{176\omega}{5} = 2200$ lb

59. F = $\omega_1\int_0^6 (8-y)(2)(6-y)\,dy + \omega_2\int_{-6}^0 (8-y)(2)(y+6)\,dy = 2\omega_1\int_0^6 48 - 14y + y^2\,dy +$

     $2\omega_2\int_{-6}^0 48 + 2y - y^2\,dy = 2\omega_1\left[48y - 7y^2 + \dfrac{y^3}{3}\right]_0^6 + 2\omega_2\left[48y + y^2 - \dfrac{y^3}{3}\right]_{-6}^0 = 216\omega_1 + 360\omega_2$

61.   a)

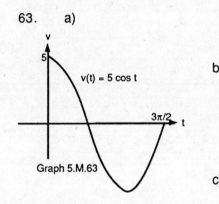

Graph 5.M.61

b) $\int_0^3 \left|3t^2 - 15t + 18\right| dt = 3\int_0^2 t^2 - 5t + 6\ dt - 3$

$\int_2^3 t^2 - 5t + 6\ dt = 3\left[\dfrac{t^3}{3} - \dfrac{5t^2}{2} + 6t\right]_0^2 - 3\left[\dfrac{t^3}{3} - \dfrac{5t^2}{2} + 6t\right]_2^3 =$

$\dfrac{29}{2}$ m

c) $\int_0^3 3t^2 - 15t + 18\ dt = 3\left[\dfrac{t^3}{3} - \dfrac{5t^2}{2} + 6t\right]_0^3 = \dfrac{27}{2}$ m

63.   a)

Graph 5.M.63

b) $\int_0^{3\pi/2} |5\cos t|\ dt = \int_0^{\pi/2} 5\cos t\ dt - \int_{\pi/2}^{5\pi/2} 5\cos t\ dt =$

$\left[5\sin t\right]_0^{\pi/2} - \left[5\sin t\right]_2^{3\pi/2} = 15$ ft

c) $\int_0^{3\pi/2} 5\cos t\ dt = \left[5\sin t\right]_0^{3\pi/2} = -5$ ft

CHAPTER 6

THE CALCULUS OF TRANSCENDENTAL FUNCTIONS

6.1 INVERSE FUNCTIONS

1. a) $f^{-1}(x) = \dfrac{x-3}{2}$

    b)
    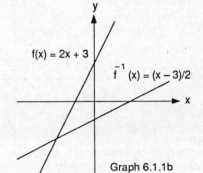
    Graph 6.1.1b

    c) $\left.\dfrac{df}{dx}\right|_{x=-1} = 2\left.\vphantom{\dfrac{df}{dx}}\right|_{x=-1} = 2$

    $\left.\dfrac{df^{-1}}{dx}\right|_{x=1} = \dfrac{1}{2}\left.\vphantom{\dfrac{df}{dx}}\right|_{x=1} = \dfrac{1}{2}$

3. a) $f^{-1}(x) = 5x - 35$

    b)
    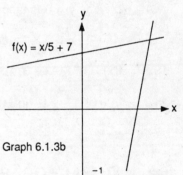
    Graph 6.1.3b

    c) $\left.\dfrac{df}{dx}\right|_{x=-1} = \dfrac{1}{5}\left.\vphantom{\dfrac{df}{dx}}\right|_{x=-1} = \dfrac{1}{5}$

    $\left.\dfrac{df^{-1}}{dx}\right|_{x=34/5} = 5\left.\vphantom{\dfrac{df}{dx}}\right|_{x=34/5} = 5$

5. $f(x) = x^2 + 1,\ x \ge 0 \Rightarrow y = x^2 + 1.\ x = y^2 + 1 \Rightarrow y = \pm\sqrt{x-1}$, and $(2,5)$ is on the graph of $y = f(x) \Rightarrow$ $(5,2)$ is on the graph of $y = f^{-1}(x) \Rightarrow f^{-1}(x) = \sqrt{x-1}$.

7. $f(x) = x^3 - 1 \Rightarrow y = x^3 - 1.\ x = y^3 - 1 \Rightarrow y = \sqrt[3]{x+1}$.

9. $f(x) = x^5 \Rightarrow y = x^5.\ x = y^5 \Rightarrow y = \sqrt[5]{x} \Rightarrow f^{-1}(x) = \sqrt[5]{x}.\ f(f^{-1}(x)) = f(\sqrt[5]{x}) = \left(\sqrt[5]{x}\right)^5 = x$ and

    $f^{-1}(f(x)) = f^{-1}(x^5) = \sqrt[5]{x^5} = x$.

11. $f(x) = x^3 + 1 \Rightarrow y = x^3 + 1.\ x = y^3 + 1 \Rightarrow y = \sqrt[3]{x-1} \Rightarrow f^{-1}(x) = \sqrt[3]{x-1}.\ f(f^{-1}(x)) = f(\sqrt[3]{x-1}) =$

    $\left(\sqrt[3]{x-1}\right)^3 + 1 = x - 1 + 1 = x$ and $f^{-1}(f(x)) = f^{-1}(x^3+1) = \sqrt[3]{(x^3+1)-1} = x$.

13. $f(x) = \dfrac{1}{x^2},\ x > 0 \Rightarrow y = \dfrac{1}{x^2}.\ x = \dfrac{1}{y^2} \Rightarrow y = \dfrac{1}{\sqrt{x}} \Rightarrow f^{-1}(x) = \dfrac{1}{\sqrt{x}},\ x > 0.\ f(f^{-1}(x)) = f\left(\dfrac{1}{\sqrt{x}}\right) = \dfrac{1}{\left(\dfrac{1}{\sqrt{x}}\right)^2} = \dfrac{1}{\dfrac{1}{x}} = x$

    and $f^{-1}(f(x)) = f^{-1}\left(\dfrac{1}{x^2}\right) = \dfrac{1}{\sqrt{\dfrac{1}{x^2}}} = \dfrac{1}{\dfrac{1}{\sqrt{x^2}}} = \sqrt{x^2} = x,\ x > 0$.

15. $f(x) = (x + 1)^2, x \geq 1 \Rightarrow y = (x + 1)^2.$ $x = (y + 1)^2 \Rightarrow y = \sqrt{x} - 1 \Rightarrow f^{-1}(x) = \sqrt{x} - 1.$ $f(f^{-1}(x)) =$
     $f(\sqrt{x} - 1) = \left[(\sqrt{x} - 1) + 1\right]^2 = \left[\sqrt{x}\right]^2 = x$ and $f^{-1}(f(x)) = f^{-1}\left((x + 1)^2\right) = \sqrt{(x + 1)^2} - 1 =$
     $|x + 1| - 1 = (x + 1) - 1 = x, \ x \geq 1.$

17. a)

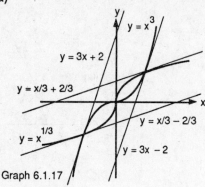

Graph 6.1.17

     b) $y = 0$ is tangent to $y = x^3$ at $x = 0.$

             $x = 0$ is tangent to $y = \sqrt[3]{x}$ at $x = 0.$

19. $f(x) = x^2 - 4x - 3, x > 2 \Rightarrow f'(x) = 2x - 4.$ $\left.\dfrac{df^{-1}}{dx}\right|_{x = -3} = \left.\dfrac{1}{2x - 4}\right|_{x = 4} = \dfrac{1}{4}.$

21. $(g \circ f)(x) = x \Rightarrow g(f(x)) = x \Rightarrow g'(f(x)) \, f'(x) = 1$

6.2 NATURAL LOGARITHMS

1.   a) $\ln 0.75 = \ln \dfrac{3}{4} = \ln 3 - \ln 4$ b) $\ln \dfrac{4}{9} = 2(\ln 2 - \ln 3)$

     c) $\ln \dfrac{1}{2} = -\ln 2$ d) $\ln \sqrt[3]{3} = \dfrac{1}{3} \ln 3$

     e) $\ln 3\sqrt{2} = \ln 3 + \dfrac{1}{2} \ln 2$ f) $\ln \sqrt{13.5} = \dfrac{1}{2}(3 \ln 3 - \ln 2)$

3. $\dfrac{d}{dx} \ln 3x = \dfrac{3}{3x} = \dfrac{1}{x}$ 5. $\dfrac{d}{dx} \ln x^3 = \dfrac{3x^2}{x^3} = \dfrac{3}{x}$

7. $\dfrac{d}{dx} \ln x^{3/2} = \dfrac{(3/2) \, x^{1/2}}{x^{3/2}} = \dfrac{3}{2x}$ 9. $\dfrac{d}{dx} \ln(x + 2) = \dfrac{1}{x + 2}$

11. $\dfrac{d}{dx} \ln(2 - \cos x) = \dfrac{\sin x}{2 - \cos x}$ 13. $\dfrac{d}{dx} \ln(\ln x) = \dfrac{1/x}{\ln x} = \dfrac{1}{x \ln x}$

15. $\dfrac{d}{dx} \left(x \sin(\ln x) + \cos(\ln x)\right) = \sin(\ln x) + x \cos(\ln x)\left(\dfrac{1}{x}\right) - \sin(\ln x)\left(\dfrac{1}{x}\right) =$
     $\sin(\ln x) + \cos(\ln x) - \dfrac{\sin(\ln x)}{x}$

17. $\dfrac{d}{dx} \ln \dfrac{1}{x\sqrt{x+1}} = -\dfrac{d}{dx}\left(\ln x + \dfrac{1}{2}\ln(x+1)\right) = -\left(\dfrac{1}{x} + \dfrac{1}{2(x+1)}\right) = -\dfrac{3x+2}{2x(x+1)}$

19. $\dfrac{d}{dx}\left(\dfrac{1+\ln x}{1-\ln x}\right) = \dfrac{\left(\frac{1}{x}\right)(1-\ln x) - \left(-\frac{1}{x}\right)(1+\ln x)}{(1-\ln x)^2} = \dfrac{2}{x(1-\ln x)^2}$

21. $\dfrac{d}{dx}\ln(\sec(\ln x))\dfrac{\sec(\ln x)\tan(\ln x)}{x\sec(\ln x)} = \dfrac{\tan(\ln x)}{x}$

23. $\dfrac{d}{dx}\ln\left[\dfrac{\left(x^2+1\right)^5}{\sqrt{1-x}}\right] = \dfrac{d}{dx}\left[5\ln\left(x^2+1\right) - \dfrac{1}{2}\ln(1-x)\right] = \dfrac{10x}{x^2+1} + \dfrac{1}{2(1-x)}$

25. $\dfrac{d}{dx}\displaystyle\int_{x^2/2}^{x^2}\ln\sqrt{t}\,dt = 2x\ln|x| - x\ln\dfrac{|x|}{\sqrt{2}}$

27. $y = \sqrt{x(x+1)} = (x(x+1))^{1/2} \Rightarrow \ln y = \dfrac{1}{2}\ln(x(x+1)) \Rightarrow 2\ln y = \ln(x) + \ln(x+1) \Rightarrow \dfrac{2y'}{y} = \dfrac{1}{x} + \dfrac{1}{x+1} \Rightarrow$

$y' = \left(\dfrac{1}{2}\right)\sqrt{x(x+1)}\left(\dfrac{1}{x} + \dfrac{1}{x+1}\right)$

29. $y = \sqrt{x+3}\sin x = (x+3)^{1/2}\sin x \Rightarrow \ln y = \left(\dfrac{1}{2}\right)\ln(x+3) + \ln(\sin x) \Rightarrow$

$y' = \sqrt{x+3}(\sin x)\left(\dfrac{1}{2(x+3)} + \cot x\right)$

31. $y = x(x+1)(x+2) \Rightarrow \ln y = \ln(x) + \ln(x+1) + \ln(x+2) \Rightarrow \dfrac{y'}{y} = \dfrac{1}{x} + \dfrac{1}{x+1} + \dfrac{1}{x+2} \Rightarrow$

$y' = x(x+1)(x+2)\left[\dfrac{1}{x} + \dfrac{1}{x+1} + \dfrac{1}{x+2}\right]$

33. $y = \dfrac{x+5}{x\cos x} \Rightarrow \ln y = \ln(x+5) - \ln(x) - \ln(\cos x) \Rightarrow \dfrac{y'}{y} = \dfrac{1}{x+5} - \dfrac{1}{x} + \dfrac{\sin x}{\cos x} \Rightarrow$

$y' = \dfrac{x+5}{x\cos x}\left[\dfrac{1}{x+5} - \dfrac{1}{x} + \tan x\right]$

35. $y = \dfrac{x\sqrt{x^2+1}}{(x+1)^{2/3}} \Rightarrow \ln y = \ln(x) + \dfrac{1}{2}\ln(x^2+1) - \dfrac{2}{3}\ln(x+1) \Rightarrow \dfrac{y'}{y} = \dfrac{1}{x} + \dfrac{x}{x^2+1} - \dfrac{2}{3(x+1)} \Rightarrow$

$y' = \dfrac{x\sqrt{x^2+1}}{(x+1)^{2/3}}\left[\dfrac{1}{x} + \dfrac{x}{x^2+1} - \dfrac{2}{3(x+1)}\right]$

37. $y = \sqrt[3]{\dfrac{x(x-2)}{x^2+1}} \Rightarrow \ln y = \dfrac{1}{3}\left[\ln(x) + \ln(x-2) - \ln(x^2+1)\right] \Rightarrow y' = \dfrac{1}{3}\sqrt[3]{\dfrac{x(x-2)}{x^2+1}}\left[\dfrac{1}{x} + \dfrac{1}{x-2} - \dfrac{2x}{x^2+1}\right]$

39. $\displaystyle\lim_{x\to\infty}\ln\left(\dfrac{1}{x}\right) = -\infty$

41. $\displaystyle\lim_{x\to\infty}\int_x^{2x}\dfrac{1}{t}\,dt = \lim_{x\to\infty}\left[\ln|t|\right]_x^{2x} = \lim_{x\to\infty}\ln\left(\dfrac{2x}{x}\right) = \ln 2$

43. $\displaystyle\lim_{\theta\to 0^+}\dfrac{\ln(\sin\theta)}{\ln(\cot\theta)} = \lim_{\theta\to 0^+}\dfrac{\cot\theta}{-\csc^2\theta/\cot\theta} = -\lim_{\theta\to 0^+}\dfrac{\cot^2\theta}{\csc^2\theta} = -\lim_{\theta\to 0^+}\dfrac{\csc^2\theta - 1}{\csc^2\theta} =$

$-\displaystyle\lim_{\theta\to 0^+}\left(1 - \sin^2\theta\right) = -1$

45. $\displaystyle\int_{-3}^{-2}\dfrac{1}{x}\,dx = \left[\ln|x|\right]_{-3}^{-2} = \ln(2) - \ln(3) = \ln\left(\dfrac{2}{3}\right)$

47. $\displaystyle\int \frac{2y}{y^2 - 25}\, dy = \ln\left|y^2 - 25\right| + C$

49. $\displaystyle\int_0^\pi \frac{\sin x}{2 - \cos x}\, dx = \Big[\ln|2 - \cos x|\Big]_0^\pi = \ln(3) - \ln(1) = \ln 3,\ \text{or}$

   $\displaystyle\int_0^\pi \frac{\sin x}{2 - \cos x}\, dx = \int_1^3 \frac{1}{u}\, du = \Big[\ln u\Big]_1^3 = \ln 3,\ \text{where } u = 2 - \cos x$

51. $\displaystyle\int_1^2 \frac{2 \ln x}{x}\, dx = \int_0^{\ln 2} 2u\, du = \Big[u^2\Big]_0^{\ln 2} = (\ln 2)^2,\ \text{where } u = \ln x$

53. $\displaystyle\int_2^4 \frac{dx}{x(\ln x)^2} = \int_2^4 (\ln x)^{-2}\, \frac{1}{x}\, dx = \Big[-(\ln x)^{-1}\Big]_2^4 = \frac{1}{\ln 4}$

55. $\displaystyle\int \frac{3 \sec^2 t}{6 + 3 \tan t}\, dt = \ln|6 + 3 \tan t| + C$

57. $\displaystyle\int_0^{\pi/2} \tan\frac{x}{2}\, dx = -2 \int_0^{\pi/2} \frac{\left(-\sin\frac{x}{2}\right)\left(\frac{1}{2}\right)}{\cos\frac{x}{2}}\, dx = -2\left[\ln\cos\left(\frac{x}{2}\right)\right]_0^{\pi/2} = \ln 2$

59. $\displaystyle\int_{\pi/2}^\pi 2 \cot\left(\frac{\theta}{3}\right) d\theta = 6 \int_{\pi/2}^\pi \frac{\left(\cos\frac{\theta}{3}\right)\left(\frac{1}{3}\right)}{\sin\frac{\theta}{3}}\, d\theta = 6\left[\ln\left(\sin\frac{\theta}{3}\right)\right]_{\pi/2}^\pi = \ln 27$

61. $\displaystyle\int \frac{dx}{2\sqrt{x} + 2x} = \int \frac{\frac{1}{2}x^{-1/2}}{1 + x^{1/2}}\, dx = \ln\left|1 + \sqrt{x}\right| + C$

63. a) If $f(x) = \ln(1 + x) \Rightarrow f'(x) = \dfrac{1}{1 + x}$, then $L(x) = f'(0)(x - 0) + f(0) = 1(x - 0) + 0 = x$. If $f(x) =$

   $\ln(1 + x) \Rightarrow f''(x) = \dfrac{-1}{(1 + x)^2}$, then $Q(x) = f(0) + f'(0)(x - 0) + \dfrac{f''(0)}{2}(x - 0)^2 = 0 + 1(x - 0) +$

   $\dfrac{-1}{2}(x - 0)^2 = x + \dfrac{1}{2}x^2.$

   b) $\left|e_1(x)\right| \leq \dfrac{1}{2}\left|1\right|(0.1)^2 = 0.005$ and $\left|e_2(x)\right| \leq \dfrac{1}{6}\left|2\right|(0.1)^3 = 0.00033$

65. $\displaystyle\int_1^5 \ln 2x - \ln x\, dx = \int_1^5 -\ln x + \ln 2 + \ln x\, dx = (\ln 2)\int_1^5 dx = \ln 16$

67. $\displaystyle V = \pi \int_0^3 \left(\frac{2}{\sqrt{y + 1}}\right)^2 dy = 4\pi \int_0^3 \frac{1}{y + 1}\, dy = 4\pi\Big[\ln|y + 1|\Big]_0^3 = 4\pi \ln 4$

69. $\displaystyle V = 2\pi \int_{1/2}^2 x\, \frac{1}{x^2}\, dx = 2\pi \int_{1/2}^2 \frac{1}{x}\, dx = 2\pi\Big[\ln x\Big]_{1/2}^2 = \pi \ln 16$

71.   a) $y = x^2/8 - \ln x \Rightarrow 1 + (y')^2 = 1 + \left(\dfrac{x}{4} - \dfrac{1}{x}\right)^2 = 1 + \left(\dfrac{x^2 - 4}{4x}\right)^2 = \left(\dfrac{x^2 + 4}{4x}\right)^2 \Rightarrow$

$L = \displaystyle\int_4^8 \sqrt{1 + (y')^2}\, dx = \int_4^8 \dfrac{x^2 + 4}{4x}\, dx = \int_4^8 \dfrac{x}{4} + \dfrac{1}{x}\, dx = \left[\dfrac{x^2}{8} + \ln x\right]_4^8 = 6 + \ln 2$

   b) $y = \sin x - \dfrac{1}{4}\ln(\sec x + \tan x) \Rightarrow y' = \cos x - \dfrac{1}{4}\dfrac{\sec x \tan x + \sec^2 x}{\sec x + \tan x} = \cos x - \dfrac{\sec x}{4} \Rightarrow$

$1 + (y')^2 = 1 + \left(\cos x - \dfrac{\sec x}{4}\right)^2 = \left(\cos x + \dfrac{\sec x}{4}\right)^2 \Rightarrow L = \displaystyle\int_0^{\pi/3} \sqrt{1 + (y')^2}\, dx =$

$\displaystyle\int_0^{\pi/3} \cos x + \dfrac{\sec x}{4}\, dx = \left[\sin x + \dfrac{1}{4}\ln(\sec x + \tan x)\right]_0^{\pi/3} = \dfrac{\sqrt{3}}{2} + \dfrac{1}{4}\ln\left(2 + \sqrt{3}\right)$

73.   a) $M_y = \displaystyle\int_1^2 x\,\dfrac{1}{x}\, dx = 1,\ M_x = \int_1^2 \dfrac{1}{2x}\dfrac{1}{x}\, dx = \dfrac{1}{2}\int_1^2 \dfrac{1}{x^2}\, dx = \left[-\dfrac{1}{2x}\right]_1^2 = \dfrac{1}{4},\ M = \int_1^2 \dfrac{1}{x}\, dx = \ln 2 \Rightarrow$

$\overline{x} = \dfrac{M_x}{M} = \dfrac{1}{\ln 2} \approx 1.44$ and $\overline{y} = \dfrac{M_y}{M} = \dfrac{1/4}{\ln 2} \approx 0.36$

   b)

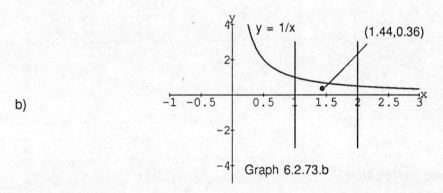

Graph 6.2.73.b

75. $\dfrac{dy}{dx} = 1 + \dfrac{1}{x}$ at $(1,3) \Rightarrow y = x + \ln(x) + C$ at $(1,3) \Rightarrow C = 2 \Rightarrow y = x + \ln(x) + 2$

77. $\dfrac{d^2s}{dt^2} = \dfrac{4}{(4-t)^2} \Rightarrow \dfrac{ds}{dt} = \displaystyle\int \dfrac{4}{(4-t)^2}\, dt = -4\int (4-t)^{-2}(-1)\, dt = \dfrac{4}{4-t} + C$ and $2 = \dfrac{4}{4-0} + C \Rightarrow$

$v = \dfrac{4}{4-t} + 1;$ distance $= \displaystyle\int_1^2 \dfrac{4}{4-t} + 1\, dt = \left[-4\ln(4-t) + t\right]_1^2 = \left(4\ln\left(\dfrac{3}{2}\right) + 1\right)$ m

79.   Those points near $x = 0$ will yield better approximations than those near 0.5.
From the graph the error in using $L(x)$ is
$\leq \left|\ln\left(\dfrac{3}{2}\right) - \dfrac{1}{2}\right| = 0.09453$ and the error in
using $Q(x)$ is $\leq \left|\ln\left(\dfrac{3}{2}\right) - \left(\dfrac{1}{2} - \dfrac{1}{8}\right)\right| = 0.2195.$

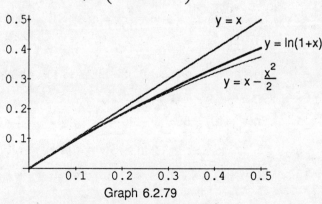

Graph 6.2.79

81.   The graph of $y = -\ln|\sin x|$ would have
       inverted arches.

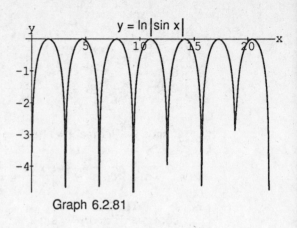

Graph 6.2.81

83.   a)

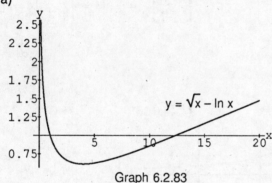

Graph 6.2.83

b)

$y = \sqrt{x} - \ln x, x \geq 0 \Rightarrow y = x^{1/2} - \ln x \Rightarrow y' = \frac{1}{2}x^{-1/2} - x^{-1} \Rightarrow y'' = -\frac{1}{4}x^{-3/2} + x^{-2} = \frac{4 - \sqrt{x}}{4x^2} \Rightarrow$

at $x = 16$ there is a point of inflection.

6.3 THE EXPONENTIAL FUNCTION

1.   a) 7   b) $\ln e = 1$ c) $\ln(e \ln e) = \ln e = 1$

3.   a) $e^{-\ln x^2} = e^{\ln\left(1/x^2\right)} = \frac{1}{x^2}$ b) $e^{(\ln x + \ln y)} = e^{\ln xy} = xy$ c) $\ln e^{1/x} = \frac{\ln e}{x} = \frac{1}{x}$

5.   a) $e^{2k} = 4 \Rightarrow 2k = \ln 4 \Rightarrow k = \ln 2$ b) $e^{5k} = \frac{1}{4} \Rightarrow 5k = -\ln 4 \Rightarrow k = -\frac{\ln 4}{5}$

      c) $e^{(k/1000)} = a \Rightarrow \frac{k}{1000} = \ln a \Rightarrow k = 1000 \ln a$

7.   a) $e^t = 1 \Rightarrow t = 0$ b) $4 e^{-0.1t} = 20 \Rightarrow -0.1t = \ln 5 \Rightarrow t = -10 \ln 5$

      c) $e^{-0.3t} = 27 \Rightarrow -0.3t = \ln 27 \Rightarrow t = -10 \ln 3$

9. $\ln ny = 2t + 4 \Rightarrow y = e^{(2t + 4)}$

11. $\ln(y - 40) = 5t \Rightarrow y - 40 = e^{5t} \Rightarrow y = e^{5t} + 40$

13. $e^{\sqrt{y}} = x^2 \Rightarrow \sqrt{y} = 2 \ln x \Rightarrow y = 4(\ln x)^2$

15. $\ln(y - 1) = x + \ln x \Rightarrow y - 1 = x e^x \Rightarrow y = x e^x + 1$

17. $\dfrac{d}{dx} e^{-5x} = -5 e^{-5x}$

19. $\dfrac{d}{dx} e^{(5-7x)} = -7 e^{(5-7x)}$

21. $\dfrac{d}{dx}\left(xe^x - e^x\right) = e^x + x e^x - e^x = x e^x$

23. $\dfrac{d}{dx}\left(x^2 - 2x + 2\right) e^x = (2x - 2) e^x + \left(x^2 - 2x + 2\right) e^x = x^2 e^x$

25. $\dfrac{d}{dx} e^x(\sin x + \cos x) = e^x(\sin x + \cos x) + e^x(\cos x - \sin x) = 2e^x \cos x$

27. $\dfrac{d}{dx} \cos\left(e^{-x^2}\right) = \left(-\sin\left(e^{-x^2}\right)\right)\left(e^{-x^2}\right)(-2x) = 2x\, e^{-x^2} \sin e^{-x^2}$

29. $\dfrac{d}{dx} e^{(\cos x + \ln x)} = \dfrac{d}{dx}\left(x\, e^{\cos x}\right) = e^{\cos x} + x\, e^{\cos x}(-\sin x) = e^{\cos x}(1 - x \sin x)$

31. $\dfrac{d}{dx}\left(\left(x^3 - 3x^2 + 6x\right)e^x - 6x + \sqrt{2}\right) = \left(3x^2 - 6x + 6\right)e^x + \left(x^3 - 3x^2 + 6x\right)e^x - 6 = \left(x^3 + 6\right)e^x - 6$

33. $\dfrac{d}{dx}\displaystyle\int_0^{\ln x} \sin e^t \, dt = \dfrac{\sin x}{x}$, the Fundamental Theorem of Calculus

35. $\ln y = x \sin x \Rightarrow \dfrac{dy}{y} = (\sin x + x \cos x)dx \Rightarrow \dfrac{dy}{dx} = (\sin x + x \cos x)e^{x \sin x}$

37. $e^{2x} = \sin(x + 3y) \Rightarrow 2 e^{2x} = (\cos(x + 3y))(1 + y') \Rightarrow 1 + 3y' = \dfrac{2e^{2x}}{\cos(x + 3y)} \Rightarrow y' = \dfrac{2e^{2x} - \cos(x + 3y)}{3 \cos(x + 3y)}$

39. $\displaystyle\int_1^{e^2} \dfrac{1}{x}\, dx = [\ln x]_1^{e^2} = 2$

41. $\displaystyle\int_{-\ln 2}^{0} e^{-x}\, dx = \left[-e^{-x}\right]_{-\ln 2}^{0} = 1$

43. $\displaystyle\int_{\ln 3}^{\ln 5} e^{2\theta}\, d\theta = \left[\dfrac{e^{2\theta}}{2}\right]_{\ln 3}^{\ln 5} = 8$

45. $\displaystyle\int_1^4 \dfrac{e^{\sqrt{r}}}{2\sqrt{r}}\, dr = \int_1^4 e^{r^{1/2}} \dfrac{1}{2} r^{-1/2}\, dr = \left[e^{\sqrt{r}}\right]_1^4 = e^2 - e$

47. $\displaystyle\int_e^{e^e} \dfrac{1}{x \ln x}\, dx = \int_e^{e^e} (\ln x)^{-1}\left(\dfrac{1}{x}\right) dx = [\ln \ln x]_e^{e^e} = \ln\left(\ln e^e\right) - \ln(\ln e) = \ln e - \ln 1 = 1 - 0 = 1$

49. $\displaystyle\int_{\ln \pi/6}^{\ln \pi/2} 2e^u \cos e^u\, du = \left[2 \sin e^u\right]_{\ln \pi/6}^{\ln \pi/2} = 2\left[\sin \dfrac{\pi}{2} - \sin \dfrac{\pi}{6}\right] = 1$

51. $\displaystyle\int \left(1 + e^{\tan \theta}\right) \sec^2\theta\, d\theta = \int \left(\sec^2\theta + e^{\tan \theta}\sec^2\theta\right) d\theta = \tan \theta + e^{\tan \theta} + C$

53. $\displaystyle\lim_{\theta \to 0} \dfrac{\cos \theta - 1}{e^\theta - \theta - 1} = \lim_{\theta \to 0} \dfrac{-\sin \theta}{e^\theta - 1} = \lim_{\theta \to 0} \dfrac{-\cos \theta}{e^\theta} = -1$

55. $\displaystyle\lim_{t \to \infty} \dfrac{e^t + t^2}{e^t - 1} = \lim_{t \to \infty} \dfrac{e^t + 2t}{e^t} = \lim_{t \to \infty} \dfrac{e^t + 2}{e^t} = \lim_{t \to \infty} \dfrac{e^t}{e^t} = 1$

57. a) If $f(x) = e^x \Rightarrow f'(x) = e^x$, then $L(x) + f'(0)(x - 0) + f(0) = x + 1$. If $f(x) = e^x \Rightarrow f''(x) = e^x$, then

   $Q(x) = f(0) + f'(0)(x - 0) + \dfrac{f''(0)}{2}(x - 0)^2 = 1 + x + \dfrac{x^2}{2}$.

   b) $\left|e_1(x)\right| \le \dfrac{1}{2}|3|(0.1)^2 = 0.015$ and $\left|e_2(x)\right| \le \dfrac{1}{6}|3|(0.1)^3 = 0.0005$, since $f'(x) = f''(x) =$

   $e^x < 3^1$ for $x \in [0,0.1]$

59. $f(x) = x^2 \ln x^{-1} \Rightarrow f'(x) = 2x \ln x^{-1} + x^2\left(\dfrac{-x^{-2}}{x^{-1}}\right) = 2x \ln\left(\dfrac{1}{x}\right) - x,\ f'(x) = 0 \Rightarrow x = 0.60653$, by

Newton's method; $f(0.60653) = 1.09861$, the maximum value

61. $y = (x-3)^2 e^x \Rightarrow y' = (x-1)(x-3)e^x \Rightarrow$ $\underset{1 \qquad\qquad 3}{++++ \mid ------ \mid ++++}$ $y' \Rightarrow$ at $x = 3$

a local minimum

63. $A = \displaystyle\int_0^{\ln 3} e^{2x} - e^x\, dx = \left[\dfrac{e^{2x}}{2} - e^x\right]_0^{\ln 3} = 2$

65. $A(t) = \displaystyle\int_0^t e^{-x}\, dx = \left[-e^{-x}\right]_0^t = 1 - \dfrac{1}{e^t},\ V(t) = \pi\int_0^t e^{-2x}\, dx = \left[\dfrac{\pi e^{-2x}}{-2}\right]_0^t = \dfrac{\pi}{2} - \dfrac{\pi}{2\,e^{2t}}$ and

$\dfrac{V(t)}{A(t)} = \dfrac{\dfrac{\pi}{2}\left[1 - \dfrac{1}{e^{2t}}\right]}{1 - \dfrac{1}{e^t}} = \dfrac{\pi\left(e^{2t} - 1\right)}{2\left(e^{2t} - e^t\right)}$ a) $\underset{t \to \infty}{\mathrm{Lim}}\ A(t) = \underset{t \to \infty}{\mathrm{Lim}}\ 1 - \dfrac{1}{e^t} = 1$

b) $\underset{t \to \infty}{\mathrm{Lim}}\ \dfrac{V(t)}{A(t)} = \dfrac{\pi}{2}\ \underset{t \to \infty}{\mathrm{Lim}}\ \dfrac{e^{2t} - 1}{e^{2t} - e^t} = \dfrac{\pi}{2}\ \underset{t \to \infty}{\mathrm{Lim}}\ \dfrac{2}{2 - \dfrac{1}{e^t}} = \dfrac{\pi}{2}$

c) $\underset{t \to 0^+}{\mathrm{Lim}}\ \dfrac{V(t)}{A(t)} = \dfrac{\pi}{2}\ \underset{t \to 0^+}{\mathrm{Lim}}\ \dfrac{e^{2t} - 1}{e^{2t} - e^t} = \dfrac{\pi}{2}\ \underset{t \to 0^+}{\mathrm{Lim}}\ \dfrac{2e^{2t}}{2e^{2t} - e^t} = \dfrac{\pi}{2}\left(\dfrac{2}{2 - 1}\right) = \pi$

rectangle.   Consequently, $\displaystyle\int_1^a \ln(x)\, dx + \int_0^{\ln a} e^y\, dy = a \ln a$.

67. $L = \displaystyle\int_0^1 \sqrt{1 + \dfrac{e^x}{4}}\, dx \Rightarrow \dfrac{dy}{dx} = \dfrac{e^{x/2}}{2} \Rightarrow y = e^{x/2} + C$ and $0 = e^{0/2} + C \Rightarrow y = e^{x/2} - 1$

69. $\dfrac{dy}{dx} = e^x \sin\left(e^x - 2\right) \Rightarrow y = -\cos\left(e^x - 2\right) + C$ and $0 = -\cos(\pi) + C \Rightarrow y = -\cos\left(e^x - 2\right) - 1$

71. $f(x) = \ln(x) - 1 \Rightarrow f'(x) = \dfrac{1}{x} \Rightarrow x_{n+1} = x_n - \dfrac{\ln(x_n) - 1}{\dfrac{1}{x_n}} \Rightarrow x_{n+1} = x_n\left[2 - \ln(x_n)\right]$. When $x_1 = 2 \Rightarrow$

$x_2 = 2.61370564,\ x_3 = 2.71624393$ and $x_5 = 2.71828183$.

73.   a) $Q(x)$ overestimates e^x when $x < 0$
and underestimates e^x when $x > 0$.
$L(x)$ always underestimates e^x.

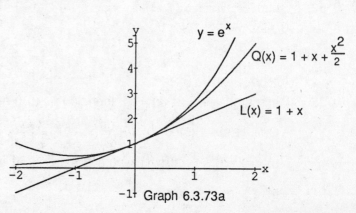

Graph 6.3.73a

b) The largest errors appear to be at
x = 1. The error in using $L(x) \leq 0.71828$
and the error in using $Q(x) \leq 0.21828$.

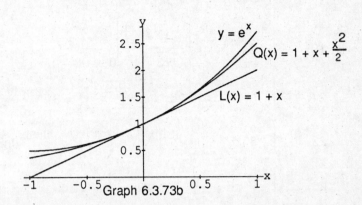

Graph 6.3.73b

75.   a) $y = e^x \Rightarrow y'' = e^x > 0$ for all $x \Rightarrow$ the graph of $y = e^x$ is always concave upward

b) area of the trapezoid ABCD $< \displaystyle\int_{\ln a}^{\ln b} e^x \, dx <$ area of the trapoezoid AEFD $\Rightarrow$

$\dfrac{1}{2}(AB + CD)(\ln b - \ln a) < \displaystyle\int_{\ln a}^{\ln b} e^x \, dx < \dfrac{e^{\ln a} + e^{\ln b}}{2}(\ln b - \ln a)$ and $\dfrac{1}{2}(AB + CD)$ is the height

of the midpoint $M = e^{(\ln a + \ln b)/2}$ since the curve containing the points B and C is linear $\Rightarrow$

$e^{(\ln a + \ln b)/2}(\ln b - \ln a) < \displaystyle\int_{\ln a}^{\ln b} e^x \, dx < \dfrac{e^{\ln a} + e^{\ln b}}{2}(\ln b - \ln a)$

c) from part b) we have $e^{(\ln a + \ln b)/2} < \dfrac{b - a}{\ln b - \ln a} < \dfrac{a + b}{2} \Rightarrow \sqrt{e^{\ln a}}\sqrt{e^{\ln b}} < \dfrac{b - a}{\ln b - \ln a} <$

$\dfrac{a + b}{2} \Rightarrow \sqrt{ab} < \dfrac{b - a}{\ln b - \ln a} < \dfrac{a + b}{2}$

6.4 OTHER EXPONENTIAL AND LOGARITHMIC FUNCTIONS

1. $y = x^\pi \Rightarrow \dfrac{dy}{dx} = \pi x^{(\pi - 1)}$ 3. $y = r^{-\sqrt{2}} \Rightarrow \dfrac{dy}{dr} = -\sqrt{2}\, r^{-\sqrt{2} - 1}$

5. $y = 2^x \Rightarrow \ln y = (\ln 2)x \Rightarrow \dfrac{y'}{y} = \ln 2 \Rightarrow \dfrac{dy}{dx} = 2^x \ln 2$

7. $y = 5^{\sqrt{s}} \Rightarrow \ln y = (\ln 5) s^{1/2} \Rightarrow \dfrac{y'}{y} = \left(\dfrac{\ln 5}{2}\right) s^{-1/2} \Rightarrow y' = \left(\dfrac{\ln 5}{2\sqrt{s}}\right) 5^{\sqrt{s}}$

9. $y = 2^{\sec x} \Rightarrow \ln y = (\ln 2)\sec x \Rightarrow \dfrac{y'}{y} = (\ln 2)(\sec x)(\tan x) \Rightarrow \dfrac{dy}{dx} = 2^{\sec x}(\ln 2)(\sec x)(\tan x)$

11. $y = x^{\ln x}, x > 0 \Rightarrow \ln y = (\ln x)^2 \Rightarrow \dfrac{y'}{y} = 2(\ln x)\left(\dfrac{1}{x}\right) \Rightarrow \dfrac{dy}{dx} = \left(x^{\ln x}\right)\left(\dfrac{\ln x^2}{x}\right)$

13. $y = (\sqrt{t})^t = t^{t/2} \Rightarrow \ln y = \left(\dfrac{t}{2}\right)(\ln t) \Rightarrow \dfrac{dy}{dt} = (\sqrt{t})^t\left(\dfrac{\ln t}{2} + \dfrac{1}{2}\right)$

15. $y = (\sin x)^{\sin x} \Rightarrow \ln y = (\sin x)\ln(\sin x) \Rightarrow \dfrac{y'}{y} = (\cos x)\ln(\sin x) + (\sin x)\left(\dfrac{\cos x}{\sin x}\right) \Rightarrow$

$y' = \left((\cos x)\left(\ln(\sin x)\right) + \cos x\right)(\sin x)^{\sin x}$

17. $y = (\ln x)^{\ln x} \Rightarrow \ln y = (\ln x)\ln(\ln x) \Rightarrow \dfrac{y'}{y} = \dfrac{\ln(\ln x)}{x} + \dfrac{1}{x} \Rightarrow y' = \left(\dfrac{\ln(\ln x) + 1}{x}\right)(\ln x)^{\ln x}$

19. $\displaystyle\int_0^1 3x^{\sqrt{3}} \, dx = \left[\dfrac{3x^{\left(\sqrt{3}+1\right)}}{\sqrt{3} + 1}\right]_0^1 = \dfrac{3}{\sqrt{3} + 1}$

21. $\displaystyle\int_0^3 (\sqrt{2} + 1) \, x^{\sqrt{2}} \, dx = \left[x^{\sqrt{2} + 1}\right]_0^3 = 3^{\sqrt{2} + 1}$

23. $\displaystyle\int_0^1 5^x \, dx = \int_0^1 e^{(\ln 5)x} \, dx = \left(\dfrac{1}{\ln 5}\right)\int_0^1 e^{(\ln 5)x}(\ln 5) \, dx = \left(\dfrac{1}{\ln 5}\right)\left[e^{(\ln 5)x}\right]_0^1 = \dfrac{4}{\ln 5}$

25. $\displaystyle\int_0^1 \dfrac{1}{2^x} \, dx = -\int_0^{-1} 2^u \, du = \int_{-1}^0 2^u \, du = \dfrac{1}{\ln 4}$, where $u = -x$. See question 20 for evaluation of the

last integral.

27. $\displaystyle\int_{-1}^0 4^{-x} \ln 2 \, dx = \int_{-1}^0 2^{-2x} \ln 2 \, dx = -\dfrac{1}{2}\int_{-1}^0 e^{(-2(\ln 2))x}(-2\ln 2) \, dx = -\dfrac{1}{2}\left[e^{\ln 2^{-2x}}\right]_{-1}^0 = \dfrac{3}{2}$

29. $\displaystyle\int_1^{\sqrt{2}} x \, 2^{x^2} \, dx = \dfrac{1}{2 \ln 2}\int_1^{\sqrt{2}} e^{(\ln 2)x^2}(2\ln 2)x \, dx = \dfrac{1}{\ln 4}\left[2^{x^2}\right]_1^{\sqrt{2}} = \dfrac{1}{\ln 2}$

31. $\displaystyle\int_2^4 x^{2x}(1 + \ln x) \, dx = \dfrac{1}{2}\int_2^4 e^{2x\ln x}(2 + 2\ln x) \, dx = \dfrac{1}{2}\left[e^{2x\ln x}\right]_2^4 = \dfrac{4^8 - 2^4}{2} = 32760$

33.   a) $\dfrac{\ln 16}{\ln 4} = \dfrac{2\ln 4}{\ln 4} = 2$ b) $\dfrac{\ln 8}{\ln 32} = \dfrac{3\ln 2}{5\ln 2} = \dfrac{3}{5}$ c) $\dfrac{\ln 2}{2\ln 4} = \dfrac{\ln 2}{4\ln 2} = \dfrac{1}{4}$

35.   a) let $z = \log_4 x \Rightarrow 2^{2z} = x \Rightarrow 2^z = \sqrt{x}$

   b) let $z = \log_3 x \Rightarrow 3^z = x \Rightarrow 9^z = x^2$

   c) $\log_2\left[e^{(\ln 2)\sin x}\right] = \log_2 2^{\sin x} = \sin x$

37.   a) $\dfrac{\log_2 x}{\log_3 x} = \dfrac{\ln x}{\ln 2} \div \dfrac{\ln x}{\ln 3} = \dfrac{\ln 3}{\ln 2}$ b) $\dfrac{\log_2 x}{\log_8 x} = \dfrac{\ln x}{\ln 2} \div \dfrac{\ln x}{3\ln 2} = 3$ c) $\dfrac{\ln a}{\ln x} \cdot \dfrac{2\ln x}{\ln a} = 2$

39. $7 + 5 = x \Rightarrow x = 12$

41. $y = \dfrac{\ln x}{\ln 4} + \dfrac{\ln x^2}{\ln 4} = \dfrac{\ln x}{\ln 4} + 2\dfrac{\ln x}{\ln 4} = 3\dfrac{\ln x}{\ln 4} \Rightarrow y' = \dfrac{3}{x \ln 4}$

43. $y = \dfrac{(\ln r)^2}{(\ln 2)(\ln 3)} \Rightarrow y' = \dfrac{2(\ln r)}{r(\ln 2)(\ln 3)}$

45. $y = \dfrac{\ln(s + 1)}{2\ln 10} \Rightarrow y' = \dfrac{1}{(\ln 100)(s + 1)}$

47. $y = \dfrac{\ln 6}{\ln t} = (\ln 6)(\ln t)^{-1} \Rightarrow y' = -(\ln 6)(\ln t)^{-2}\left(\dfrac{1}{t}\right) = \dfrac{-\ln 6}{t(\ln t)^2}$

49. $y = \dfrac{(\ln 2)(\ln x)}{\ln(\ln 2)} \Rightarrow y' = \dfrac{\ln 2}{x \ln(\ln 2)}$

51. $y = \dfrac{\theta \ln e}{\ln 5} = \dfrac{\theta}{\ln 5} \Rightarrow y' = \dfrac{1}{\ln 5}$

53. $y = \dfrac{\ln\left[\dfrac{x^2 e^x}{2\sqrt{x+1}}\right]}{\ln 2} = \dfrac{2\ln x + x - \ln 2 - \frac{1}{2}\ln x + 1}{\ln 2} \Rightarrow y' = \dfrac{2}{x\ln 2} + \dfrac{1}{\ln 2} - \dfrac{1}{2x\ln 2} = \dfrac{3 + 2x}{2x\ln 2}$

55. $y = 3\dfrac{\ln\left(\log_2 t\right)}{\ln 8} = \dfrac{3\ln\left(\dfrac{\ln t}{\ln 2}\right)}{3\ln 2} \Rightarrow y' = \dfrac{\dfrac{\ln 2}{\ln t}\left(\dfrac{1}{t\ln 2}\right)}{\ln 2} = \dfrac{1}{t(\ln 2)(\ln t)}$

57. $\displaystyle\int \dfrac{\log_{10} x}{x}\,dx = \dfrac{1}{\ln 10}\int (\ln x)\left(\dfrac{1}{x}\right)dx = \dfrac{1}{\ln 10}\left[\dfrac{(\ln x)^2}{2}\right] + C$

59. $\displaystyle\int_1^4 \dfrac{(\ln 2)(\log_2 x)}{x}\,dx = \int_1^4 \ln x\left(\dfrac{1}{x}\right)dx = \left[\dfrac{(\ln x)^2}{2}\right]_1^4 = \ln 4$

61. $\displaystyle\int_0^2 \dfrac{\log_2(x+2)}{x+2}\,dx = \dfrac{1}{\ln 2}\int_0^2 \ln(x+2)\left(\dfrac{1}{x+2}\right)dx = \dfrac{1}{\ln 2}\left[\dfrac{\ln(x+2)^2}{2}\right]_0^2 = \dfrac{3\ln 2}{2}$

63. $\displaystyle\int_0^9 \dfrac{2\log_{10}(x+1)}{x+1}\,dx = \dfrac{2}{\ln 10}\int_0^9 \ln(x+1)\left(\dfrac{1}{x+1}\right)dx = \dfrac{2}{\ln 10}\left[\dfrac{(\ln x+1)^2}{2}\right]_0^9 = \ln 10$

65. $\displaystyle\int \dfrac{1}{x\log_{10}x}\,dx = (\ln 10)\int \dfrac{1/x}{\ln x}\,dx = (\ln 10)\ln(\ln x) + C$

67. $\displaystyle\lim_{x \to 0} \dfrac{3^{\sin x} - 1}{x} = \lim_{x \to 0} \dfrac{3^{\sin x}(\ln 3)\cos x}{1} = \ln 3$

69. $\displaystyle\lim_{x \to 1^+} x^{1/(1-x)} = e$, let $y = x^{1/(1-x)} \Rightarrow \ln y = \dfrac{\ln x}{1-x} \Rightarrow \lim_{x \to 1^+}\dfrac{\ln x}{1-x} = \lim_{x \to 1^+}\dfrac{1/x}{-1} = -1 \Rightarrow$

$\displaystyle\lim_{x \to 1^+} x^{1/(1-x)} = e^{-1} = \dfrac{1}{e}$

71. $\displaystyle\lim_{x \to 0^+} x^x = 1$, let $y = x^x \Rightarrow \ln y = x\ln x = \dfrac{\ln x}{1/x} \Rightarrow \lim_{x \to 0^+}\dfrac{\ln x}{1/x} = \lim_{x \to 0^+}\dfrac{1/x}{-1/x^2} = \lim_{x \to 0^+} -x = 0 \Rightarrow$

$\displaystyle\lim_{x \to 0^+} x^x = e^0 = 1$

73. $\displaystyle\lim_{x \to \infty}(\ln x)^{1/x} = \lim_{x \to \infty} e^{\ln(\ln x)/x} = \lim_{x \to \infty} e^{1/x\ln x} = e^0 = 1$

75. a) If $f(x) = \log_3 x = \dfrac{\ln x}{\ln 3}$, then $f'(x) = \dfrac{1}{(\ln 3)x}$ and $f''(x) = \dfrac{-1}{(\ln 3)x^2}$. $\therefore L(x) = f'(3)(x-3) + f(3) =$

$\dfrac{x}{\ln 27} - \dfrac{i}{\ln 3} + 1 = 0.30341x + 0.08976$ and $Q(x) = f(3) + f'(3)(x-3) + \dfrac{f''(3)}{2}(x-3)^2 = 1 +$

$\dfrac{1}{\ln 27}(x-3) - \dfrac{1}{18\ln 3}(x-3)^2 = 1 + 0.30341x - 0.05057(x-3)^2$.

b) $\left|e_1(x)\right| \leq \dfrac{1}{2}\left|\dfrac{-1}{(\ln e^{-1})(2.9)^2}\right|(0.1)^2 = 0.00059$, since $\left|f''(x)\right| \leq f'(e^{-1})$ for all $x \in [2.9, 3.1]$;

$\left|e_2(x)\right| \leq \dfrac{1}{6}\left|\dfrac{2}{(\ln e^{-1})(2.9)^3}\right|(0.1)^3 = 0.00001$, since $\left|f'''(x)\right| \leq f'''(e^{-1})$ for all $x \in [2.9, 3.1]$

77. $f(x) = 2^x - x^2 \Rightarrow f'(x) = \dfrac{2^x}{\ln 2} - 2x \Rightarrow x_{n+1} = x_n - \dfrac{2^{x_n} - (x_n)^2}{\dfrac{2^{x_n}}{\ln 2} - 2(x_n)}$. If $x_0 = -0.5$, then $x_1 = -0.726274865$,

$x_2 = -0.759393388$, and $x_{13} = -0.766664696$ $\therefore$ the coordinates

are $(-0.766664696, 0.587774756)$

79.   a) If $x = \left[H_3O^+\right]$ and $S - x = \left[OH^-\right]$, then $x(S - x) = 10^{-14} \Rightarrow S(x) = x + \dfrac{10^{-14}}{x}$.

$\therefore S'(x) = 1 - \dfrac{10^{-14}}{x^2}$ and $S''(x) = \dfrac{10^{-14}}{x^3} > 0 \Rightarrow$ a minimum at $x = 10^{-7}$

       b) $pH = -\log_{10}\left[10^{-7}\right] = 7$

       c) $\dfrac{\left[OH^-\right]}{\left[H_3O^+\right]} = \dfrac{S - x}{x} = \dfrac{x + \dfrac{10^{-14}}{x} - x}{x} = \dfrac{10^{-14}}{x^2} \Rightarrow$ at $x = 10^{-7}$ $\dfrac{\left[OH^-\right]}{\left[H_3O^+\right]} = 1$

81.   Let O = original sound level = $10 \log_{10}(I \times 10^{12})$ db. Solving $O + 10 = 10 \log_{10}(kI \times 10^{12})$ for $k \Rightarrow$

$10 \log_{10}(I \times 10^{12}) + 10 = 10 \log_{10}(kI \times 10^{12}) \Rightarrow 1 = \log_{10} k \Rightarrow k = 10$.

83.   a) We should asign 1 to the value of $(\sin x)^x$ at $x = 0$.

       b) $\underset{x \to 0^+}{\text{Lim}} (\sin x)^x = \underset{x \to 0^+}{\text{Lim}} e^{\frac{\ln(\sin x)}{1/x}} =$

       $\underset{x \to 0^+}{\text{Lim}} e^{\frac{x^2}{\tan x}} = \underset{x \to 0^+}{\text{Lim}} e^{\frac{2x}{\sec^2 x}} = e^0 = 1$

       c) The maximum value is at $x \approx 1.55$

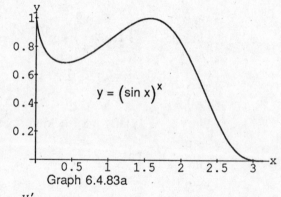

$y = (\sin x)^x$

Graph 6.4.83a

       d) The root in question is near 1.57.

$y' = \left(\ln(\sin x) - x \cot x\right)(\sin x)^x$

Graph 6.4.83d

       e) $y' = 0 \Rightarrow \left(\ln(\sin x) - x \cot x\right)(\sin x)^x = 0 \Rightarrow \ln(\sin x) - x \cot x = 0 \Rightarrow x \approx 1.570796327$,

by Newton's method

       f)

x	1.55	1.57	1.57070796327
$(\sin x)^x$	0.999664854	0.999999502	1

85.

a)

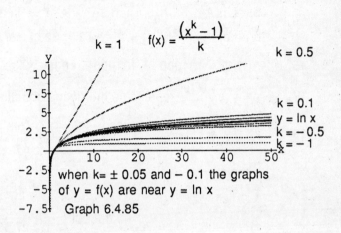

$f(x) = \dfrac{(x^k - 1)}{k}$

when k = ± 0.05 and − 0.1 the graphs of y = f(x) are near y = ln x

Graph 6.4.85

b) $\displaystyle\lim_{k \to 0} \frac{x^k - 1}{k} = \lim_{k \to 0} \frac{x^k \ln x}{1} = \ln x$

6.5 GROWTH AND DECAY

1. a) $k = \dfrac{\ln .99}{1000} \approx -0.00001$

   b) $0.9 = e^{kt} \Rightarrow kt = \ln(0.9) \Rightarrow t = \dfrac{\ln(0.9)}{k} \approx 10536$ yr

   c) $y = y_o e^{(20000)k} \approx y_o(0.82) \Rightarrow 82\%$

3. $\dfrac{dy}{dt} = -0.6y \Rightarrow y = y_o e^{-0.6t}$. If $y_o = 100$, t = 1 and $y = y_o e^{-0.6t}$, then $y = 100e^{-0.6} \approx 54.88$ grams.

5. $L(x) = L_o e^{-kx} \Rightarrow \dfrac{L_o}{2} = L_o e^{-18k} \Rightarrow \ln\dfrac{1}{2} = -18k \Rightarrow k = \dfrac{\ln 2}{18} \approx 0.0385 \ \therefore L(x) = L_o e^{-0.0385} \Rightarrow$

   $\dfrac{L_o}{10} = L_o e^{-0.0385x} \Rightarrow \ln 10 = 0.0385x \Rightarrow x \approx 59.8$ ft

7. $y = y_o e^{kt}, \ y_o = 1 \Rightarrow y = e^{kt}. \ \therefore$ when y = 2 at t = 0.5, $2 = e^{0.5t} \Rightarrow \ln 2 = 0.5k \Rightarrow k = \dfrac{\ln 2}{0.5} = \ln 4 \Rightarrow$

   $y = e^{(\ln 4)t} \Rightarrow y(24) = e^{24 \ln 4} = 4^{24} \approx 2.8147497 \times 10^{14}$

9. a) $10000e^{k(1)} = 7500 \Rightarrow e^k = 0.75 \Rightarrow k = \ln 0.75 \ \therefore y = 10000e^{(\ln 0.75)t}$. Now $1000 = $

   $10000e^{(\ln 0.75)t} \Rightarrow \ln 0.1 = (\ln 0.75)t \Rightarrow t = \dfrac{\ln 0.1}{\ln 0.75} \approx 8.00$ (to the nearest hundredth of a year).

   b) $1 = 10000e^{(\ln 0.75)t} \Rightarrow \ln 0.0001 = (\ln 0.75)t \Rightarrow t = \dfrac{\ln 0.0001}{\ln 0.75} \approx 32.02$ (to the nearest

   hundredth of a year)

11. For $P = \dfrac{1}{5}P_o, \ \dfrac{P_o}{5} = P_o e^{-0.1t} \Rightarrow \dfrac{1}{5} = e^{-0.1t} \Rightarrow \ln 5 = 0.1t \Rightarrow t = \dfrac{\ln 5}{0.1} \approx 16.09$ yr

13.   a)   The amount of money after t years. A(t), is given by $A(t) = A_o e^t$.

b)   Let $A(t) = 3A_o$. Then $3A_o = A_o e^t \Rightarrow \ln 3 = t$ or $t \approx 1.099$ years.

c)   Let $t = 1 \Rightarrow A(1) = A_o e \approx 2.71828 A_o$ or more than 2.7 times the original amount, A_o.

15. $A(100) = 131000 \Rightarrow 131000 = 1000 e^{100r} \Rightarrow \ln 131 = 100r \Rightarrow r = \dfrac{\ln 131}{100} \approx 0.04875$ or 4.875%

17. $y = y_o e^{-0.18t}$ represents the decay equation; solving $(0.9)y_o = y_o e^{-0.18t} \Rightarrow t = \dfrac{\ln(0.9)}{-0.18} \approx 0.585$ days

19. $y_o = y_o e^{-kt} = y_o e^{-(k)(3/k)} = y_o e^{-3} = \dfrac{y_o}{e^3} < \dfrac{y_o}{20} = 5\% \Rightarrow$ after three mean lifetimes less than 5%

remains

21. $T - T_s = (T_o - T_s)e^{-kt}$, $T_o = 90°C$, $T_s = 20°C \Rightarrow 60 - 20 = 70 e^{-10k} \Rightarrow k = \dfrac{\ln(7/4)}{10} \approx 0.05596$

a) $35 - 20 = 70 e^{-0.05596t} \Rightarrow t \approx 27.5$ min the total time; it will take $27.5 - 10 = 17.5$ min

to reach to 35°C

b) $T - T_s = (T_o - T_s)e^{-kt}$, $T_o = 90°C$, $T_s = -15°C \Rightarrow 35 + 15 = 105 e^{-0.05596t} \Rightarrow$

$t \approx 13.26$ min

23. $T - T_s = (T_o - T_s)e^{-kt}$, $39 - T_s = (46 - T_s)e^{-10k}$ and $33 - T_s = (46 - T_s)e^{-20k} \Rightarrow$

$\dfrac{39 - T_s}{46 - T_s} = e^{-10k}$ and $\dfrac{33 - T_s}{46 - T_s} = e^{-20k} = \left(e^{-10k}\right)^2 \Rightarrow \dfrac{33 - T_s}{46 - T_s} = \left(\dfrac{39 - T_s}{46 - T_s}\right)^2 \Rightarrow$

$(33 - T_s)(46 - T_s) = (39 - T_s)^2 \Rightarrow T_s = -3°C$

25. $\dfrac{c_o}{2} = c_o e^{-k(5700)} \Rightarrow k = \dfrac{\ln 2}{5700} \approx 0.0001216 \Rightarrow c(t) = c_o e^{-0.0001216t} \Rightarrow (0.445)c_o = c_o e^{-0.0001216t} \Rightarrow$

$t \approx 6658.3$ yr

27.   from exercise 25 we have $c(t) = c_o e^{-0.0001216t} \Rightarrow (0.995)c_o = c_o e^{-0.0001216t} \Rightarrow t \approx 41.22$ yr

6.6 THE RATE AT WHICH FUNCTIONS GROW

1.   a)   Yes, $\displaystyle \lim_{x \to \infty} \frac{x + 3}{e^x} = \lim_{x \to \infty} \frac{1}{e^x} = 0$

b)   Yes, $\displaystyle \lim_{x \to \infty} \frac{x^3 - 3x + 1}{e^x} = \lim_{x \to \infty} \frac{3x^2 - 3}{e^x} = \lim_{x \to \infty} \frac{6x}{e^x} = \lim_{x \to \infty} \frac{6}{e^x} = 0$

c)   Yes, $\displaystyle \lim_{x \to \infty} \frac{\sqrt{x}}{e^x} = \lim_{x \to \infty} \frac{(1/2)x^{(-1/2)}}{e^x} = 0$ d)   No, $\displaystyle \lim_{x \to \infty} \frac{4^x}{e^x} = \lim_{x \to \infty} \left(\frac{4}{e}\right)^x = \infty$

e)   Yes, $\displaystyle \lim_{x \to \infty} \frac{\left(\frac{5}{2}\right)^x}{e^x} = \lim_{x \to \infty} \left(\frac{5}{2e}\right)^x = 0$ f)   Yes, $\displaystyle \lim_{x \to \infty} \frac{\ln x}{e^x} = \lim_{x \to \infty} \frac{1}{xe^x} = 0$

g) Yes, $\displaystyle\lim_{x \to \infty} \frac{\log_{10}x}{e^x} = \left(\frac{1}{\ln 10}\right)\lim_{x \to \infty}\frac{\ln x}{e^x} = \left(\frac{1}{\ln 10}\right)(0) = 0$

h) Yes, $\displaystyle\lim_{x \to \infty}\frac{e^{-x}}{e^x} = \lim_{x \to \infty}\frac{1}{e^{2x}} = 0$

i) No, $\displaystyle\lim_{x \to \infty}\frac{e^{x+1}}{e^x} = \lim_{x \to \infty} e = e \neq 0$

j) No, $\displaystyle\lim_{x \to \infty}\frac{(1/2)e^x}{e^x} = \lim_{x \to \infty}\frac{1}{2} = \frac{1}{2} \neq 0$

3. a) Yes, $\displaystyle\lim_{x \to \infty}\frac{\log_3 x}{\ln x} = \lim_{x \to \infty}\frac{\ln x}{(\ln 3)(\ln x)} = \frac{1}{\ln 3}$

    b) Yes, $\displaystyle\lim_{x \to \infty}\frac{\log_2 x^2}{\ln x} = \lim_{x \to \infty}\frac{2\ln x}{(\ln 2)(\ln x)} = \frac{2}{\ln 2}$

    c) Yes, $\displaystyle\lim_{x \to \infty}\frac{\log_{10}\sqrt{x}}{\ln x} = \lim_{x \to \infty}\frac{\ln x}{2(\ln 10)(\ln x)} = \frac{1}{\ln 100}$

    d) No, $\displaystyle\lim_{x \to \infty}\frac{1/x}{\ln x} = \lim_{x \to \infty} -\frac{1}{x} = 0$

    e) No, $\displaystyle\lim_{x \to \infty}\frac{1/\sqrt{x}}{\ln x} = \lim_{x \to \infty} -\frac{1}{2\sqrt{x}} = 0$

    f) No, $\displaystyle\lim_{x \to \infty}\frac{e^{-x}}{\ln x} = \lim_{x \to \infty}\frac{1}{e^x \ln x} = 0$

    g) No, $\displaystyle\lim_{x \to \infty}\frac{x}{\ln x} = \lim_{x \to \infty} x = \infty$

    h) Yes, $\displaystyle\lim_{x \to \infty}\frac{5\ln x}{\ln x} = 5$

    i) No, $\displaystyle\lim_{x \to \infty}\frac{2}{\ln x} = 0$

    j) No, $\displaystyle 0 = \lim_{x \to \infty}\frac{-1}{\ln x} \leq \lim_{x \to \infty}\frac{\sin x}{\ln x} < \lim_{x \to \infty}\frac{1}{\ln x} = 0 \Rightarrow \lim_{x \to \infty}\frac{\sin x}{\ln x} = 0$

5. $\displaystyle\lim_{x \to \infty}\frac{\sqrt{10x+1}}{\sqrt{x}} = \sqrt{\lim_{x \to \infty}\frac{10x+1}{x}} = \sqrt{10}$ and $\displaystyle\lim_{x \to \infty}\frac{\sqrt{x+1}}{\sqrt{x}} = \sqrt{\lim_{x \to \infty}\frac{x+1}{x}} = \sqrt{1} = 1$. Since

    the growth rate is transitive ,we may conclude that $\sqrt{10x+1}$ and $\sqrt{x+1}$ have the same growth rate.

7. a) False b) False c) True   d) True   e) True   f) True   g) False h) True

9. $\displaystyle\lim_{x \to \infty}\frac{x^n}{e^x} = \lim_{x \to \infty}\frac{n\,x^{n-1}}{e^x} = \cdots = \lim_{x \to \infty}\frac{n!}{e^x} = 0 \Rightarrow x^n = 0(e^x)$ for any non-negative integer n

11. a) $\displaystyle\lim_{x \to \infty}\frac{x^{1/n}}{\ln x} = \lim_{x \to \infty}\frac{x^{(1-n)/n}}{n\,(1/x)} = \left(\frac{1}{n}\right)\lim_{x \to \infty} x^{1/n} = \infty \Rightarrow \ln x = o(x^{1/n})$ for any positive integer n

    b) $\displaystyle \ln\left(e^{17000000}\right) = 17000000 < \left(e^{17 \times 10^6}\right)^{1/10^6} = e^{17} \approx 24154952.75$

    c) $x \approx 3.430631121 \times 10^{15}$

    d) In the interval $[3.41 \times 10^{15}, 3.45 \times 10^{15}]$
    we have $\ln x = 10 \ln \ln x$.

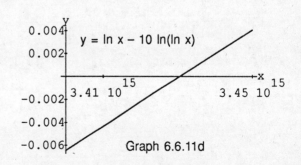

Graph 6.6.11d

13. $\text{Lim}_{x \to \infty} \dfrac{x}{x \log_2 x} = \text{Lim}_{x \to \infty} \dfrac{\ln 2}{\ln x} = 0 \Rightarrow n \log_2 n$ grows faster than n, $\text{Lim}_{x \to \infty} \dfrac{x}{x^2} = \text{Lim}_{x \to \infty} \dfrac{1}{x} = 0 \Rightarrow n^2$ grows

   faster than n, $\text{Lim}_{x \to \infty} \dfrac{x}{x \left(\log_2 x\right)^2} = (\ln 2)^2 \left(\text{Lim}_{x \to \infty} \dfrac{1}{(\ln x)^2} \right) = 0 \Rightarrow n \left(\log_2 n\right)^2$ grows faster than n

   $\therefore$ n is the best choice

15. If $\text{Lim}_{x \to \infty} \dfrac{f(x)}{g(x)} = L \neq 0$, then $\text{Lim}_{x \to \infty} \dfrac{g(x)}{f(x)} = \dfrac{1}{L}$ and by definition $f = O(g)$ and $g = O(f)$.

17. The error in the trapezoidal rule $= \left|E_t\right| \leq \left(\dfrac{b-a}{12}\right) h^2 M_1$, the error in the Simpson's rule =

   $\left|E_s\right| \leq \left(\dfrac{b-a}{180}\right) h^4 M_2$, where $h = \dfrac{b-a}{n}$; $\text{Lim}_{h \to 0} \dfrac{\left|E_t\right|}{h^2} \leq \text{Lim}_{h \to 0} \dfrac{(b-a) M_1}{12} = \dfrac{(b-a) M_1}{12} \Rightarrow \left|E_t\right|$ is $O(h^2)$,

   $\text{Lim}_{h \to 0} \dfrac{\left|E_s\right|}{h^4} \leq \text{Lim}_{h \to 0} \dfrac{(b-a) M_2}{180} = \dfrac{(b-a) M_2}{180} \Rightarrow \left|E_s\right|$ is $O(h^4)$, when $h \to 0 \Rightarrow n \to \infty \Rightarrow \text{Lim}_{n \to \infty} \left|\dfrac{E_s}{E_t}\right| =$

   $\text{Lim}_{n \to \infty} \left(\dfrac{(b-a)^5 M_2}{180 \, n^4} \dfrac{12 \, n^2}{(b-a)^3 M_1} \right) = \left(\dfrac{M_2 (b-a)^2}{15 \, M_1} \right) \left(\text{Lim}_{n \to \infty} \dfrac{1}{n^2} \right) = 0 \Rightarrow E_t$ grows faster than E_s

6.7 THE INVERSE TRIGONOMETRIC FUNCTIONS

1. a) $\dfrac{\pi}{4}$ b) $-\dfrac{\pi}{3}$ c) $\dfrac{\pi}{6}$

3. a) $-\dfrac{\pi}{6}$ b) $\dfrac{\pi}{4}$ c) $-\dfrac{\pi}{3}$

5. a) $\dfrac{\pi}{3}$ b) $\dfrac{3\pi}{4}$ c) $\dfrac{\pi}{6}$

7. a) $\dfrac{3\pi}{4}$ b) $\dfrac{\pi}{6}$ c) $\dfrac{2\pi}{3}$

9. a) $\dfrac{\pi}{4}$ b) $-\dfrac{\pi}{3}$ c) $\dfrac{\pi}{6}$

11. a) $\dfrac{3\pi}{4}$ b) $\dfrac{\pi}{6}$ c) $\dfrac{2\pi}{3}$

13. $\cos \alpha = \dfrac{\sqrt{3}}{2}$, $\tan \alpha = \dfrac{1}{\sqrt{3}}$, $\sec \alpha = \dfrac{2}{\sqrt{3}}$, $\csc \alpha = 2$

15. $\sin\left[\dfrac{\pi}{12} + \cos^{-1} \dfrac{\sqrt{2}}{2}\right] = \sin\left[\dfrac{\pi}{12} + \dfrac{\pi}{4}\right] = \dfrac{\sqrt{3}}{2}$

17. $\sin^{-1}\left[\dfrac{1}{2} \sec\left[\cos^{-1} \dfrac{1}{2}\right]\right] = \sin^{-1}\left[\dfrac{1}{2} \sec\left[\dfrac{\pi}{3}\right]\right] = \sin^{-1}\left[\dfrac{1}{2} \, 2\right] = \dfrac{\pi}{2}$

19. $\csc\left[\sec^{-1} 2\right] + \cos\left[\tan^{-1}\left(-\sqrt{3}\right)\right] = \csc\left[\dfrac{\pi}{3}\right] + \cos\left[-\dfrac{\pi}{3}\right] = \dfrac{2}{\sqrt{3}} + \cos\left[\dfrac{\pi}{3}\right] = \dfrac{4 + \sqrt{3}}{2\sqrt{3}}$

21. $\sec^{-1}\left(\sqrt{2}\left(\cos\left(\cot^{-1} 1\right)\right)\right) = \sec^{-1}\left(\sqrt{2}\left(\cos\left(\frac{\pi}{4}\right)\right)\right) = \sec^{-1}\left(\sqrt{2}\,\frac{1}{\sqrt{2}}\right) = \sec^{-1} 1 = 0$

23. $\sec^{-1}\left[\sec\left[-\frac{\pi}{6}\right]\right] = \sec^{-1}\left[\frac{2}{\sqrt{3}}\right] = \frac{\pi}{6}$

25. $\sin\left(\tan^{-1} 2.4\right) = \frac{12}{13}$

27. $\sin\left[\tan^{-1} \frac{x}{2}\right] = \frac{x}{\sqrt{x^2 + 4}}$

29. $\cot\left[\sin^{-1} \frac{3}{\sqrt{4x^2 + 9}}\right] = \frac{2|x|}{3}$

31. $\sin\left(\tan^{-1} \frac{x}{\sqrt{1 - x^2}}\right) = x$

33. $\sec^{-1}\left[\sqrt{2}\cos\left[\frac{13\pi}{12} - \sin^{-1}\left[-\frac{1}{2}\sin\left[\sec^{-1}\sqrt{3} + \csc^{-1}\sqrt{3}\right]\right]\right]\right] =$

$\sec^{-1}\left[\sqrt{2}\cos\left[\frac{13\pi}{12} - \sin^{-1}\left[-\frac{1}{2}\sin\left[\frac{\pi}{2}\right]\right]\right]\right] = \sec^{-1}\left[\sqrt{2}\cos\left[\frac{13\pi}{12} - \sin^{-1}\left[-\frac{1}{2}\right]\right]\right] =$

$\sec^{-1}\left[\sqrt{2}\cos\left[\frac{13\pi}{12} + \frac{\pi}{6}\right]\right] = \sec^{-1}\left[\sqrt{2}\cos\left[\frac{5\pi}{4}\right]\right] = \sec^{-1}\left[\sqrt{2}\left(-\frac{1}{\sqrt{2}}\right)\right] = \sec^{-1}[-1] = \pi$

35. a) $\lim\limits_{x \to 1^-} \sin^{-1} x = \frac{\pi}{2}$ b) $\lim\limits_{x \to -1^+} 2\cos^{-1} x = \pi$

37. a) $\lim\limits_{x \to \infty} \sec^{-1}\sqrt{x} = \frac{\pi}{2}$ b) $\lim\limits_{x \to -\infty} \sec^{-1}(x + 5) = \frac{\pi}{2}$

39. The angle α = the large angle between the wall and the right end of the blackboard − the small angle between the left end of the blackboard and the wall = $\cot^{-1}\left(\frac{x}{15}\right) - \cot^{-1}\left(\frac{x}{3}\right)$

41. $\frac{ds}{dt} = \cos^2 \pi s \Rightarrow \int \sec^2 \pi s \, ds = \int dt \Rightarrow \frac{\tan(\pi s)}{\pi} = t + C$ and $\frac{\tan(0)}{\pi} = 0 + C \Rightarrow C = 0 \therefore \pi t = \tan(\pi s)$.

If $s = \frac{1}{4}$, then $\pi t = \tan\left(\frac{\pi}{4}\right) \Rightarrow t = \frac{1}{\pi}$ sec. If $s \to \frac{\pi}{2}$, then $\pi t = \tan\left(\frac{\pi}{2}\right) \Rightarrow t \to \infty$ and s will never reach $\frac{1}{2}$ because it would take an infinite amount of time.

43. $\alpha = 65° - \beta = 65° - \tan^{-1}\left(\frac{21}{50}\right) \approx 65° - 22.78° \approx 42.23°$

45. a) as $\to \infty \Rightarrow \sec^{-1}(-x) \to \left(\frac{\pi}{2}\right)^+$ and $\sec^{-1}(x) \to \left(\frac{\pi}{2}\right)^-$

b) $\cos^{-1}(-x) = \pi - \cos^{-1}(x)$, where $-1 \le x \le 1 \Rightarrow \cos^{-1}\left(-\frac{1}{x}\right) = \pi - \cos^{-1}\left(\frac{1}{x}\right)$, where $x \ge 1$ or

$x \le -1 \Rightarrow \sec^{-1}(-x) = \pi - \sec^{-1}(x)$

47.

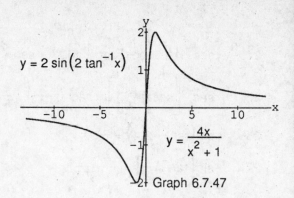

The graphs are identical for $y = 2 \sin\left(2 \tan^{-1} x\right) =$

$4 \left(\sin\left(\tan^{-1} x\right)\right)\left(\cos\left(\tan^{-1} x\right)\right) =$

$4 \left(\dfrac{x}{\sqrt{x^2 + 1}}\right)\left(\dfrac{1}{\sqrt{x^2 + 1}}\right) = \dfrac{4x}{x^2 + 1}$.

Graph 6.7.47

49.   a) Domain, all reals except those having the form
       $\dfrac{\pi}{2} + k\pi$ where k is an integer; Range $(-\pi/2, \pi/2)$; No,

       many computer graphs will not be correct

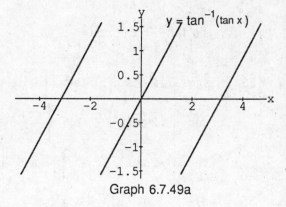

Graph 6.7.49a

   b) Domain, all reals; Range, all reals; Yes, many
       computer graphs will be correct

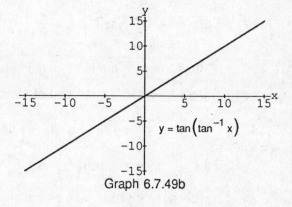

Graph 6.7.49b

51.   a) Domain, all real; Range, $[0,\pi]$; Yes, many
           computer graphs will be correct

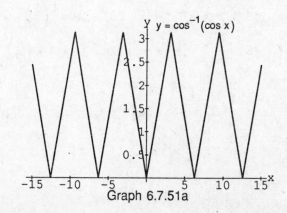

Graph 6.7.51a

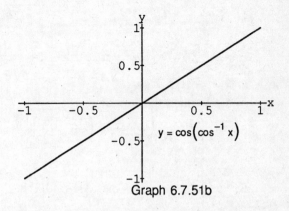

Graph 6.7.51b

b) Domain, $[-1,1]$; Range, $[-1,1]$; No, many computer graphs will not be correct

6.8 DERIVATIVES OF INVERSE TRIGONOMETRIC FUNCTIONS; RELATED INTEGRALS

1. $y = \cos^{-1}\left(x^2\right) \Rightarrow \dfrac{dy}{dx} = \dfrac{-2x}{\sqrt{1-x^4}}$

3. $y = 5\tan^{-1}(3x) \Rightarrow \dfrac{dy}{dx} = \dfrac{15}{1+9x^2}$

5. $y = \sin^{-1}\left(\dfrac{x}{2}\right) \Rightarrow \dfrac{dy}{dx} = \dfrac{1}{\sqrt{4-x^2}}$

7. $y = \sec^{-1}(5x) \Rightarrow \dfrac{dy}{dx} = \dfrac{1}{|x|\sqrt{25x^2-1}}$

9. $y = \csc^{-1}\left(x^2+1\right) \Rightarrow \dfrac{dy}{dx} = \dfrac{-2x}{\left(x^2+1\right)\sqrt{x^4+2x^2}}$

11. $y = \csc^{-1}\left(x^{1/2}\right) + \sec^{-1}\left(x^{1/2}\right) \Rightarrow \dfrac{dy}{dx} = \dfrac{-\frac{1}{2}x^{-1/2}}{\sqrt{x}\sqrt{x-1}} + \dfrac{\frac{1}{2}x^{-1/2}}{\sqrt{x}\sqrt{x-1}} = 0$

13. $y = \cot^{-1}\left((x-1)^{1/2}\right) \Rightarrow \dfrac{dy}{dx} = \dfrac{-1}{2x\sqrt{x-1}}$

15. $y = \left(x^2-1\right)^{1/2} - \sec^{-1}(x) \Rightarrow \dfrac{dy}{dx} = \left(\dfrac{1}{2}\right)\left(x^2-1\right)^{-1/2}(2x) - \dfrac{1}{|x|\sqrt{x^2-1}} = \dfrac{x\,|x|-1}{|x|\sqrt{x^2-1}}$

17. $y = x\sin^{-1}(x) + \sqrt{1-x^2} \Rightarrow \dfrac{dy}{dx} = \sin^{-1}x + \dfrac{x}{\sqrt{1-x^2}} - \dfrac{x}{\sqrt{1-x^2}} = \sin^{-1}x$

19. If $y = \ln x - \dfrac{1}{2}\ln\left(1+x^2\right) - \dfrac{\tan^{-1}x}{x} + C$, then $dy = \left(\dfrac{1}{x} - \dfrac{x}{1+x^2} - \dfrac{\frac{x}{1+x^2}-\tan^{-1}x}{x^2}\right)dx =$

$\left(\dfrac{1}{x} - \dfrac{x}{1+x^2} - \dfrac{1}{x\left(1+x^2\right)} + \tan^{-1}x\right)dx = \dfrac{x\left(1+x^2\right)-x^3-x+\left(\tan^{-1}x\right)\left(1+x^2\right)}{x^2\left(1+x^2\right)}dx =$

$\dfrac{\tan^{-1}x}{x^2}dx$, which verifies the formula.

21. If $y = x\left(\sin^{-1}x\right)^2 - 2x + 2\sqrt{1-x^2}\,\sin^{-1}x + C$, then $dy =$

$$\left[\left(\sin^{-1}x\right)^2 + \frac{2x\left(\sin^{-1}x\right)}{\sqrt{1-x^2}} - 2 + \frac{-2x}{\sqrt{1-x^2}}\sin^{-1}x + 2\sqrt{1-x^2}\,\frac{1}{\sqrt{1-x^2}}\right]dx = \left(s din^{-1}x0\right)^2 dx, \text{ which}$$

verifies the formula.

23. $\displaystyle\int_0^{1/2} \frac{1}{\sqrt{1-x^2}}\,dx = \left[\sin^{-1}(x)\right]_0^{1/2} = \frac{\pi}{6}$

25. $\displaystyle\int_{\sqrt{2}}^{2} \frac{1}{x\sqrt{x^2-1}}\,dx = \left[\sec^{-1}(x)\right]_{\sqrt{2}}^{2} = \frac{\pi}{12}$

27. $\displaystyle\frac{1}{2}\int_0^{\sqrt{2}/2} \frac{2y}{\sqrt{1-y^4}}\,dy = \left(\frac{1}{2}\right)\left[\sin^{-1}\left(y^2\right)\right]_0^{\sqrt{2}/2} = \frac{\pi}{12}$

29. $\displaystyle\int_0^2 \frac{1}{1+(x-1)^2}\,dx = \int_{-1}^1 \frac{1}{1+u^2}\,du = \left[\tan^{-1}(u)\right]_{-1}^1 = \frac{\pi}{2}$, where $u = x - 1$

31. $\displaystyle\int_{-\pi/2}^{\pi/2} \frac{2\cos\theta}{1+\sin^2\theta}\,d\theta = \int_{-1}^1 \frac{2}{1+u^2}\,du = 2\left[\tan^{-1}(u)\right]_{-1}^1 = \pi$, where $u = \sin\theta$

33. $\displaystyle\int_1^3 \frac{1}{\sqrt{y}(1+y)}\,dy = \int_1^{\sqrt{3}} \frac{2u}{u(1+u^2)}\,du = 2\int_1^{\sqrt{3}} \frac{1}{1+u^2}\,du = 2\left[\tan^{-1}u\right]_1^{\sqrt{3}} = \frac{\pi}{6}$, where $u = \sqrt{y}$

35. $\displaystyle\int_{1/\sqrt{3}}^{1} \frac{2}{2x\sqrt{4x^2-1}}\,dx = \left[\sec^{-1}(2x)\right]_{1/\sqrt{3}}^{1} = \frac{\pi}{6}$

37. $\displaystyle\int \frac{1}{x\sqrt{x^4-1}}\,dx = \frac{1}{2}\int \frac{2x}{x^2\sqrt{\left(x^2\right)^2-1}}\,dx = \frac{1}{2}\sec^{-1}\left(x^2\right) + C$

39. $\displaystyle\int_2^4 \frac{1}{2y\sqrt{y-1}}\,dy = \int_{\sqrt{2}}^2 \frac{2u}{2u^2\sqrt{u^2-1}}\,du = \int_{\sqrt{2}}^2 \frac{1}{u\sqrt{u^2-1}}\,du = \left[\sec^{-1}(u)\right]_{\sqrt{2}}^2 = \frac{\pi}{12}$, where $u^2 = y$

41. $\displaystyle\int_{-2/3}^{-\sqrt{2}/3} \frac{1}{t\sqrt{9t^2-1}}\,dt = \int_{-2/3}^{-\sqrt{2}/3} \frac{3}{3t\sqrt{(3t)^2-1}}\,dt = \left[\sec^{-1}|3t|\right]_{-2/3}^{-\sqrt{2}/3} = -\frac{\pi}{12}$

43. $\displaystyle\lim_{x\to 0} \frac{\sin^{-1}(x)}{x} = \lim_{x\to 0} \frac{1}{\sqrt{1-x^2}} = 1$

45. $V = \displaystyle\pi\int_{-\sqrt{3}/3}^{\sqrt{3}} \left(\frac{1}{\sqrt{1+x^2}}\right)^2 dx = \pi\int_{-\sqrt{3}/3}^{\sqrt{3}} \frac{1}{1+x^2}\,dx = \pi\left[\tan^{-1}(x)\right]_{-\sqrt{3}/3}^{\sqrt{3}} = \frac{\pi^2}{2}$

47. $\alpha(x) = \cot^{-1}\left(\dfrac{x}{15}\right) - \cot^{-1}\left(\dfrac{x}{3}\right)$, $x > 0 \Rightarrow \alpha'(x) = \dfrac{-15}{225+x^2} + \dfrac{3}{9+x^2} = \dfrac{-15(9+x^2)+3(225+x^2)}{(225+x^2)(9+x^2)}$;

solving $\alpha'(x) = 0 \Rightarrow -135 - 15x^2 + 675 + 3x^2 = 0 \Rightarrow x = 3\sqrt{5}$, $\alpha'(x) > 0$ when $0 < x < 3\sqrt{5}$ and

$\alpha'(x) < 0$ for $x > 3\sqrt{5} \Rightarrow$ at $3\sqrt{5}$ ft there is a maximum

49. $\theta(x) = \pi - \cot^{-1}(x) - \cot^{-1}(2-x) \Rightarrow \theta'(x) = \dfrac{4(1-x)}{(1+x^2)(1+(2-x)^2)}$, solving $\theta'(x) = 0 \Rightarrow x = 1$;

$\theta'(x) > 0$ for $0 < x < 1$ and $\theta'(x) < 0$ for $x > 1 \Rightarrow$ at $x = 1$ there is a maximum of $\theta(1) = \dfrac{\pi}{2}$

51. $\dfrac{dy}{dx} = \dfrac{1}{x\sqrt{x^2-1}}$ at $(2,\pi) \Rightarrow y = \sec^{-1}|x| + C = \cos^{-1}\left|\dfrac{1}{x}\right| + C$ at $(2,\pi) \Rightarrow \pi - \dfrac{\pi}{3} = C \Rightarrow C = \dfrac{2\pi}{3} \Rightarrow$

$y = \sec^{-1}|x| + \dfrac{2\pi}{3}$, but $x > 0 \Rightarrow y = \sec^{-1}x + \dfrac{2\pi}{3}$

53. $\dfrac{dy}{dx} = \dfrac{-1}{\sqrt{1-x^2}}$ at $(-\sqrt{2}/2, \pi/2) \Rightarrow y = \cos^{-1}(x) + C$ at $(-\sqrt{2}/2, \pi/2) \Rightarrow C = -\dfrac{\pi}{4} \Rightarrow y = \cos^{-1}(x) - \dfrac{\pi}{4}$

55. The Calculus Tool Kit yields 0.643517104.

57. Yes, $\sin^{-1}(x)$ and $\cos^{-1}(x)$ differ by the constant $\dfrac{\pi}{2}$.

59. $y = \cos^{-1}u \Rightarrow \cos y = u \Rightarrow (-\sin y)\,y' = u' \Rightarrow y' = -\dfrac{u'}{\sin y}$. Whereas $0 < y < \pi$, we have

$\sin y = \sqrt{1-\cos^2 y} \Rightarrow y' = -\dfrac{u'}{\sqrt{1-u^2}}$

61. $y = \cot^{-1}u = \dfrac{\pi}{2} - \tan^{-1}u \Rightarrow y' = -\dfrac{u'}{1+u^2}$

63.
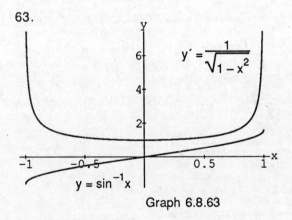
Graph 6.8.63

65.
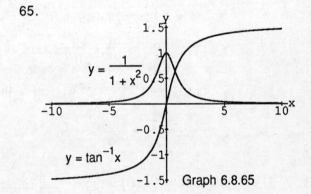
Graph 6.8.65

6.9 HYPERBOLIC FUNCTIONS

1. $\cosh x = \dfrac{5}{4}, \tanh x = -\dfrac{3}{5}, \coth x = -\dfrac{5}{3}, \operatorname{sech} x = \dfrac{4}{5}, \operatorname{csch} = -\dfrac{4}{3}$

3. $\sinh x = \dfrac{8}{15}, \tanh x = \dfrac{8}{17}, \coth x = \dfrac{17}{8}, \operatorname{sech} x = \dfrac{15}{17}, \operatorname{csch} = \dfrac{15}{8}$

5. $2\cosh(\ln x) = 2\left(\dfrac{e^{\ln x} + e^{-\ln x}}{2}\right) = x + \dfrac{1}{x}$

7. $\cosh(5x) + \sinh(5x) = \dfrac{e^{5x} + e^{-5x}}{2} + \dfrac{e^{5x} - e^{-5x}}{2} = e^{5x}$

9. $\cosh(3x) - \sinh(3x) = \dfrac{e^{3x} + e^{-3x}}{2} - \dfrac{e^{3x} - e^{-3x}}{2} = e^{-3x}$

11. a) $\sinh(2x) = \sinh(x+x) = \sinh(x)\cosh(x) + \cosh(x)\sinh(x) = 2\sinh(x)\cosh(x)$

    b) $\cosh(2x) = \cosh(x+x) = \cosh(x)\cosh(x) + \sinh(x)\sin(x) = \cosh^2(x) + \sinh^2(x)$

13. $y = \sinh(3x) \Rightarrow \dfrac{dy}{dx} = 3\cosh(3x)$ 15. $y = 2\tanh\left(\dfrac{x}{2}\right) \Rightarrow \dfrac{dy}{dx} = \operatorname{sech}^2\left(\dfrac{x}{2}\right)$

17. $y = \ln(\operatorname{sech}(x)) \Rightarrow \dfrac{dy}{dx} = -\dfrac{\operatorname{sech}(x)\tanh(x)}{\operatorname{sech}(x)} = -\tanh(x)$

19. $y = \ln(\operatorname{csch}(x) + \coth(x)) \Rightarrow \dfrac{dy}{dx} = \dfrac{-\operatorname{csch}(x)\coth(x) - \operatorname{csch}^2(x)}{\operatorname{csch}(x) + \coth(x)} = -\operatorname{csch}(x)$

21. $y = \left(\dfrac{1}{2}\right)\ln|\tanh(x)| \Rightarrow \dfrac{dy}{dx} = \dfrac{\operatorname{sech}^2 x}{2\tanh x} = \dfrac{\cosh(x)}{2\sinh(x)\cosh^2(x)} = \dfrac{1}{2\sinh(x)\cosh(x)} = \dfrac{1}{\sinh(2x)} = \operatorname{csch}(2x)$

23. a) $y = \cosh^2(x) \Rightarrow \dfrac{dy}{dx} = 2\cosh(x)\sinh(x) = \sinh(2x)$

    b) $y = \sinh^2(x) \Rightarrow \dfrac{dy}{dx} = 2\sinh(x)\cosh(x) = \sinh(2x)$

    c) $y = \left(\dfrac{1}{2}\right)\cosh(2x) \Rightarrow \dfrac{dy}{dx} = \left(\dfrac{1}{2}\right)(\sinh(2x))(2) = \sinh(2x)$

25. $y = \sinh^{-1}(2x) \Rightarrow \dfrac{dy}{dx} = \dfrac{2}{\sqrt{1+4x^2}}$

27. $y = (1-x)\tanh^{-1}(x) \Rightarrow \dfrac{dy}{dx} = (-1)\tanh^{-1}(x) + (1-x)\left(\dfrac{1}{1-x^2}\right) = \dfrac{1}{1+x} - \tanh^{-1}(x)$

29. $y = x\operatorname{sech}^{-1}(x) \Rightarrow \dfrac{dy}{dx} = \operatorname{sech}^{-1}(x) + \dfrac{(x)(-1)}{x\sqrt{1-x^2}} = \operatorname{sech}^{-1}(x) - \dfrac{1}{\sqrt{1-x^2}}$

31. $y = \sinh^{-1}(\tan(x)) \Rightarrow \dfrac{dy}{dx} = \dfrac{\sec^2 x}{\sqrt{1+\tan^2 x}} = \dfrac{\sec^2 x}{\sec x} = \sec x$

33. $y = \tanh^{-1}(\sin x)$, where $-\dfrac{\pi}{2} < x < \dfrac{\pi}{2} \Rightarrow \dfrac{dy}{dx} = \dfrac{\cos x}{1-\sin^2 x} = \dfrac{\cos x}{\cos^2 x} = \sec x$

35. $y = \operatorname{sech}^{-1}(\sin x)$, where $0 < x < \dfrac{\pi}{2} \Rightarrow \dfrac{dy}{dx} = \dfrac{-\cos x}{(\sin x)\sqrt{1+\sin^2 x}} = \dfrac{-1}{\sin x} = -\csc x$

37. a) If $y = \tan^{-1}(\sinh x) + C$, then $dy = \dfrac{\cosh x}{1+\sinh^2 x}\,dx = \dfrac{\cosh x}{\cosh^2 x}\,dx = \operatorname{sech} x\,dx$, which verifies

the formula.

    b) If $y = \sin^{-1}(\tanh x) + C$, then $dy = \dfrac{\operatorname{sech}^2 x}{\sqrt{1-\tanh^2 x}}\,dx = \dfrac{\operatorname{sech}^2 x}{\operatorname{sech} x}\,dx = \operatorname{sech} x\,dx$, which verifies

the formula.

39.   If $y = x \operatorname{sech}^{-1}x + \sin^{-1}x + C$, then $dy = \left(\operatorname{sech}^{-1}x + x\left(\dfrac{-1}{x\sqrt{1-x^2}}\right) + \dfrac{1}{\sqrt{1-x^2}}\right)dx = \operatorname{sech}^{-1}x\,dx$,

which verifies the formula.

41.   If $y = \dfrac{x^2-1}{2}\coth^{-1}x + \dfrac{x}{2} + C$, then $dy = \left(x\coth^{-1}x + \left(\dfrac{x^2-1}{2}\right)\left(\dfrac{1}{1-x^2}\right) + \dfrac{1}{2}\right)dx = x\coth^{-1}x\,dx$,

which verifies the formula.

43.   If $y = \dfrac{2x^2-1}{4}\cosh^{-1}x - \dfrac{x}{4}\sqrt{x^2-1} + C$, then $dy =$

$\left(x\cosh^{-1}x + \left(\dfrac{2x^2-1}{4}\right)\left(\dfrac{1}{\sqrt{x^2-1}}\right) - \dfrac{1}{4}\sqrt{x^2-1} - \left(\dfrac{x}{4}\right)\left(\dfrac{x}{\sqrt{x^2-1}}\right)\right)dx =$

$\left(x\cosh^{-1}x + \dfrac{2x^2-1-\left(x^2-1\right)-x^2}{4\sqrt{x^2-1}}\right)dx = \left(x\cosh^{-1}x + \dfrac{2x^2-1-x^2+1-x^2}{4\sqrt{x^2-1}}\right)dx =$

$\left(x\cosh^{-1}x\right)dx$, which verifies the formula.

45. $\displaystyle\int_{-1}^{1}\cosh(5x)\,dx = \dfrac{1}{5}\int_{-1}^{1}\cosh(5x)\,5\,dx = \left[\dfrac{\sinh(5x)}{5}\right]_{-1}^{1} = \left(\dfrac{2}{5}\right)\sinh 5$

47. $\displaystyle\int_{-3}^{3}\sinh(x)\,dx = 0$, $\sinh(x)$ is odd

49. $\displaystyle\int_{0}^{1/2}4\,e^x\cosh(x)\,dx = \int_{0}^{1/2}4\,e^x\left(\dfrac{e^x+e^{-x}}{2}\right)dx = \int_{0}^{1/2}e^{2x}(2)+2\,dx = \left[e^{2x}+2x\right]_{0}^{1/2} = e$

51. $\displaystyle\int\dfrac{\cosh(\ln x)}{x}\,dx = \sinh(\ln x)+C$

53. $\displaystyle\int_{0}^{\ln 3}\operatorname{sech}^2(x)\,dx = \left[\tanh(x)\right]_{0}^{\ln 3} = \dfrac{4}{5}$

55. $2\displaystyle\int_{1}^{4}\dfrac{\cosh(\sqrt{x})}{2\sqrt{x}}\,dx = 2\left[\sinh\left(\sqrt{x}\right)\right]_{1}^{4} = 2(\sinh(2)-\sinh(1))$

57. $\displaystyle\int_{-\ln 2}^{0}\cosh^2 3x\,dx = \dfrac{1}{2}\int_{-\ln 2}^{0}\cosh(6x)+1\,dx = \dfrac{1}{2}\left[\dfrac{\sinh 6x}{6}+x\right]_{-\ln 2}^{0} =$

$\dfrac{\sinh(\ln 64)}{12} + \ln\sqrt{2} \approx 3.012589$

59. $\displaystyle\int\operatorname{sech}^3 5x\tanh 5x\,dx = \int(\operatorname{sech}5x)^2(\operatorname{sech}5x\tanh 5x)\,dx = \dfrac{(\operatorname{sech}5x)^3}{3}+C$

61. $\sinh^{-1}\left(\dfrac{-5}{12}\right) = \ln\left(-\dfrac{5}{12}+\sqrt{\dfrac{25}{144}+1}\right) = \ln\left(\dfrac{2}{3}\right)$

63. $\cosh^{-1}\left(\dfrac{5}{3}\right) = \ln\left(\dfrac{5}{3}+\sqrt{\dfrac{25}{9}-1}\right) = \ln 3$

65. $\tanh^{-1}\left(-\dfrac{1}{2}\right) = \dfrac{1}{2}\ln\left(\dfrac{1-1/2}{1+1/2}\right) = -\dfrac{\ln 3}{2}$

67. $\coth^{-1}\left(\dfrac{5}{4}\right) = \left(\dfrac{1}{2}\right)\ln\left(\dfrac{\frac{9}{4}}{\frac{1}{4}}\right) = \left(\dfrac{1}{2}\right)\ln 9 = \ln 3$

69. $\operatorname{sech}^{-1}\left(\dfrac{3}{5}\right) = \ln\left(\dfrac{1 + \sqrt{1 - 9/25}}{3/5}\right) = \ln 3$

71. $\operatorname{csch}^{-1}\left(-\dfrac{1}{\sqrt{3}}\right) = \ln\left(-\sqrt{3} + \dfrac{\sqrt{4/3}}{1/\sqrt{3}}\right) = \ln(-\sqrt{3} + 2)$

73. a) $\displaystyle\int_0^1 \dfrac{1}{\sqrt{1 + x^2}}\,dx = \left[\sinh^{-1}(x)\right]_0^1 = \sinh^{-1}(1) - \sinh^{-1}(0) = \sinh^{-1}(1)$

    b) $\sinh^{-1}(1) = \ln(1 + \sqrt{2})$

75. a) $\displaystyle\int_{5/4}^{5/3} \dfrac{1}{\sqrt{x^2 - 1}}\,dx = \left[\cosh^{-1}(x)\right]_{5/4}^{5/3} = \cosh^{-1}\left(\dfrac{5}{3}\right) - \cosh^{-1}\left(\dfrac{5}{4}\right)$

    b) $\cosh^{-1}\left(\dfrac{5}{3}\right) - \cosh^{-1}\left(\dfrac{5}{4}\right) = \ln\left(5/3 + \sqrt{25/9 - 1}\right) - \ln\left(5/4 + \sqrt{25/16 - 1}\right) = \ln\left(\dfrac{3}{2}\right)$

77. a) $\displaystyle\int_{5/4}^2 \dfrac{1}{1 - x^2}\,dx = \left[\coth^{-1}(x)\right]_{5/4}^2 = \coth^{-1}(2) - \coth^{-1}\left(\dfrac{5}{4}\right)$

    b) $\coth^{-1}(2) - \coth^{-1}\left(\dfrac{5}{4}\right) = \left(\dfrac{1}{2}\right)\left[\ln(3) - \ln\left(\dfrac{9/4}{1/4}\right)\right] = \left(\dfrac{1}{2}\right)\ln\left(\dfrac{1}{3}\right)$

79. a) $\displaystyle\int_1^2 \dfrac{1}{x\sqrt{4 + x^2}}\,dx = -\dfrac{1}{2}\int_{1/2}^1 \dfrac{-1}{u\sqrt{1 + u^2}}\,du = -\dfrac{1}{2}\left[\operatorname{csch}^{-1}(u)\right]_{1/2}^1 =$

    $\left(\dfrac{1}{2}\right)\left(\operatorname{csch}^{-1}\left(\dfrac{1}{2}\right) - \operatorname{csch}^{-1}(1)\right)$, where $u = \dfrac{x}{2}$

    b) $\left(\dfrac{1}{2}\right)\left(\operatorname{csch}^{-1}\left(\dfrac{1}{2}\right) - \operatorname{csch}^{-1}(1)\right) = \left(\dfrac{1}{2}\right)\left[\ln\left(2 + \dfrac{\sqrt{5/4}}{1/2}\right) - \ln(1 + \sqrt{2})\right] = \left(\dfrac{1}{2}\right)\ln\left(\dfrac{2 + \sqrt{5}}{1 + \sqrt{2}}\right)$

81. $V = \pi\displaystyle\int_0^2 \cosh^2(x) - \sinh^2(x)\,dx = \pi\int_0^2 1\,dx = 2\pi$

83. $M_x = \displaystyle\int_{-\ln\sqrt{3}}^{\ln\sqrt{3}} \dfrac{\operatorname{sech}^2(x)}{2}\,dx = \dfrac{1}{2}[\tanh(x)]\,\Big|_{-\ln\sqrt{3}}^{\ln\sqrt{3}} = \left(\dfrac{1}{2}\right)\left(\dfrac{1}{2} - \left(-\dfrac{1}{2}\right)\right) = \dfrac{1}{2}$, $M = \displaystyle\int_{-\ln\sqrt{3}}^{\ln\sqrt{3}} \operatorname{sech}(x)\,dx =$

$\left[\sin^{-1}(\tanh(x))\right]_{-\ln\sqrt{3}}^{\ln\sqrt{3}} = \dfrac{\pi}{3} \Rightarrow \overline{y} = \dfrac{M_x}{M} = \dfrac{3}{2\pi}$, and $\overline{x} = 0$, by symmetry

85. a) $y = \cosh x \Rightarrow ds = \sqrt{(dx)^2 + (dy)^2} = \sqrt{(dx)^2 + \sinh^2 x\,(dx)^2} = \cosh x\,dx$, $M_x = \displaystyle\int_{-\ln 2}^{\ln 2} y\,ds =$

$\displaystyle\int_{-\ln 2}^{\ln 2} \cosh x\,ds = \int_{-\ln 2}^{\ln 2} \cosh^2 x\,dx = \int_0^{\ln 2} \cosh 2x + 1\,dx = \left[\dfrac{\sinh 2x}{2} + x\right]_0^{\ln 2} =$

$\dfrac{15}{16} + \ln 2$, $M = 2\displaystyle\int_0^{\ln 2} \sqrt{1 + \sinh^2 x}\,dx = 2\int_0^{\ln 2} \cosh x\,dx = 2[\sinh x]_0^{\ln 2} = \dfrac{3}{2}$

$\therefore \overline{y} = \dfrac{M_x}{M} = \dfrac{\dfrac{15}{16} + \ln 2}{\dfrac{3}{2}} = \dfrac{5}{8} + \dfrac{\ln 4}{3}$ and $\overline{x} = 0$ by symmetry

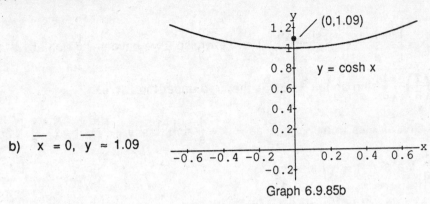

Graph 6.9.85b

b) $\overline{x} = 0$, $\overline{y} \approx 1.09$

87.   a) $f(x) = \dfrac{f(x) + f(-x)}{2}$ because $\dfrac{f(x) - f(-x)}{2} = 0$

      b) $f(x) = \dfrac{f(x) - f(-x)}{2}$ because $\dfrac{f(x) + f(-x)}{2} = 0$

89. $\dfrac{dy}{dx} = \dfrac{-1}{x\sqrt{1-x^2}} + \dfrac{x}{\sqrt{1-x^2}} \Rightarrow \displaystyle\int dy = \int \dfrac{-1}{x\sqrt{1-x^2}}\,dx + \int \dfrac{x}{\sqrt{1-x^2}}\,dx \Rightarrow$

$y = \text{sech}^{-1}(x) - \sqrt{1-x^2} + C$ at $(1,0) \Rightarrow C = 0 \Rightarrow y = \text{sech}^{-1}(x) - \sqrt{1-x^2}$

91.   a)   Let the point located at $(\cosh x, 0)$ be called T. $A(u)$ = area of the triangle $\triangle$ OTP – the area

      under the curve $y = \sqrt{x^2 - 1}$ from A to T $\Rightarrow A(u) = \dfrac{1}{2}\cosh u \sinh u - \displaystyle\int_{1}^{\cosh u} \sqrt{x^2 - 1}\,dx$.

      b) $A(u) = \dfrac{1}{2}\cosh u \sinh u - \displaystyle\int_{1}^{\cosh u} \sqrt{x^2 - 1}\,dx \Rightarrow A'(u) = \dfrac{1}{2}\left(\cosh^2 u + \sinh^2 u\right) -$

$\sqrt{\cosh^2 u - 1}\,(\sinh u) = \dfrac{1}{2}\cosh^2 u + \dfrac{1}{2}\sinh^2 u - \sinh^2 u = \dfrac{1}{2}\left(\cosh^2 u - \sinh^2 u\right) = \left(\dfrac{1}{2}\right)(1) = \dfrac{1}{2}$

      c) $A'(u) = \dfrac{1}{2} \Rightarrow A(u) = \dfrac{u}{2} + C$, from part a we have $A(0) = 0 \Rightarrow C = 0$ and $A(u) = \dfrac{u}{2} \Rightarrow u = 2A$

6.10 HANGING CABLES

1.   a) $y = \dfrac{1}{2}\cosh 2x \Rightarrow 1 + (y')^2 = 1 + \sinh^2 2x = \cosh^2 2x \Rightarrow L = \displaystyle\int_{0}^{\ln\sqrt{5}} \sqrt{\cosh^2 2x}\,dx =$

$\displaystyle\int_{0}^{\ln\sqrt{5}} \cosh 2x\,dx = \left[\dfrac{\sinh 2x}{2}\right]_{0}^{\ln\sqrt{5}} = \dfrac{6}{5}$

      b) $y = \dfrac{1}{a}\cosh ax \Rightarrow 1 + (y')^2 = 1 + \sinh^2 ax \Rightarrow \cosh^2 ax \Rightarrow L = \displaystyle\int_{0}^{b} \sqrt{\cosh^2 ax}\,dx =$

$\displaystyle\int_{0}^{b} \cosh ax\,dx = \left[\dfrac{\sinh ax}{a}\right]_{0}^{b} = \dfrac{\sinh ab}{a}$

3. $A = \int_0^b \dfrac{1}{a} \cosh ax\, dx = \left[\dfrac{\sinh ax}{a^2}\right]_0^b = \dfrac{\sinh ab}{a^2}$. From exercise 2 we have $s = \dfrac{1}{a}\sinh ab \Rightarrow$ the area

of the rectangle is $\left(\dfrac{1}{a}\right)\left(\dfrac{1}{a}\sinh ab\right) = \dfrac{\sinh ab}{a^2}$, the area under the curve.

5.   a) $s = \dfrac{1}{a}\sinh ax \Rightarrow \sinh ax = as \Rightarrow ax = \sinh^{-1} as \Rightarrow x = \dfrac{1}{a}\sinh^{-1} as;\ y = \dfrac{1}{a}\cosh ax = \dfrac{1}{a}\sqrt{\cosh^2 ax} =$

$\dfrac{1}{a}\sqrt{\sinh^2 ax + 1} = \dfrac{1}{a}\sqrt{a^2 s^2 + 1} = \sqrt{s^2 + \dfrac{1}{a^2}}$.

b) $dx = \dfrac{1}{a}\dfrac{1}{\sqrt{1+a^2 s^2}}\,a\,ds \Rightarrow (dx)^2 = \dfrac{1}{1+a^2 s^2}(ds)^2,\ dy = \dfrac{1}{\sqrt{s^2 + \dfrac{1}{a^2}}}\,s\,ds \Rightarrow (dy)^2 = \dfrac{1}{s^2 + \dfrac{1}{a^2}}s^2(ds)^2$

$\therefore (dx)^2 + (dy)^2 = \left(\dfrac{1}{1+a^2 s^2} + \dfrac{s^2}{s^2 + \dfrac{1}{a^2}}\right)(ds)^2 = \left(\dfrac{1}{1+a^2 s^2} + \dfrac{a^2 s^2}{a^2 s^2 + 1}\right)(ds)^2 = \left(\dfrac{1+a^2 s^2}{1+a^2 s^2}\right)(ds)^2 = (ds)^2$

6.M MISCELLANEOUS EXERCISES

1. $y = \ln\left(\cos^{-1}x\right) \Rightarrow y' = \dfrac{\dfrac{-1}{\sqrt{1-x^2}}}{\cos^{-1}x} = \dfrac{-1}{\sqrt{1-x^2}\,\cos^{-1}x}$

3. $y = \log_2 x^2 = \dfrac{2\ln x}{\ln 2} \Rightarrow \dfrac{dy}{dx} = \dfrac{2}{(\ln 2)x}$

5. $y = \left(1+x^2\right)e^{\tan^{-1}x} \Rightarrow y' = 2xe^{\tan^{-1}x} + \left(1+x^2\right)\left(\dfrac{e^{\tan^{-1}x}}{1+x^2}\right) = 2xe^{\tan^{-1}x} + e^{\tan^{-1}x}$

7. $y = (x+2)^{x+2} \Rightarrow \ln y = (x+2)\ln(x+2) \Rightarrow \dfrac{dy}{y} = \left(\ln(x+2) + 1\right)dx \Rightarrow$

$y' = \left(\ln(x+2) + 1\right)(x+2)^{x+2}$

9. $y = \sin^{-1}\left(\sqrt{1-x}\right) = \sin^{-1}\left((1-x)^{1/2}\right) \Rightarrow \dfrac{dy}{dx} = \dfrac{(1/2)(1-x)^{-1/2}(-1)}{\sqrt{1-(1-x)}} = \dfrac{-1}{2\sqrt{x-x^2}}$

11. $y = x\tan^{-1}(x) - (1/2)\ln x \Rightarrow \dfrac{dy}{dx} = \tan^{-1}(x) + x\left(\dfrac{1}{1+x^2}\right) - \left(\dfrac{1}{2}\right)\left(\dfrac{1}{x}\right) = \tan^{-1}(x) + \dfrac{x}{1+x^2} - \dfrac{1}{2x}$

13. $y = 2\sqrt{x-1}\,\sec^{-1}\sqrt{x} = 2(x-1)^{1/2}\sec^{-1}(x)^{1/2} \Rightarrow \dfrac{dy}{dx} =$

$2\left[\left(\dfrac{1}{2}\right)(x-1)^{-1/2}\sec^{-1}(x)^{1/2} + (x-1)^{1/2}\left(\dfrac{(1/2)x^{-1/2}}{\sqrt{x}\sqrt{x-1}}\right)\right] = \dfrac{\sec^{-1}\sqrt{x}}{\sqrt{x-1}} + \dfrac{1}{x}$

15. If $y = t\tan^{-1}t - \dfrac{1}{2}\ln\left(1+t^2\right) + C$, then $dy = \left(\tan^{-1}t + \dfrac{t}{1+t^2} - \dfrac{t}{1+t^2}\right)dt = \tan^{-1}t\,dt$, which

verifies the formula.

17. $x^{\ln y} = 2 \Rightarrow (\ln y)(\ln x) = \ln 2 \Rightarrow \dfrac{dy}{y}\ln x + (\ln y)\dfrac{dx}{x} = 0 \Rightarrow \dfrac{dy}{dx} = -\dfrac{y\ln y}{x\ln x}$

19. $y = s - \coth s \Rightarrow y' = 1 + \operatorname{csch}^2 s = \coth^2 s$

21. $y = \ln(\operatorname{csch} s) + s \coth s \Rightarrow y' = \dfrac{-\operatorname{csch} s \coth s}{\operatorname{csch} s} + \coth s - s\operatorname{csch}^2 s = -s\operatorname{csch}^2 s$

23. $y = \sin^{-1}(\tanh s) \Rightarrow y' = \dfrac{\operatorname{sech}^2 s}{\sqrt{1 - \tanh^2 s}} = \operatorname{sech} s$

25. $y = \sqrt{1 + v^2}\,\sinh^{-1} v \Rightarrow y' = \dfrac{v}{\sqrt{1 + v^2}}\sinh^{-1} v + \sqrt{1 + v^2}\,\dfrac{1}{\sqrt{1 + v^2}} = \dfrac{v \sinh^{-1} v}{\sqrt{1 + v^2}} + 1$

27. $y = 1 - \tanh^{-1}\left(v^{-1}\right), |v| > 1 \Rightarrow y' = \left(\dfrac{-1}{1 - v^{-2}}\right)\left(-v^{-2}\right) = \dfrac{1}{v^2 - 1}$

29. $y = \operatorname{sech}^{-1}(\cos 2v),\ 0 < v < \pi/4 \Rightarrow y'\dfrac{-(-\sin 2v)(2)}{(\cos 2v)\sqrt{1 - \cos^2 2v}} = \dfrac{2\sin 2v}{(\cos 2v)(\sin 2v)} = 2\sec 2v$

31. $\displaystyle\int_0^{\ln 22} \dfrac{1}{3 - 2e^{-x}}\,dx = \dfrac{1}{3}\int_0^{\ln 22} \dfrac{3e^x}{3e^x - 2}\,dx = \dfrac{1}{3}\Big[\ln\big|3e^x - 2\big|\Big]_0^{\ln 22} = \ln 4$

33. $\displaystyle\int_1^{\exp(\pi/3)} \dfrac{\tan(\ln t)}{t}\,dt = \int_0^{\pi/3} \tan u\,du = -\int_0^{\pi/3}\dfrac{\sin u}{\cos u}\,du = -\Big[\ln|\cos u|\Big]_0^{\pi/3} = \ln 2,\ \text{where } u = \ln t$

35. $\displaystyle\int_1^{e}\dfrac{2\ln t}{t\ln^2 t + 31t}\,dt = \int_0^{1}\dfrac{2u}{u^2 + 31}\,du = \Big[\ln\big(u^2 + 31\big)\Big]_0^1 = \ln\left(\dfrac{32}{31}\right),\ \text{where } u = \ln t$

37. $\displaystyle\int_1^{e}\dfrac{\pi\cos(\pi\ln z)}{z}\,dz = \int_0^{\pi}\cos u\,du = [\sin u]_0^{\pi} = 0,\ \text{where } u = \pi\ln z$

39. $\displaystyle\int_1^{e}\dfrac{8(\ln 3)\left(\log_3 x\right)}{x}\,dx = \int_1^{e}\dfrac{8(\ln 3)(\ln x)}{x(\ln 3)}\,dx = 8\int_1^{e}(\ln x)\left(\dfrac{1}{x}\right)dx = 4\Big[(\ln x)^2\Big]_1^{e} = 4$

41. $\displaystyle\int_0^{\pi/4} 2^{\tan\theta}\sec^2\theta\,d\theta = \left[\dfrac{2^{\tan\theta}}{\ln 2}\right]_0^{\pi/4} = \dfrac{1}{\ln 2}$

43. $\displaystyle\int \dfrac{3}{x\sqrt{\ln x - \ln^2 x}}\,dx = \int\dfrac{3}{x\sqrt{\left(\frac{1}{2}\right)^2 - \left(\ln(x) - \frac{1}{2}\right)^2}}\,dx = \int\dfrac{3}{\sqrt{1 - u^2}}\,du =$

$3\sin^{-1}u + C = 3\sin^{-1}\left(\ln(x) - \dfrac{1}{2}\right) + C,\ \text{where } \dfrac{u}{2} = \ln(x) - \dfrac{1}{2}$

45. $\displaystyle\int_1^{3\sqrt 3}\dfrac{1}{y^{2/3} + y^{4/3}}\,dy = 3\int_1^{3\sqrt 3}\dfrac{(1/3)y^{-2/3}}{1 + \left(y^{1/3}\right)^2}\,dy = 3\Big[\tan^{-1}\left(y^{1/3}\right)\Big]_1^{3\sqrt 3} = \dfrac{\pi}{4}$

47. $\displaystyle\int\dfrac{1}{x\sqrt{x^2 - 1}\left(\sec^{-1}x\right)}\,dx = \int\dfrac{1}{u}\,du = \ln u + C = \ln\left|\sec^{-1}x\right| + C,\ \text{where } x > 1 \text{ and } u = \sec^{-1}x$

49. $\displaystyle\int_0^{\ln 2} 4e^x\cosh x\,dx = 4\int_0^{\ln 2} e^x\left(\dfrac{e^x + e^{-x}}{2}\right)dx = \int_0^{\ln 2} 2e^{2x} + 2\,dx = \Big[e^{2x} + 2x\Big]_0^{\ln 2} = 3 + \ln 4$

50. $\displaystyle\int_0^{\ln 2}\dfrac{\sinh t}{1 + \cosh t}\,dt = \Big[\ln(1 + \cosh t)\Big]_0^{\ln 2} = \ln\left(\dfrac{9}{8}\right)$

51. $\displaystyle\int_{-\ln 3}^{\ln 3} 3\sqrt{\cosh 2x + 1}\, dx = \int_{-\ln 3}^{\ln 3} 3\sqrt{\cosh^2 x + \sinh^2 x + \cosh^2 x - \sinh^2}\, dx =$

$3\sqrt{2}\displaystyle\int_{-\ln 3}^{\ln 3} \cosh x\, dx = 3\sqrt{2}[\sinh x]_{-\ln 3}^{\ln 3} = 8\sqrt{2}$

53. $\displaystyle\int 10\,\text{csch}^2 s\,\coth s\, ds = -10\int (\text{csch}\, s)(-\text{csch}\, s\,\coth s)\, ds = -5\,\text{csch}^2 s + C$

55. a) $\displaystyle\int_0^{\pi/2} \frac{\sin x}{\sqrt{1 + \cos^2 x}}\, dx = -\left[\sinh^{-1}(\cos x)\right]_0^{\pi/2} = \sinh^{-1}(1)$ b) $\sinh^{-1}(1) = \ln(1 + \sqrt{2})$

57. a) $\displaystyle\int_{1/5}^{1/2} \frac{4\tanh^{-1}x}{1 - x^2}\, dx = 4\left[\frac{(\tanh^{-1}x)^2}{2}\right]_{1/5}^{1/2} = 2\left[\left(\tanh^{-1}(1/2)\right)^2 - \left(\tanh^{-1}(1/5)\right)^2\right]$

  b) $2\left[\left(\tanh^{-1}(1/2)\right)^2 - \left(\tanh^{-1}(1/5)\right)^2\right] = 2\left[\left(\frac{1}{2}\ln\left(\frac{3/2}{1/2}\right)\right)^2 - \left(\frac{1}{2}\ln\left(\frac{6/5}{4/5}\right)\right)^2\right] = \ln\left(\frac{9}{2}\right)\ln\sqrt{2}$

59. a) $\displaystyle\int_{3/5}^{4/5} \frac{2\,\text{sech}^{-1}x}{x\sqrt{1-x^2}}\, dx = -2\int_{3/5}^{4/5} \left(\text{sech}^{-1}x\right)\left(\frac{-1}{x\sqrt{1-x^2}}\right)dx =$

  $-\left[\left(\text{sech}^{-1}x\right)^2\right]_{3/5}^{4/5} = \left(\text{sech}^{-1}\frac{3}{5}\right)^2 - \left(\text{sech}^{-1}\frac{4}{5}\right)^2$

  b) $\left(\text{sech}^{-1}\frac{3}{5}\right)^2 - \left(\text{sech}^{-1}\frac{4}{5}\right)^2 = \left[\ln\left(\frac{1 + \sqrt{16/25}}{3/5}\right)\right]^2 - \left[\ln\left(\frac{1 + \sqrt{9/25}}{4/5}\right)\right]^2 = (\ln 6)\left(\ln\frac{3}{2}\right)$

61. $\dfrac{1}{y+1}\dfrac{dy}{dx} = -\dfrac{1}{x^2} \Rightarrow \displaystyle\int \frac{1}{y+1}\, dy = \int -\frac{1}{x^2}\, dx \Rightarrow \ln|y+1| + \frac{1}{x} + C \text{ and } \ln|0+1| = 1 + C \Rightarrow$

$\ln|y+1| = \dfrac{1}{x} - 1$

63. a) $e^{2y} = x^2 \Rightarrow 2y = 2\ln x \Rightarrow y = \ln x$

  b) $3^y = 2^{y+1} \Rightarrow y(\ln 3) = (y+1)(\ln 2) \Rightarrow y = \dfrac{\ln 2}{\ln\left(\frac{3}{2}\right)}$

65. $\displaystyle\lim_{x \to 0} \frac{2^{\sin x} - 1}{e^x - 1} = \lim_{x \to 0} \frac{2^{\sin x}(\ln 2)\cos x}{e^x} = \ln 2$

67. $\displaystyle\lim_{x \to 4} \frac{\sin^2(\pi x)}{e^{x-4} + 3 - x} = \lim_{x \to 4} \frac{2\pi\left(\sin(\pi x)\right)(\cos \pi x)}{e^{x-4} - 1} = \lim_{x \to 4} \frac{\pi\sin(2\pi x)}{e^{x-4} - 1} = \lim_{x \to 4} \frac{2\pi^2\cos(2\pi x)}{e^{x-4}} = 2\pi^2$

69. $\displaystyle\lim_{x \to \infty} \left(1 + \frac{3}{x}\right)^x,$ let $f(x) = \left(1 + \frac{3}{x}\right)^x \Rightarrow \ln f(x) = \dfrac{\ln\left(1 + 3x^{-1}\right)}{x^{-1}} \Rightarrow \displaystyle\lim_{x \to \infty} \ln f(x) =$

$\displaystyle\lim_{x \to \infty} \frac{\ln\left(1 + 3x^{-1}\right)}{x^{-1}} = \lim_{x \to \infty} \frac{\dfrac{-3x^{-2}}{1 + 3x^{-1}}}{-x^{-2}} = \lim_{x \to \infty} \frac{3}{1 + 3/x} = 3 \Rightarrow \lim_{x \to \infty} \left(1 + \frac{3}{x}\right)^x = e^3$

71. $\displaystyle \lim_{x \to \infty} \left[\frac{1}{n+1} + \frac{1}{n+2} \cdots \frac{1}{2n} \right] =$

$\displaystyle \lim_{x \to \infty} \left[\left(\frac{1}{n}\right)\left(\frac{1/n}{1+(1/n)}\right) + \left(\frac{1}{n}\right)\left(\frac{1/n}{1+2(1/n)}\right) + \cdots + \left(\frac{1}{n}\right)\left(\frac{1/n}{1+n(1/n)}\right) \right]$

which can be interpreted as a Riemann sum with regular partitioning.

$\displaystyle \therefore \lim_{x \to \infty} \left[\frac{1}{n+1} + \frac{1}{n+2} \cdots \frac{1}{2n} \right] = \int_0^1 \frac{1}{1+x}\, dx = b\Big[\ln(1+x)\Big]_0^1 = \ln 2$

73. $\displaystyle \frac{df^{-1}}{dx}\bigg|_{x=2+\ln 2} = \frac{1}{df/dx}\bigg|_{x=\ln 2} = \frac{1}{e^x+1}\bigg|_{x=\ln 2} = \frac{1}{3}$

75. $\displaystyle A = \int_1^e \frac{2\ln x}{x}\, dx = 2\int_1^e (\ln x)\left(\frac{1}{x}\right)dx = \Big[(\ln x)^2\Big]_1^e = 1$

77. a) $\displaystyle \int_{10}^{20} \frac{1}{x}\, dx = \int_1^2 \frac{1}{10\,u}\, 10\, du = \int_1^2 \frac{1}{u}\, du = \int_1^2 \frac{1}{x}\, dx,\ \text{where } x = 10u$

b) $\displaystyle \int_{ka}^{kb} \frac{1}{x}\, dx = \Big[\ln|x|\Big]_{ka}^{kb} = \ln(kb) - \ln(ka) = \ln\left(\frac{b}{a}\right), \text{while } \int_a^b \frac{1}{x}\, dx = \Big[\ln|x|\Big]_a^b = \ln(b) - \ln(b) = \ln\left(\frac{b}{a}\right)$

79. $\displaystyle x = \left(\frac{y}{4}\right)^2 - 2\ln\left(\frac{y}{4}\right) \Rightarrow \frac{dx}{dy} = \frac{y}{8} - 2\frac{1/4}{y/4} = \frac{y}{8} - \frac{2}{y} = \frac{y^2-16}{8y} \Rightarrow 1 + \left(\frac{dx}{dy}\right)^2 = 1 + \left(\frac{y^2-16}{8y}\right)^2 =$

$\displaystyle \left(\frac{y^2+16}{8y}\right)^2, L = \int_4^{12} \sqrt{1+\left(\frac{dx}{dy}\right)^2}\, dy = \int_4^{12} \frac{y^2+16}{8y}\, dy = \frac{1}{8}\int_4^{12} y + \frac{16}{y}\, dy =$

$\displaystyle \frac{1}{8}\left[\frac{y^2}{2} + 16\ln y\right]_4^{12} = 8 + \ln 9$

81. $\displaystyle \lim_{x \to 0} \frac{x}{6^x - 5^x} = \lim_{x \to 0} \frac{1}{(\ln 6)6^x - (\ln 5)5^x} = \frac{1}{\ln 6 - \ln 5} = \frac{1}{\ln(6/5)}$

83. a) $\displaystyle V = \pi\int_{1/4}^4 \left(\frac{1}{2\sqrt{x}}\right)^2 dx = \frac{\pi}{4}\int_{1/4}^4 \frac{1}{x}\, dx = \frac{\pi}{4}\Big[\ln|x|\Big]_{1/4}^4 = \frac{\pi\ln 4}{2}$

b) $\displaystyle M_y = \int_{1/4}^4 x\left(\frac{1}{2\sqrt{x}}\right)dx = \frac{1}{2}\int_{1/4}^4 x^{1/2}\, dx = \left[\frac{1}{3}x^{3/2}\right]_{1/4}^4 = \frac{63}{24},\ M_x = \int_{1/4}^4 \frac{1}{4\sqrt{x}}\,\frac{1}{2\sqrt{x}}\, dx =$

$\displaystyle \frac{1}{8}\int_{1/4}^4 \frac{1}{x}\, dx = \frac{1}{8}[\ln x]_{1/4}^4 = \frac{\ln 4}{4},\ M = \int_{1/4}^4 \frac{1}{2\sqrt{x}}\, dx = \frac{1}{2}\int_{1/4}^4 x^{-1/2}\, dx = \Big[\sqrt{x}\Big]_{1/4}^4 = \frac{3}{2}.\ \therefore \overline{x} =$

$\displaystyle \frac{M_y}{M} = \frac{63/24}{3/2} = \frac{7}{8}\ \text{and}\ \overline{y} = \frac{M_x}{M} = \frac{(\ln 4)/4}{3/2} = \frac{\ln 4}{12}$

85. $\displaystyle \text{erf}(x) = \frac{2}{\sqrt{\pi}}\int_0^x e^{-t^2}\, dt,\ \text{erf}(0) = 0;\ \text{erf}'(x) = e^{-x^2},\ \text{erf}'(0) = 1;\ \text{erf}''(x) = -2xe^{-x^2},\ \text{erf}''(0) = 0.$

$\displaystyle \therefore Q(x) = \text{erf}(0) + \text{erf}'(0)(x-0) + \frac{\text{erf}''(0)}{2}(x-0)^2 = x$

87. If $f(x) = e^{g(x)}$, then $f'(x) = e^{g(x)} g'(x)$ where $g'(x) = \frac{x}{1+x^4}$. $\therefore f'(2) = e^0 \frac{2}{1+16} = \frac{2}{17}$.

89.

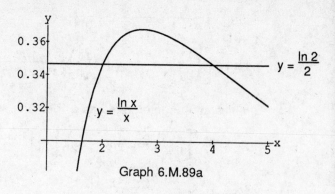

Graph 6.M.89a

a) No, there are two intersections.

b) Yes, there is only one intersection.

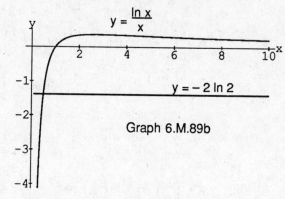

Graph 6.M.89b

91.   a) $\log_{10} 5 = \dfrac{\ln 5}{\ln 10} \approx 0.698970004$ b) $\log_2 3 = \dfrac{\ln 3}{\ln 2} \approx 1.584962501$

      c) $\log_7 2 = \dfrac{\ln 2}{\ln 7} \approx 0.356207187$

93. $A_1 = \displaystyle\int_1^e \dfrac{2 \log_2 x}{x}\,dx = \dfrac{2}{\ln 2}\int_1^e \dfrac{\ln x}{x}\,dx = \left[\dfrac{(\ln x)^2}{\ln 2}\right]_1^e = \dfrac{1}{\ln 2}\,,\ A_2 = \displaystyle\int_1^e \dfrac{2 \log_4 x}{x}\,dx =$

      $\dfrac{2}{\ln 4}\displaystyle\int_1^e \dfrac{\ln x}{x}\,dx = \left[\dfrac{(\ln x)^2}{2 \ln 2}\right]_1^e = \dfrac{1}{2 \ln 2} \Rightarrow A_1 : A_2 = 2:1$

95.   From the figure we have $\dfrac{\ln \pi}{\pi} < \dfrac{\ln e}{e} \Rightarrow e \ln \pi < \pi \ln e \Rightarrow \ln \pi^e < \ln e^\pi \Rightarrow \pi^e < e^\pi.$

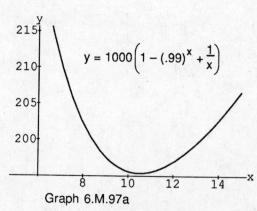

Graph 6.M.97a

97.   a) About x = 11 there is a minimum.

b) There is no maximum, however the curve is asymptotic to $y = 1000$. The curve is near 1000 when $x \geq 643$.

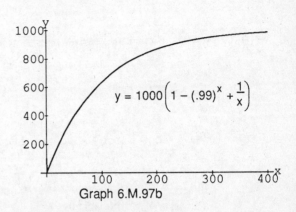

$$y = 1000\left(1 - (.99)^x + \frac{1}{x}\right)$$

Graph 6.M.97b

99. $C(t) = C_o \exp\left[-\frac{\ln 2}{5700}t\right] \Rightarrow (0.1)C_o = C_o\exp\left[-\frac{\ln 2}{5700}t\right] \Rightarrow \ln(0.1) = -\frac{\ln 2}{5700}t \Rightarrow$

$t = -\frac{5700\ \ln(0.1)}{\ln 2} \approx 18935\ yr$

101. a) $\frac{dy}{dt} = \frac{kA(c-y)}{V} \Rightarrow -\int\frac{-1}{c-y}dy = \int\frac{kA}{V}dt \Rightarrow -\ln|c-y| = \frac{kA}{V}t + C_1 \Rightarrow c - y = C_2\exp\left[-\frac{kA}{V}t\right] \Rightarrow$

$y(t) = c - C_2\exp\left[-\frac{kA}{V}t\right] \Rightarrow y(0) = y_o = c - C_2\exp\left[-\frac{kA}{V}0\right] \Rightarrow y_o = c - C_2 \Rightarrow C_2 = c - y_o \Rightarrow$

$y(t) = c - \left(c - y_o\right)\exp\left[-\frac{kA}{V}t\right] = c - \left(c - y_o\right)e^{-\frac{kA}{V}t}$

b) $\lim_{t\to\infty} y(t) = c - \left(c - y_o\right)\lim_{t\to\infty}e^{-\frac{kA}{V}t} = c - \left(c - y_o\right)(0) = c$

103. a) $\lim_{x\to\infty}\frac{\ln 2x}{\ln x^2} = \lim_{x\to\infty}\frac{\ln 2x}{2\ln x} = \lim_{x\to\infty}\frac{\ln 2 + \ln x}{2\ln x} = 0 + \frac{1}{2} = \frac{1}{2}$, $\lim_{x\to\infty}\frac{\ln 2x}{\ln(x+2)} =$

$\lim_{x\to\infty}\frac{1/x}{1/(x+2)} = \lim_{x\to\infty}\frac{x+2}{x} = 1$, $\lim_{x\to\infty}\frac{\ln x^2}{\ln(x+2)} = \lim_{x\to\infty}\frac{1/x}{1/(x+2)} = \lim_{x\to\infty}\frac{x+2}{x} = 1$

$\therefore$ all the growth rates are the same

b) $\lim_{x\to\infty}\frac{x^{\ln x}}{x^{\log_2 x}} = \lim_{x\to\infty}\frac{1}{x^{(\ln x)(1/\ln 2 - 1)}} = 0 \Rightarrow x^{\log_2 x}$ grows faster than $x^{\ln x}$

c) $\lim_{x\to\infty}\frac{\left(\frac{1}{3}\right)^x}{\left(\frac{1}{2}\right)^x} = \lim_{x\to\infty}\left(\frac{2}{3}\right)^x = 0 \Rightarrow \left(\frac{1}{2}\right)^x$ grows faster than $\left(\frac{1}{3}\right)^x$

105. a) $\lim_{x\to\infty}\frac{\ln x}{x} = \lim_{x\to\infty}\frac{1/x}{1} = 0 \Rightarrow$ true

b) $\lim_{x\to\infty}\frac{\ln\ln x}{\ln x} = \frac{\frac{1}{\ln x}\frac{1}{x}}{\frac{1}{x}} = \lim_{x\to\infty}\frac{1}{\ln x} = 0 \Rightarrow$ true

c) $\lim_{x\to\infty}\frac{x}{x + \ln x} = \lim_{x\to\infty}\frac{1}{1 + 1/x} = 1 \Rightarrow$ the same growth rate $\Rightarrow$ false

107. $y = \tan^{-1}x + \tan^{-1}\dfrac{1}{x} \Rightarrow y' = \dfrac{1}{1+x^2} + \dfrac{-\dfrac{1}{x^2}}{1+\dfrac{1}{x^2}} = \dfrac{1}{1+x^2} - \dfrac{1}{1+x^2} = 0 \Rightarrow \tan^{-1}x + \tan^{-1}\dfrac{1}{x}$ is a constant,

the constant is $\dfrac{\pi}{2}$

109. $\theta = \pi - \cot^{-1}\left(\dfrac{x}{60}\right) - \cot^{-1}\left(\dfrac{5}{3} - \dfrac{x}{30}\right), \ 0 < x < 50 \Rightarrow \theta' = \dfrac{1/60}{1+\left(\dfrac{x}{60}\right)^2} + \dfrac{-1/30}{1+\left(\dfrac{50-x}{30}\right)^2} =$

$30\left[\dfrac{2}{60^2+x^2} - \dfrac{1}{30^2+(50-x)^2}\right]$, solving $\theta' = 0 \Rightarrow x^2 - 200x + 3200 = 0 \Rightarrow x = 100 \pm 20\sqrt{17}$,

but $100 + 20\sqrt{17}$ is not in the domain, $\theta' > 0$ for $x < 20\left(5 - \sqrt{17}\right)$, and $\theta' < 0$ for

$20\left(5 - \sqrt{17}\right) < x < 50 \Rightarrow$ at $x = 20\left(5 - \sqrt{17}\right) \approx 17.54$ m is a maximum

111. Suppose g is an even function, h is an odd function and $f(x) = g(x) + h(x) = 0$ for all x. Now $f(x) +$

$f(-x) = 0 \Rightarrow g(x) + h(x) + g(-x) + h(-x) = g(x) + h(x) + g(x) - h(x) = 2\,g(x) = 0 \Rightarrow g(x) = 0$ for

all x. Hence, $f(x) = h(x)$ and $f(x) - f(-x) = 0 \Rightarrow h(x) - h(-x) = h(x) + h(x) = 2h(x) = 0 \Rightarrow$

$h(x) = 0$ for all x.

113. When solving $x\dfrac{d^2y}{dx^2} = \sqrt{1+\left(\dfrac{dy}{dx}\right)^2}$ let $p = \dfrac{dy}{dx} \Rightarrow x\dfrac{dp}{dx} = \sqrt{1+p^2} \Rightarrow \displaystyle\int \dfrac{1}{\sqrt{1+p^2}}\,dp = \int \dfrac{1}{x}\,dx \Rightarrow$

$\sinh^{-1}p = \ln x + C$ and $\sinh^{-1}0 = \ln(1) + C \Rightarrow \sinh^{-1}p = \ln x \Rightarrow \displaystyle\int dy = \int \sinh(\ln x)\,dx \Rightarrow$

$y = \displaystyle\int x - \dfrac{1}{x}\,dx = \dfrac{1}{2}\left[\dfrac{x^2}{2} - \ln x\right] + C$ and $0 = \dfrac{1}{2}\left[\dfrac{1}{2} - \ln 1\right] + C \Rightarrow y = \dfrac{x^2}{4} - \ln\sqrt{x} - \dfrac{1}{4}$

115. $y = x^2 \Rightarrow 1 + (y')^2 = 1 + (2x)^2 \Rightarrow A = 2\pi\displaystyle\int_0^1 x^2\sqrt{1+(2x)^2}\,dx = 2\pi\int_0^1 x^2\sqrt{1+4x^2}\,dx =$

$4\pi\displaystyle\int_0^1 x^2\sqrt{\left(\dfrac{1}{2}\right)^2 + x^2}\,dx = 4\pi\left[\dfrac{x\left(1/4 + 2x^2\right)\sqrt{1/4 + x^2}}{8} - \dfrac{1}{128}\sinh^{-1}2x\right]_0^1 \approx 3.809730$, formula 22

CHAPTER 7

TECHNIQUES OF INTEGRATION

7.1 BASIC INTEGRATION FORMULAS

1. $\int \frac{16x}{\sqrt{8x^2+1}}dx = \int u^{-1/2}\,du = 2u^{1/2} + C = 2\sqrt{8x^2+1} + C$, where $u = 8x^2 + 1$

3. $\int_0^{\pi/2} 3\sqrt{\sin v}\cos v\,dv = 2\left[(\sin v)^{3/2}\right]_0^{\pi/2} = 2$

5. $\int_0^1 \frac{16x}{8x^2+2}dx = \int_2^{10}\frac{1}{u}\,du = [\ln u]_2^{10} = \ln 5$, where $u = 8x^2 + 2$

7. $\int \frac{1}{x-\sqrt{x}}dx = \int \frac{2u}{u^2-u}\,du = 2\int\frac{1}{u-1}\,du = 2\ln|u-1| + C = 2\ln\left(\sqrt{x}-1\right) + C$, where $u = \sqrt{x}$

9. $\int_{-\pi}^{\pi}\sec\left(\frac{t}{3}\right)dx = 3\int_{-\pi/3}^{\pi/3}\sec u\,du = 3[\ln|\sec u + \tan u|]_{-\pi/3}^{\pi/3} = \ln\left(2+\sqrt{3}\right)^6$, where $u = \frac{t}{3}$

11. $\int_{3\pi/2}^{7\pi/4}\csc(s-\pi)\,ds = \int_{\pi/2}^{3\pi/4}\csc u\,du = -[\ln|\csc u + \cot u|]_{\pi/2}^{3\pi/4} = -\ln\left(\sqrt{2}-1\right) = \ln\left(\sqrt{2}+1\right)$,

    where $u = s - \pi$

13. $\int_0^{\sqrt{\ln 2}} 2xe^{x^2}\,dx = \left[e^{x^2}\right]_0^{\ln 2} = 1$

15. $\int_0^{\pi/3} e^{\tan v}\sec^2 v\,dv = \left[e^{\tan v}\right]_0^{\pi/3} = e^{\sqrt{3}} - 1$

17. $\int_{-1}^0 3^{(x+1)}\,dx = \int_0^1 3^u\,du = \left[\frac{3^u}{\ln 3}\right]_0^1 = \frac{2}{\ln 3}$, where $u = x + 1$

19. $\int \frac{2^{\sqrt{w}}}{2\sqrt{w}}\,dw = \int 2^u\,du = \frac{2^u}{\ln 2} + C = \frac{2^{\sqrt{w}}}{\ln 2} + C$, where $u = \sqrt{w}$

21. $\int_0^{\sqrt{3}/3}\frac{9}{1+(3u)^2}\,du = 3\left[\tan^{-1}3u\right]_0^{\sqrt{3}/3} = \pi$

23. $\int \frac{6}{\sqrt{y}(1+y)}\,dy = 12\int\frac{1}{1+u^2}\,du = 12\tan^{-1}(u) + C = 12\tan^{-1}(\sqrt{y}) + C$, where $u = \sqrt{y}$

25. $\int_0^{1/6}\frac{1}{\sqrt{1-9x^2}}\,dx = \frac{1}{3}\int_0^{1/2}\frac{1}{\sqrt{1-u^2}}\,du = \frac{1}{3}\left[\sin^{-1}(u)\right]_0^{1/2} = \frac{\pi}{18}$, where $u = 3x$

27. $\int_0^{1/\sqrt{2}}\frac{2s}{\sqrt{1-\left(s^2\right)^2}}\,ds = \left[\sin^{-1}s^2\right]_0^{1/\sqrt{2}} = \frac{\pi}{6}$

29. $\int \dfrac{6}{x\sqrt{25x^2-1}}\,dx = 6\int \dfrac{1}{u\sqrt{u^2-1}}\,du = 6\sec^{-1}|u| + C = 6\sec^{-1}|5x| + C$, where $u = 5x$

31. $\int \dfrac{dy}{\sqrt{e^{2y}-1}} = \int \dfrac{e^y}{e^y\sqrt{(e^y)^2-1}}\,dy = \sec^{-1}(e^y) + C$

33. $\int_{1}^{e^{\pi/3}} \dfrac{1}{x\cos(\ln x)}\,dx = \int_{0}^{\pi/3} \sec u\,du = \Big[\ln|\sec u + \tan u|\Big]_{0}^{\pi/3} = \ln(2+\sqrt{3})$, where $u = \ln x$

35. $\int \dfrac{\ln x}{x + 4x\ln^2 x}\,dx = \int \dfrac{u}{1+4u^2}\,du = \dfrac{1}{8}\int \dfrac{8u}{1+4u^2}\,dx = \dfrac{1}{8}\ln(1+4u^2) + C = \dfrac{1}{8}\ln(1+4\ln^2 x) + C,$

where $u = \ln x$

37. $\int \dfrac{1}{\sqrt{-x^2+4x-3}}\,dx = \int \dfrac{1}{\sqrt{1-(x-2)^2}}\,dx = \int \dfrac{1}{\sqrt{1-u^2}}\,du = \Big[\sin^{-1}(u)\Big] + C =$

$\Big[\sin^{-1}(x-2)\Big] + C$, where $u = x - 2$

39. $\int_{1}^{2} \dfrac{8}{x^2-2x+2}\,dx = 8\int_{1}^{2} \dfrac{1}{1+(x-1)^2}\,dx = 8\int_{0}^{1} \dfrac{1}{1+u^2}\,du = 8\Big[\tan^{-1}(u)\Big]_{0}^{1} = 2\pi$, where $u = x - 1$

41. $\int \dfrac{1}{(x+1)\sqrt{x^2+2x}}\,dx = \int \dfrac{1}{(x+1)\sqrt{(x+1)^2-1}}\,dx = \int \dfrac{1}{u\sqrt{u^2-1}}\,du =$

$\Big[\sec^{-1}|u|\Big] + C = \Big[\sec^{-1}|x+1|\Big] + C$, when $|x+1| > 1$ and $u = x + 1$

43. $\int_{\pi/4}^{3\pi/4} (\csc x - \cot x)^2\,dx = \int_{\pi/4}^{3\pi/4} 2\csc^2 x - 1 - 2\csc x \cot x\,dx =$

$\Big[-2\cot(x) - x + 2\csc(x)\Big]_{\pi/4}^{3\pi/4} = 4 - \dfrac{\pi}{2}$

45. $\int_{\pi/6}^{\pi/3} (\csc x - \sec x)(\sin x + \cos x)\,dx = \int_{\pi/6}^{\pi/3} \dfrac{\cos x}{\sin x} + \dfrac{-\sin x}{\cos x}\,dx = \Big[\ln|\sin x| + \ln|\cos x|\Big]_{\pi/6}^{\pi/3} = 0$

47. $\int \dfrac{x}{x+1}\,dx = \int 1 - \dfrac{1}{x+1}\,dx = x - \ln|x+1| + C$

49. $\int_{\sqrt{2}}^{3} \dfrac{2x^3}{x^2-1}\,dx = \int_{\sqrt{2}}^{3} 2x + \dfrac{2x}{x^2-1}\,dx = \Big[x^2 + \ln\big|x^2-1\big|\Big]_{\sqrt{2}}^{3} = 7 + \ln 8$

51. $\int_{0}^{\sqrt{3}/2} \dfrac{1-x}{\sqrt{1-x^2}}\,dx = \int_{0}^{\sqrt{3}/2} \left(\dfrac{1}{\sqrt{1-x^2}} - (1-x^2)^{-1/2}(x)\right)\,dx = \Big[\sin^{-1}(x) + \sqrt{1-x^2}\Big]_{0}^{\sqrt{3}/2} = \dfrac{2\pi-3}{6}$

53. $\int_{0}^{\pi/4} \dfrac{1+\sin x}{\cos^2 x}\,dx = \int_{0}^{\pi/4} \sec^2 x + \sec x \tan x\,dx = \Big[\tan x + \sec x\Big]_{0}^{\pi/4} = \sqrt{2}$

55. $y = \ln(\cos x) \Rightarrow 1 + (y')^2 = 1 + \left(\dfrac{-\sin x}{\cos x}\right)^2 = 1 + \tan^2 x = \sec^2 x \Rightarrow L = \int_{0}^{\pi/3} \sqrt{1+(y')^2}\,dx =$

$\int_{0}^{\pi/3} \sqrt{\sec^2 x}\,dx = \int_{0}^{\pi/3} \sec x\,dx = \Big[\ln|\sec x + \tan x|\Big]_{0}^{\pi/3} = \ln(2+\sqrt{3})$

57. $M_x = \int_{-\pi/4}^{\pi/4} \frac{\sec^2 x}{2} \, dx = \int_0^{\pi/4} \sec^2 x \, dx = \left[\tan x\right]_0^{\pi/4} = 1, \; M = 2\int_0^{\pi/4} \sec x \, dx = 2\left[\ln|\sec x + \tan x|\right]_0^{\pi/4} =$

$2\ln\left(\sqrt{2} + 1\right) = \ln\left(2\sqrt{2} + 3\right) \Rightarrow \overline{y} = \frac{M_x}{M} = \frac{1}{\ln\left(2\sqrt{2} + 3\right)}$, and $\overline{x} = 0$ by symmetry

59. $M_x = \int_{\pi/6}^{5\pi/6} \left(\frac{2 + \csc x}{2}\right)(2 - \csc x) \, dx = \frac{1}{2}\int_{\pi/6}^{5\pi/6} 4 - \csc^2 x \, dx = \frac{1}{2}\left[4x + \cot x\right]_{\pi/6}^{5\pi/6} = \frac{4\pi}{3} - \sqrt{3},$

$M = \int_{\pi/6}^{5\pi/6} 2 - \csc x \, dx = \left[2x + \ln|\csc x + \cot x|\right]_{\pi/6}^{5\pi/6} = \frac{4\pi}{3} + \ln\left(7 - 4\sqrt{3}\right) \Rightarrow \overline{x} = \frac{\pi}{2}$, by symmetry,

and $\overline{y} = \frac{M_x}{M} = \frac{\frac{4\pi}{3} - \sqrt{3}}{\frac{4\pi}{3} + \ln\left(7 - 4\sqrt{3}\right)} \approx 1.58$

61. $\int \left[(s^2 - 1)(x + 1)\right]^{-2/3} dx = \int \left[(x - 1)(x + 1)^2\right]^{-2/3} dx = \frac{1}{2}\int \left(\frac{x - 1}{x + 1}\right)^{-2/3} \frac{2 \, dx}{(x + 1)^2} =$

$\frac{3}{2}\left(\frac{x - 1}{x + 1}\right)^{1/3} + C$, where $d\left(\frac{x - 1}{x + 1}\right) = \frac{2 \, dx}{(x + 1)^2}$.

a) If $u = \frac{1}{x + 1}$, then $x = +1 = \frac{1}{u}$, $dx = -\frac{du}{u^2}$, $\frac{x - 1}{x + 1} = \frac{\frac{1}{u} - 2}{\frac{1}{u}} = 1 - 2u$ and $(x + 1)^2 = \frac{1}{u^2}$.

$\therefore \int \left[(x^2 - 1)(x + 1)\right]^{-2/3} dx = \frac{1}{2}\int (1 - 2u)^{-2/3}(-2 \, du) = \frac{3}{2}(1 - 2u)^{1/3} + C = \frac{3}{2}\sqrt[3]{1 - \frac{2}{x + 1}} + C$

b) If $u = \left(\frac{x - 1}{x + 1}\right)^k$, $k \neq 0$, then $u^{1/k} = \frac{x - 1}{x + 1}$, $\frac{u^{1/k - 1}}{k} du = \frac{2 \, dx}{(x + 1)^2}$, $x = \frac{1 + u^{1/k}}{1 - u^{1/k}}$ and with replacing

each x in $\frac{x - 1}{x + 1}$ with $x = \frac{1 + u^{1/k}}{1 - u^{1/k}}$ we get $u^{1/k}$. $\therefore \int \left[(x^2 - 1)(x + 1)\right]^{-2/3} dx =$

$\frac{1}{2}\int \frac{\left(u^{1/k}\right)^{-2/3} u^{1/k - 1}}{k} du = \frac{1}{2k}\int u^{1/k - 1} du = \frac{3u^{1/3k}}{2} + C = \frac{3}{2}\sqrt[3]{u^{1/k}} + C = \frac{3}{2}\sqrt[3]{\left(\frac{x - 1}{x + 1}\right)} + C \Rightarrow$

for all indicated values of k the corresponding substitution will work, as well as for many others.

c) If $x = \tan u$, then $dx = \sec^2 u \, du$. $\therefore \int \left[(x^2 - 1)(x + 1)\right]^{-2/3} dx =$

$\frac{1}{2}\int \left(\frac{\tan u - 1}{\tan u + 1}\right)^{-2/3} \frac{2\sec^2 u \, du}{(\tan u + 1)^2} = \frac{3}{2}\left(\frac{\tan u - 1}{\tan u + 1}\right)^{1/3} + C = \frac{3}{2}\left(\frac{x - 1}{x + 1}\right)^{1/3} + C$,

where $d\left(\frac{\tan u - 1}{\tan u + 1}\right) = \frac{2\sec^2 u \, du}{(\tan u + 1)^2}$.

d) If $x = \tan^2 u$, then $dx = 2 \tan u \sec^2 u \, du$. $\therefore \int \left[(x^2 - 1)(x + 1) \right]^{-2/3} dx =$

$\dfrac{1}{2} \int \left(\dfrac{\tan^2 u - 1}{\tan^2 u + 1} \right)^{-2/3} \dfrac{4 \tan u \sec^2 u \, du}{\left(\tan^2 u + 1 \right)^2} dx = \dfrac{3}{2} \left(\dfrac{\tan^2 u - 1}{\tan^2 u + 1} \right)^{1/3} + C = \dfrac{3}{2} \left(\dfrac{x - 1}{x + 1} \right)^{1/3} + C$, where

$d \left(\dfrac{\tan^2 u - 1}{\tan^2 + 1} \right) = \dfrac{4 \tan u \sec^2 u \, du}{\left(\tan^2 u + 1 \right)^2}$.

e) If $x = 2 \tan u + 1$, then $dx = 2 \sec^2 u \, du$. $\therefore \int \left[(x^2 - 1)(x + 1) \right]^{-2/3} dx =$

$\dfrac{1}{2} \int \left(\dfrac{2 \tan u}{2 \tan u + 2} \right)^{-2/3} \dfrac{4 \sec^2 u \, du}{(2 + 2 \tan u)^2} = \dfrac{3}{2} \left(\dfrac{2 \tan u}{2 \tan u + 2} \right)^{1/3} + C = \dfrac{3}{2} \left(\dfrac{x - 1}{x + 1} \right)^{1/3} + C$, where

$d \left(\dfrac{2 \tan u}{2 \tan u + 2} \right) = \dfrac{4 \sec^2 u \, du}{(2 + 2 \tan u)^2}$.

f) If $x = \cos u$, then $dx = - \sin u \, du$. $\therefore \int \left[(x^2 - 1)(x + 1) \right]^{-2/3} dx =$

$\dfrac{1}{2} \int \left(\dfrac{\cos u - 1}{\cos u + 1} \right)^{-2/3} \left(\dfrac{- 2 \sin u \, du}{(\cos u + 1)^2} \right) dx = \dfrac{3}{2} \left(\dfrac{\cos u - 1}{\cos u + 1} \right)^{1/3} + C = \dfrac{3}{2} \left(\dfrac{x - 1}{x + 1} \right)^{1/3} + C$, where

$d \left(\dfrac{\cos u - 1}{\cos u + 1} \right) = \dfrac{- 2 \sin u \, du}{(\cos u + 1)^2}$.

g) If $x = \cosh u$, then $dx = \sinh u \, du$. $\therefore \int \left[(x^2 - 1)(x + 1) \right]^{-2/3} dx =$

$\dfrac{1}{2} \int \left(\dfrac{\cosh u - 1}{\cosh u + 1} \right)^{-2/3} 2 \sinh u \, du = \dfrac{3}{2} \left(\dfrac{\cosh u - 1}{\cosh u + 1} \right)^{1/3} + C = \dfrac{3}{2} \left(\dfrac{x - 1}{x + 1} \right)^{1/3} + C$, where

$d \left(\dfrac{\cosh u - 1}{\cosh u + 1} \right) = \dfrac{2 \sinh u \, du}{\cosh u + 1}$.

63. $25 \left[\ln|\sec x + \tan x| \right]_{30°}^{45°} = 25 \ln \left(\dfrac{\sqrt{2} + 1}{\sqrt{3}} \right) \approx 8.30$ cm

7.2 INTEGRATION BY PARTS

1.   Let $v = x$ and $du = \sin x \Rightarrow dv = dx$ and $u = -\cos x$
$$\int x \sin x \, dx = -x \cos x + \int \cos x \, dx = -x \cos x + \sin x + C$$

3. $\int t^2 \cos t \, dt = t^2 \sin t + 2t \cos t - 2 \sin t + C$, tabular integration

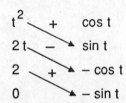

5.   Let $v = \ln x$ and $du = x \, dx \Rightarrow dv = \dfrac{1}{x}$ and $u = \dfrac{x^2}{2}$
$$\int_1^2 x \ln x \, dx = \left[\frac{x^2 \ln x}{2}\right]_1^2 - \frac{1}{2}\int_1^2 x \, dx = \left[\frac{x^2 \ln x}{2} - \frac{x^2}{4}\right]_1^2 = \ln(4) - \frac{3}{4}$$

7.   Let $v = \tan^{-1}(x)$ and $du = dx \Rightarrow dv = \dfrac{dx}{1 + x^2}$ and $u = x$
$$\int \tan^{-1}(x) \, dx = x \tan^{-1}(x) - \frac{1}{2}\int \frac{2x}{1 + x^2} \, dx = x \tan^{-1}(x) - \ln\sqrt{1 + x^2} + C$$

9.   Let $v = x$ and $du = \sec^2 x \, dx \Rightarrow dv = dx$ and $u = \tan x$
$$\int x \sec^2 x \, dx = x \tan x + \int \frac{-\sin x}{\cos x} \, dx = x \tan x + \ln|\cos x| + C$$

11. $\int x^3 e^x \, dx = \left(x^3 - 3x^2 + 6x - 6\right) e^x + C$, tabular integration

13. $\int \left(x^2 - 5x\right) e^x \, dx = \left[\left(x^2 - 5x\right) - (2x - 5) + (2)\right] e^x + C = \left(x^2 - 7x + 7\right) e^x + C$, tabular integration

$$\begin{array}{lcl}
x^2 - 5x & + & e^x \\
2x - 5 & - & e^x \\
2 & + & e^x \\
0 & & e^x
\end{array}$$

15. $\int x^5 e^x \, dx = \left(x^5 - 5x^4 + 20x^3 - 60x^2 + 120x - 120\right) e^x + C$, tabular integration

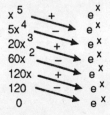

17. $\int_0^{\pi/2} \theta^2 \sin 2\theta \, d\theta = \left[-\frac{\theta^2 \cos 2\theta}{2} + \frac{\theta \sin 2\theta}{2} + \frac{\cos 2\theta}{4} \right]_0^{\pi/2} = \frac{\pi^2 - 4}{8}$, tabular integration

$\theta^2 \quad + \quad \sin 2\theta$

$2\theta \quad - \quad (-\cos 2\theta)/2$

$2 \quad + \quad (-\sin 2\theta)/4$

$0 \quad\quad\quad (\cos 2\theta)/8$

19. Let $v = \sec^{-1} t$ and $du = t \, dt \Rightarrow dv = \dfrac{dt}{t\sqrt{t^2 - 1}}$ and $u = \dfrac{t^2}{2} \Rightarrow$

$\int_{2/\sqrt{3}}^{2} t \sec^{-1} t \, dt = \left[\dfrac{t^2 \sec^{-1} t}{2} \right]_{2/\sqrt{3}}^{2} - \dfrac{1}{4} \int_{2/\sqrt{3}}^{2} \left(t^2 - 1 \right)^{-1/2} 2t \, dt =$

$\left[\dfrac{t^2 \sec^{-1} t}{2} \right]_{2/\sqrt{3}}^{2} - \dfrac{1}{2} \left[\left(t^2 - 1 \right)^{1/2} \right]_{2/\sqrt{3}}^{2} = \dfrac{5\pi - 3\sqrt{3}}{9}$

21. Let $u = e^\theta$ and $dv = \sin \theta \, d\theta \Rightarrow du = e^\theta d\theta$ and $v = -\cos \theta \Rightarrow \int e^\theta \sin \theta \, d\theta =$

$-e^\theta \cos \theta + \int e^\theta \cos \theta \, d\theta.$ Now let $u = e^\theta$ and $dv = \cos \theta \, d\theta \Rightarrow du = e^\theta d\theta$ and

$v = \sin \theta \Rightarrow \int e^\theta \sin \theta \, d\theta = -e^\theta \cos \theta + \int e^\theta \cos \theta \, d\theta = -e^\theta \cos \theta + e^\theta \sin \theta -$

$\int e^\theta \sin \theta \, d\theta \Rightarrow 2 \int e^\theta \sin \theta \, d\theta = -e^\theta \cos \theta + e^\theta \sin \theta + C$

$\therefore \int e^\theta \sin \theta \, d\theta = \dfrac{1}{2} \left[-e^\theta \cos \theta + e^\theta \sin \theta \right] + C.$

23. Let $u = e^{2x}$ and $dv = \cos 3x \, dx \Rightarrow du = 2e^{2x} dx$ and $v = \dfrac{\sin 3x}{3} \Rightarrow \int e^{2x} \sin 3x \, dx = \dfrac{e^{2x} \sin 3x}{3} -$

$\dfrac{2}{3} \int e^{2x} \sin 3x \, dx,$ now let $u = e^{2x}$ and $dv = \sin 3x \, dx \Rightarrow du = 2e^{2x} dx$ and $v = -\dfrac{\cos 3x}{3} \Rightarrow$

$\int e^{2x} \sin 3x \, dx = \dfrac{e^{2x} \sin 3x}{3} + \dfrac{2}{3} \left[\dfrac{e^{2x} \cos 3x}{3} - \int 2e^{2x} \dfrac{\cos 3x}{3} \, dx \right] \Rightarrow$

$\dfrac{13}{9} \int e^{2x} \cos 3x \, dx = \dfrac{e^{2x} \sin 3x}{3} + \dfrac{2e^{2x} \cos 3x}{9} + C \Rightarrow \int e^{2x} \cos 3x \, dx = \dfrac{e^{2x}}{13} (3 \sin 3x + 2 \cos 3x) + C$

25. If $x = \sqrt{3s + 9}$, then $x^2 + 3s + 9$ and $\dfrac{2x \, dx}{3} = ds.$ $\therefore \int e^{\sqrt{3s + 9}} ds = \dfrac{2}{3} \int e^x dx = \dfrac{2}{3} \left(x e^x - e^x \right) + C =$

$\dfrac{2}{3} \left(\sqrt{3s + 9} \, e^{\sqrt{3s + 9}} - e^{\sqrt{3s + 9}} \right) + C$

27. $\int_0^{\pi/3} x\tan^2 x\, dx = \int_0^{\pi/3} x\sec^2 x - x\, dx = \int_0^{\pi/3} x\sec^2 x\, dx - \left[\dfrac{x^2}{2}\right]_0^{\pi/3}$. Let $v = x$ and

$du = \sec^2 x\, dx \Rightarrow dv = dx$ and $u = \tan x$. Now $\int x\sec^2 x\, dx = x\tan x + \int \dfrac{-\sin x}{\cos x}\, dx =$

$x\tan x + \ln|\cos x| + C$. $\therefore \int_0^{\pi/3} x\tan^2 x\, dx = \int_0^{\pi/3} x\sec^2 x\, dx - \left[\dfrac{x^2}{2}\right]_0^{\pi/3} =$

$\left[x\tan x + \ln|\cos x|\right]_0^{\pi/3} - \left[\dfrac{x^2}{2}\right]_0^{\pi/3} = \dfrac{\pi\sqrt{3}}{3} - \ln(2) - \dfrac{\pi^2}{18}$

29. If $\theta = \ln x$, then $d\theta = \dfrac{dx}{x}$, $e^\theta d\theta = dx$ and $\int \sin(\ln x)\, dx = \int e^\theta \sin\theta\, d\theta$. Now let $u = e^\theta$ and

$dv = \sin\theta\, d\theta \Rightarrow du = e^\theta d\theta$ and $v = -\cos\theta$, $\int e^\theta \sin\theta\, d\theta = -e^\theta \cos\theta + \int e^\theta \cos\theta\, d\theta$. Now

let $u = e^\theta$ and $dv = \cos\theta\, d\theta \Rightarrow du = e^\theta d\theta$ and $v = \sin\theta \Rightarrow \int e^\theta \sin\theta\, d\theta = -e^\theta \cos\theta +$

$\int e^\theta \cos\theta\, d\theta = -e^\theta \cos\theta + e^\theta \sin\theta - \int e^\theta \sin\theta\, d\theta \Rightarrow 2\int e^\theta \sin\theta\, d\theta =$

$-e^\theta \cos\theta + e^\theta \sin\theta + C$ $\therefore \int e^\theta \sin\theta\, d\theta = \dfrac{1}{2}\left[-e^\theta \cos\theta + e^\theta \sin\theta\right] + C = \int x\sin(\ln x)\, dx =$

$\dfrac{1}{2}\left[-x\cos(\ln x) + x\sin(\ln x)\right] + C$.

31. a) Let $v = x$ and $du = \sin x \Rightarrow dv = dx$ and $u = -\cos x$, $A = \int_0^{\pi} x\sin x\, dx = -x\cos x + \int_0^{\pi} \cos x\, dx =$

$\left[-x\cos x + \sin x\right]_0^{\pi} = \pi$

b) $A = -\int_{\pi}^{2\pi} x\sin x\, dx = \left[x\cos x - \sin x\right]_{\pi}^{2\pi} = 3\pi$

33. $V = 2\pi \int_0^1 x e^{-x}\, dx = 2\pi\left[-x e^{-x} - e^{-x}\right]_0^1 = 2\pi - \dfrac{4\pi}{e}$

35. Let $z = \ln x \Rightarrow e^z = x$ and $e^z dz = dx$,

a) $M_x = \int_1^e \dfrac{\ln^2 x}{2}\, dx = \dfrac{1}{2}\int_0^1 z^2 e^z\, dz = \dfrac{1}{2}\left[\left(z^2 - 2z + 2\right)e^z\right]_0^1 = \dfrac{e-2}{2}$, tabular integration.

$$
\begin{array}{ccc}
z^2 & + & e^z \\
2z & - & e^z \\
2 & + & e^z \\
0 & & e^z
\end{array}
$$

$M_y = \int_1^e x\ln x\, dx = \int_0^1 z e^{2z}\, dz = \left[\left(\dfrac{z}{2} - \dfrac{1}{4}\right)e^{2z}\right]_0^1 = \dfrac{e^2 + 1}{4}$, tabular integration.

$$
\begin{array}{ccc}
z & + & e^{2z} \\
1 & - & e^{2z}/2 \\
0 & & e^{2z}/4
\end{array}
$$

$$M = \int_1^e \ln x \, dx = \left[(z-1)e^z\right]_0^1 = 1, \therefore \; \overline{x} = \frac{M_y}{M} = \frac{e^2+1}{4} \text{ and } \overline{y} = \frac{M_x}{M} = \frac{e-2}{2}$$

b)

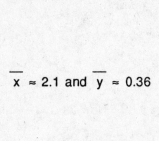

Graph 7.2.35b

$\overline{x} \approx 2.1$ and $\overline{y} \approx 0.36$

37. $M_y = \int_0^\pi x(1+x)\sin x \, dx = \left[\left(x+x^2\right)(-\cos x) + (1+2x)(\sin x) + (2)(\cos x)\right]_0^\pi = \pi^2 + \pi - 4,$

tabular integration

$$
\begin{array}{lcl}
x + x^2 & + & \sin x \\
1 + 2x & - & -\cos x \\
2 & + & -\sin x \\
0 & & \cos x
\end{array}
$$

39. $\int \sin^{-1} x \, dx = x \sin^{-1} x - \int \sin y \, dy = x \sin^{-1} x + \cos y + C = x \sin^{-1} x + \cos\left(\sin^{-1} x\right) + C$

41. $\int \sec^{-1} x \, dx = x \sec^{-1} x - \int \sec y \, dy = x \sec^{-1} x - \ln\left|\sec y + \tan y\right| + C = x \sec^{-1} x -$

$\ln\left|\sec\left(\sec^{-1} x\right) + \tan(\sec^{-1} x)\right| + C = x \sec^{-1} x - \ln\left|x + \sqrt{x^2 - 1}\right| + C$

43. Yes, $\cos^{-1} x$ is the angle whose cosine is x which implies $\sin(\cos^{-1} x) = \sqrt{1 - x^2}$.

45. a) $\int \sinh^{-1} x \, dx = x \sinh^{-1} x - \int \sinh y \, dy = x \sinh^{-1} x - \cosh y + C = x \sinh^{-1} x - \cosh\left(\sinh^{-1} x\right) +$

C; check, $d\left[x \sinh^{-1} x - \cosh\left(\sinh^{-1} x\right) + C\right] = \left[\sinh^{-1} x + \dfrac{x}{\sqrt{1+x^2}} - \sinh\left(\sinh^{-1} x\right)\dfrac{1}{\sqrt{1+x^2}}\right] dx =$

$\sinh^{-1} x \, dx$

b) $\int \sinh^{-1} x \, dx = x \sinh^{-1} x - \int x \dfrac{1}{\sqrt{1+x^2}} dx = x \sinh^{-1} x - \dfrac{1}{2}\int \left(1+x^2\right)^{-1/2} 2x \, dx = x \sinh^{-1} x +$

$\left(1+x^2\right)^{1/2} + C$; check, $d\left[x \sinh^{-1} x + \left(1+x^2\right)^{1/2} + C\right] = \left[\sinh^{-1} x + \dfrac{x}{\sqrt{1+x^2}} - \dfrac{x}{\sqrt{1+x^2}}\right] dx =$

$\sinh^{-1} x \, dx$

7.3 TRIGONOMETRIC INTEGRALS

1. $\displaystyle\int_0^{\pi/2} \sin^5 x\, dx = \int_0^{\pi/2}\left(1 - \cos^2 x\right)^2 \sin x\, dx = -\left[\cos x - \frac{2\cos^3 x}{3} + \frac{\cos^5 x}{5}\right]_0^{\pi/2} = \frac{8}{15}$

3. $\displaystyle\int_{-\pi/2}^{\pi/2} \cos^3 x\, dx = 2\int_0^{\pi/2}\left(1 - \sin^2 x\right)\cos x\, dx = 2\left[\sin x - \frac{\sin^3 x}{3}\right]_0^{\pi/2} = \frac{4}{3}, \cos^3 x$ is even

5. $\displaystyle\int \sin^7 y\, dy = \int \left[1 - \cos^2 y\right]^3 \sin y\, dy = -\cos y + \cos^3 y - \frac{3\cos^5 y}{5} + \frac{\cos^7 y}{7} + C$

7. $\displaystyle\int 8\sin^4 x\, dx = 8\int\left(\frac{1 - \cos 2x}{2}\right)^2 dx = \int 3 - 4\cos 2x + \cos 4x\, dx =$

$3x - 2\sin 2x + \dfrac{\sin 4x}{4} + C$

9. $\displaystyle\int_{-\pi/4}^{\pi/4} \frac{4\sin^4 x}{\cos^2 x}\, dx = 8\int_0^{\pi/4}\left(\tan^2 x\right)\left(1 - \cos^2 x\right) dx = 8\int_0^{\pi/4} \tan^2 x - \sin^2 x\, dx =$

$\displaystyle 8\int_0^{\pi/4} \sec^2 x - 1 - \frac{1 - \cos 2x}{2}\, dx = 8\int_0^{\pi/4} \sec^2 x - \frac{3}{2} + \frac{\cos 2x}{2}\, dx = 8\left[\tan x - \frac{3x}{2} + \frac{\sin 2x}{4}\right]_0^{\pi/4} =$

$10 - 3\pi$

11. $\displaystyle\int_{\pi/6}^{\pi/2} \frac{3\cos^3 t}{\sqrt{\sin^3 t}}\, dt = 3\int_{\pi/6}^{\pi/2}\left(\sin^{-3/2} t\right)\left(1 - \sin^2 t\right)\cos t\, dt = 3\int_{\pi/6}^{\pi/2}\left(\sin^{-3/2} t - \sin^{1/2}\right)\cos t\, dt =$

$\displaystyle 3\left[-2\sin^{-1/2} t - \frac{2}{3}\sin^{3/2} t\right]_{\pi/6}^{\pi/2} = \frac{13\sqrt{2} - 16}{2}$

13. $\displaystyle\int_0^{2\pi} \sin^2 2\theta - \sin^4 2\theta\, d\theta = \int_0^0 \left(u^2 - u^4\right)\left(\frac{1}{2}\right) du = 0$, where $u = \sin 2\theta$

15. $\displaystyle\int_{-\pi/2}^{\pi/4} 16\sin^2 x \cos^2 x\, dx = 4\int_{-\pi/2}^{\pi/4}\left(2\sin x \cos x\right)^2 dx = 4\int_{-\pi/2}^{\pi/4}\left(\sin 2x\right)^2 dx =$

$\displaystyle 2\int_{-\pi/2}^{\pi/4} 1 - \cos 4x\, dx = 2\left[x - \frac{\sin 4x}{4}\right]_{-\pi/2}^{\pi/4} = \frac{3\pi}{2}$

17. $\displaystyle\int 35\sin^4 x \cos^3 x\, dx = \int 35\sin^4 x\left(1 - \sin^2 x\right)\cos x\, dx = 35\left[\frac{\sin^5 x}{5} - \frac{\sin^7}{7}\right] + C$

19. $\displaystyle\int \frac{4\sin^2(\ln x)\cos^2(\ln x)}{x}\, dx = \int 4\sin^2 u \cos^2 u\, du = \int \sin^2 2u\, du = \frac{1}{2}\int 1 - \cos 4u\, du =$

$\displaystyle \frac{1}{2}\left[u - \frac{\sin 4u}{4}\right] + C = \frac{1}{2}\left[\ln x - \frac{\sin\left(\ln x^4\right)}{4}\right] + C$, where $u = \ln x$

21. $\displaystyle\int_0^{2\pi} \sqrt{\frac{1 - \cos\theta}{2}}\, d\theta = \int_0^{2\pi}\left|\sin\frac{\theta}{2}\right| d\theta = 2\int_0^{2\pi}\left(\frac{1}{2}\right)\sin\frac{\theta}{2}\, d\theta = -2\left[\cos\frac{\theta}{2}\right]_0^{2\pi} = 4$

23. $\displaystyle\int_0^{\pi} \sqrt{1 - \sin^2 t}\, dt = \int_0^{\pi} |\cos t|\, dt = \int_0^{\pi/2}\cos t\, dt - \int_{\pi/2}^{\pi}\cos t\, dt = 2$

25. $\displaystyle\int_{-\pi/4}^{\pi/4} \sqrt{1 + \tan^2 x}\, dx = \int_{-\pi/4}^{\pi/4} |\sec x|\, dx = 2\int_{0}^{\pi/4} \sec x\, dx = 2\Big[\ln|\sec x + \tan x|\Big]_{0}^{\pi/4} = \ln\big(3 + 2\sqrt{2}\big)$

27. $\displaystyle\int \theta\sqrt{1 - \cos 2\theta}\, d\theta = \sqrt{2}\int \theta\sqrt{\frac{1 - \cos 2\theta}{2}}\, d\theta = \sqrt{2}\int \theta \sin\theta\, d\theta =$

$\sqrt{2}\Big[-\theta\cos\theta + \sin\theta\Big] + C$

29. $\displaystyle\int_{0}^{\pi} \sqrt{1 + \sin x}\, dx = \int_{0}^{\pi} \sqrt{1 + \cos\left(x - \frac{\pi}{2}\right)}\, dx = \int_{-\pi/2}^{\pi/2} \sqrt{1 + \cos u}\, du =$

$\displaystyle\int_{-\pi/2}^{\pi/2} \sqrt{1 + \cos u}\, \frac{\sqrt{1 - \cos u}}{\sqrt{1 - \cos u}}\, du = \int_{-\pi/2}^{\pi/2} (1 - \cos u)^{-1/2}|\sin u|\, du = 2\int_{0}^{\pi/2} (1 - \cos u)^{-1/2}\sin u\, du =$

$\displaystyle 4\Big[(1 - \cos u)^{-1/2}\Big]_{0}^{\pi/2} = 4, \text{ where } u = x - \frac{\pi}{2}$

31. $\displaystyle\int (\cos x)\ln\left(\frac{1 - \cos 2x}{2}\right) dx = \int (\cos x)\ln\left(\sin^2 x\right) dx = 2\int (\cos x)\ln(\sin x)\, dx = 2\int u\, e^u\, du =$

$2(u - 1)\, e^u + C = 2(\ln(\sin x) - 1)(\sin x) + C, \text{ where } u = \ln(\sin x)$

33. Let $v = \sec x$ and $du = \sec^2 x\, dx \Rightarrow dv = \sec x \tan x\, dx$ and $u = \tan x$, $\displaystyle\int \sec^3 x\, dx =$

$\displaystyle \sec x \tan x - \int \sec x \left(\sec^2 x - 1\right) dx = \sec x \tan x - \int \sec^3 x\, dx + \int \sec x\, dx \Rightarrow \int \sec^3 x\, dx =$

$\displaystyle \frac{1}{2}(\sec x \tan x + \ln|\sec x + \tan x|) + C \quad \therefore \int_{-\pi/3}^{0} 2\sec^3 x\, dx = \Big[\sec x \tan x + \ln|\sec x + \tan x|\Big]_{-\pi/3}^{0} =$

$2\sqrt{3} - \ln\big(2 - \sqrt{3}\big)$

35. $\displaystyle\int_{\pi/2}^{\pi} \csc^3 \frac{x}{2}\, dx = 2\int_{\pi/4}^{\pi/2} \csc^3 z\, dz, \text{ where } z = \frac{x}{2}.$ Let $v = \csc z$ and $du = \csc^2 z\, dz \Rightarrow dv =$

$\displaystyle -\csc z \cot z\, dz$ and $u = -\cot z$, $\displaystyle\int_{\pi/4}^{\pi/2} \csc^3 z\, dz = \Big[-\csc z \cot z\Big]_{\pi/4}^{\pi/2} - \int_{\pi/4}^{\pi/2} (\csc z)\left(\csc^2 z - 1\right) dz =$

$\displaystyle\sqrt{2} - \int_{\pi/4}^{\pi/2} \csc^3 z\, dz + \int_{\pi/4}^{\pi/2} \csc z\, dz = \sqrt{2} - \int_{\pi/4}^{\pi/2} \csc^3 z\, dz - \Big[\ln|\csc z - \cot z|\Big]_{\pi/4}^{\pi/2} =$

$\displaystyle\sqrt{2} - \int_{\pi/4}^{\pi/2} \csc^3 z\, dz - \ln\big(\sqrt{2} - 1\big) \Rightarrow \int_{\pi/4}^{\pi/2} \csc^3 z\, dz = \sqrt{2} + \ln\big(\sqrt{2} - 1\big)$

37. $\displaystyle\int_{0}^{\pi/4} \tan^2 x \sec x\, dx = \int_{0}^{\pi/4} \left(\sec^2 x - 1\right)\sec x\, dx = \int_{0}^{\pi/4} \sec^3 x\, dx - \int_{0}^{\pi/4} \sec x\, dx;$ let $v = \sec x$ and

$\displaystyle du = \sec^2 x\, dx \Rightarrow dv = \sec x \tan x\, dx, \ u = \tan x$ and $\int \sec^3 x\, dx = \sec x \tan x -$

$\displaystyle\int \sec x \left(\sec^2 x - 1\right) dx = \sec x \tan x - \int \sec^3 x\, dx + \int \sec x\, dx \Rightarrow \int \sec^3 x\, dx =$

$\displaystyle\frac{1}{2}(\sec x \tan x + \ln|\sec x + \tan x|) + C. \quad \therefore \int_{0}^{\pi/4} \sec^3 x\, dx - \int_{0}^{\pi/4} \sec x\, dx =$

$\displaystyle\Big[\sec x \tan x - \ln|\sec x + \tan x|\Big]_{0}^{\pi/4} = \frac{\sqrt{2} - \ln\big(\sqrt{2} + 1\big)}{2}.$

39. $\int_0^{\pi/4} \sec^4\theta \, d\theta = \int_0^{\pi/4} \left(1 + \tan^2\theta\right)\left(\sec^2\theta\right) d\theta = \left[\tan\theta + \frac{\tan^3\theta}{3}\right]_0^{\pi/4} = \frac{4}{3}$

41. $\int \csc^4\theta \, d\theta = \int \left(1 + \cot^2\theta\right) \csc^2\theta \, d\theta = -\cot\theta + \frac{\cot^3\theta}{3} + C$

43. $\int_0^{\pi/4} 4\tan^3 x \, dx = 4\int_0^{\pi/4} \left(\sec^2 x - 1\right) \tan x \, dx = 4\int_0^{\pi/4} \tan x \sec^2 x - \frac{\sin x}{\cos x} \, dx =$

$4\left[\frac{\tan^2 x}{2} + \ln|\cos x|\right]_0^{\pi/4} = 2 - \ln 4$

45. $\int_{\pi/4}^{\pi/2} 8\cot^4 t \, dt = 8\int_{\pi/4}^{\pi/2} \left(\cot^2 t\right)\left(\csc^2 t - 1\right) dt = 8\int_{\pi/4}^{\pi/2} \left(\cot^2 t\right)\left(\csc^2 t\right) - \left(\csc^2 t - 1\right) dt =$

$8\left[-\frac{\cot^3 t}{3} + \cot t + t\right]_{\pi/4}^{\pi/2} = \frac{6\pi - 16}{3}$

47. Let $u = \ln(\cos t)$ and $dv = \sec^2 t \, dt \Rightarrow du = -\tan t$ and $v = \tan t$; $\int \left(\sec^2 t\right)\left(\ln(\cos t)\right) dt =$

$(\tan t) \ln(\cos t) + \int \tan^2 t \, dt = (\tan t)(\ln(\cos t)) + \int \sec^2 t - 1 \, dt = (\tan t) \ln(\cos t) + \tan t - t + C$

49. $\int \tan^3 x (\sec x)^{3/2} \, dx = \int \left(\sec^2 x - 1\right)(\sec x)^{1/2}(\sec x \tan x) \, dx =$

$\int \left(\sec^{5/2} - \sec^{1/2}\right) (\sec x \tan x) \, dx = \frac{2}{7}\sec^{7/2} - \frac{2}{3}\sec^{3/2} + C$

51. $\int_{-\pi}^0 \sin 3x \cos 2x \, dx = \frac{1}{2}\int_{-\pi}^0 \sin x + \sin 5x \, dx = -\frac{1}{2}\left[\cos x + \frac{\cos 5x}{5}\right]_{-\pi}^0 = -\frac{6}{5}$

53. $\int_{\pi/12}^{\pi/6} 8\sin 4x \sin 2x \, dx = 4\int_{\pi/12}^{\pi/6} [\cos 2x - \cos 6x] \, dx = 4\left[\frac{\sin 2x}{2} - \frac{\cos 6x}{6}\right]_{\pi/12}^{\pi/6} = \frac{3\sqrt{3} - 1}{3}$

55. $\int_{2\pi}^{4\pi} \cos\frac{x}{3} \cos\frac{x}{4} \, dx = \frac{1}{2}\int_{2\pi}^{4\pi} \cos\frac{x}{12} \cos\frac{7x}{12} \, dx = \frac{1}{2}\left[12\sin\frac{x}{12} + \frac{12}{7}\sin\frac{7x}{12}\right]_{2\pi}^{4\pi} = \frac{24\sqrt{3} - 18}{7}$

57. The integrands for a, b, c, e, g and i are odd and are integrated over a symmetric interval about zero, which implies that the corresponding integrals are zero. The integrands for d, f and h are nonnegative and even which implies that the corresponding integrals are not zero.

59. The integrals in parts d, f and h have even integrands. $\int_{-\pi \div 2}^{\pi/2} x\sin x \, dx = 2[-x\cos x + \sin x]_0^{\pi/2} = 2$,

$\int_{-\pi/2}^{\pi/2} \cos^3 x \, dx = 2\int_0^{\pi/2} \left(1 - \sin^2\right) \cos x \, dx = 2\left[\sin x - \frac{\sin^3 x}{3}\right]_0^{\pi/2} = \frac{4}{3}$, $\int_{-\pi/2}^{\pi/2} \sin x \sin 2x \, dx =$

$4\int_0^{\pi/2} \sin^2 x \cos x \, dx = 4\left[\frac{\sin^3 x}{3}\right]_0^{\pi/2} = \frac{4}{3}$

61. Let $v = \csc x$ and $du = \csc^2 x\, dx \Rightarrow dv = -\csc x \cot x\, dx$ and $u = -\cot^2 x$

$\int \csc^3 dx = -\csc x \cot x - \int \cot^2 x \csc x\, dx = -\csc x \cot x - \int \left(\csc^2 x - 1 \right) \csc x\, dx =$

$-\csc x \cot x - \int \csc^3 x\, dx + \int \csc x\, dx \Rightarrow 2\int \csc^3 x\, dx = -\csc x \cot x - \ln|\csc x + \cot x| + C \Rightarrow$

$\int \csc^3 x\, dx = -\frac{1}{2} \csc x \cot x - \frac{1}{2}\ln|\csc x + \cot x| + C$

63. a) Let $f(x)$ be an odd function and $h(x) = \frac{1}{f(x)}$. Now $h(-x) = \frac{1}{f(-x)} = -\frac{1}{f(-x)} = -\frac{1}{f(x)} = -h(x) \Rightarrow$

$h(x)$ is odd. $\therefore$ the reciprocal of an odd function is odd. Likewise the reciprocal of an even function

is also even.

b) Let $f(x)$ be an odd function , $g(x)$ an even function and $h(x) = \frac{f(x)}{h(x)}$. Now $h(-x) = \frac{f(-x)}{g(-x)} =$

$\frac{-f(x)}{g(x)} = -h(x) \Rightarrow h(x)$ is odd. $\therefore$ the quotient of an odd function and an even function is odd.

7.4 TRIGONOMETRIC SUBSTITUTIONS

1. $\displaystyle\int_{-2}^{2} \frac{1}{4 + x^2}\, dx = 2\int_{0}^{2} \frac{1}{2^2 + x^2}\, dx = 2\left[\frac{1}{2} \tan^{-1} \frac{x}{2} \right]_0^2 = \frac{\pi}{4}$

3. $\displaystyle\int_{0}^{3/2} \frac{1}{\sqrt{9 - x^2}}\, dx = \int_{0}^{\pi/6} \frac{3\cos\theta}{|3\cos\theta|}\, dx = \frac{\pi}{6}$

5. $\displaystyle\int \frac{1}{\sqrt{x^2 - 4}}\, dx = \int \frac{2\sec\theta\tan\theta}{\sqrt{4\left(\sec^2\theta - 1\right)}}\, d\theta = \int \sec\theta\, d\theta = \ln\left|\sec\theta + \tan\theta\right| + C =$

$\ln\left|x + \sqrt{x^2 - 4}\right| + C$, where $x = 2\sec\theta$

7. $\displaystyle\int \frac{2}{\sqrt{t} + 4t\sqrt{t}}\, dt = 2\int \frac{2}{1 + (2u)^2}\, du = 2\tan^{-1} 2u + C = 2\tan^{-1} 2\sqrt{t} + C$, where $u = \sqrt{t}$

9. $\displaystyle\int_{0}^{3\sqrt{2}/4} \frac{1}{\sqrt{9 - 4x^2}}\, dx = \frac{3}{2}\int_{0}^{\pi/4} \frac{\cos\theta}{3\cos\theta}\, d\theta = \frac{\pi}{8}$, where $2x = 3\sin\theta$

11. $\displaystyle\int_{1/\sqrt{3}}^{1} \frac{2}{z\sqrt{4z^2 - 1}}\, dz = 2\int_{1/\sqrt{3}}^{1} \frac{2}{2z\sqrt{(2z)^2 - 1}}\, dz = 2\left[\sec^{-1}|2z| \right]_{1/\sqrt{3}}^{1} = \frac{\pi}{3}$

13. $\displaystyle\int \frac{1}{y\sqrt{4 + \ln^2 y}}\, dy = \int \frac{1}{\sqrt{4 + u^2}}\, du = \int \frac{1/2}{\sqrt{1 + \left(\frac{u}{2}\right)^2}}\, du = \sinh^{-1}\left(\frac{u}{2}\right) + C = \sinh^{-1}\left(\frac{\ln y}{2}\right) + C,$

where $u = \ln y$

15. $\displaystyle\int_1^2 \frac{6}{\sqrt{4-(x-1)^2}}\,dx = \int_0^{\pi/6} \frac{6(2\cos\theta)}{2\,|\cos\theta|}\,d\theta = \pi$, where $x - 1 = 2\sin\theta$

17. $\displaystyle\int_1^3 \frac{1}{y^2 - 2y + 5}\,dy = \int_1^3 \frac{1}{(y-1)^2 + 2^2}\,dy = \frac{1}{2}\left[\tan^{-1}\left(\frac{y-1}{2}\right)\right]_1^3 = \frac{\pi}{8}$

19. $\displaystyle\int_1^{3/2} \frac{x-1}{\sqrt{2x - x^2}}\,dx = \int_1^{3/2} \frac{x-1}{\sqrt{1-(x-1)^2}}\,dx = \int_0^{1/2} \left(1 - u^2\right)^{-1/2} u\,du =$

$-\left[1 - u^2\right]_0^{1/2} = \frac{2 - \sqrt{3}}{2}$, where $u = x - 1$

21. $\displaystyle\int \frac{1}{\sqrt{x^2 - 2x}}\,dx = \int \frac{1}{\sqrt{(x-1)^2 - 1}}\,dx = \int \frac{\sec\theta\tan\theta}{\tan\theta}\,d\theta = \ln\left|\sec\theta + \tan\theta\right| + C =$

$\ln\left|(x-1) + \sqrt{x^2 - 2x}\right| + C$, where $x - 1 = \sec\theta$

23. $\displaystyle\int_{-2}^2 \frac{x+2}{\sqrt{x^2 + 4x + 13}}\,dx = \int_{-2}^2 \frac{x+2}{\sqrt{(x+2)^2 + 3^2}}\,dx = \int_0^{\arctan 4/3} \frac{3\tan\theta\,3\sec^2\theta}{3\sec\theta}\,d\theta =$

$3\left[\sec\theta\right]_0^{\arctan 4/3} = 2$, where $x + 2 = 3\tan\theta$

25. $\displaystyle\int \frac{1}{\sqrt{s^2 - 2s + 5}}\,ds = \int \frac{1/2}{\sqrt{\left(\frac{s-1}{2}\right)^2 + 1}}\,ds = \sinh^{-1}\left(\frac{s-1}{2}\right) + C$

27. $\displaystyle\int \frac{3}{9x^2 - 6x + 5}\,dx = \int \frac{3}{(3x-1)^2 + 4}\,dx = \frac{1}{2}\tan^{-1}\left(\frac{3x-1}{2}\right) + C$

29. $\displaystyle\int \frac{1}{\sqrt{r^2 - 2r - 3}}\,dr = \int \frac{1/2}{\sqrt{\left(\frac{r-1}{2}\right)^2 - 1}}\,dr = \cosh^{-1}\left(\frac{r-1}{2}\right) + C$

31. $\displaystyle\int_{-2}^3 \frac{2\theta}{\theta^2 + 4\theta + 5}\,d\theta = \int_0^5 \frac{2(z-2)}{z^2 + 1}\,dz = \left[\ln\left(z^2 + 1\right) - 4\tan^{-1}z\right]_0^5 = \ln(26) - 4\tan^{-1}5,$

where $z = \theta + 2$

33. $\displaystyle\int \frac{1}{\sqrt{9x^2 - 6x + 5}}\,dx = \frac{1}{3}\int \frac{\frac{3}{2}}{\sqrt{\left(\frac{3x-1}{2}\right)^2 + 1}}\,dx = \frac{1}{3}\sinh^{-1}\left(\frac{3x-1}{2}\right) + C$

35. $\displaystyle\int_{-2}^1 \frac{dx}{\sqrt{x^2 + 4x + 13}} = \int_{-2}^1 \frac{\frac{1}{3}\,dx}{\sqrt{\left(\frac{x+2}{3}\right)^2 + 1}} = \left[\sinh^{-1}\left(\frac{x+2}{3}\right)\right]_{-2}^1 = \sinh^{-1}(1)$

37. $\displaystyle\int_5^6 \frac{1}{\sqrt{t^2 - 2t - 8}}\,dt = \int_5^6 \frac{\frac{1}{3}\,dt}{\sqrt{\left(\frac{t-1}{3}\right)^2 - 1}} = \left[\cosh^{-1}\left(\frac{t-1}{3}\right)\right]_5^6 = \cosh^{-1}\left(\frac{5}{3}\right) - \cosh^{-1}\left(\frac{4}{3}\right)$

39. $\displaystyle\int \frac{r\,dr}{\sqrt{r^2 + 4r + 5}} = \int \frac{z - 2}{\sqrt{z^2 + 1}}\,dz = \left(z^2 + 1\right)^{1/2} - 2\sinh^{-1}z + C = \sqrt{r^2 + 4r + 5} - 2\sinh^{-1}(r+2) + C,$

where $z = r + 2$

41. $\displaystyle\int \frac{4x^2}{\left(1 - x^2\right)^{3/2}}\,dx = \int \frac{4\sin^2\theta\cos\theta}{\cos^3\theta}\,d\theta = 4\int \sec^2\theta - 1\,d\theta = 4\tan\theta - 4\theta + C =$

$\displaystyle\frac{4x}{\sqrt{1 - x^2}} - 4\sin^{-1}x + C$, where $x = \sin\theta$

43. $\displaystyle A = \int_0^3 \frac{\sqrt{9 - x^2}}{3}\,dx = \frac{1}{3}\int_0^{\pi/2} 3\cos\theta\,3\cos\theta\,d\theta = \frac{3}{2}\int_0^{\pi/2} 1 + \cos 2\theta\,d\theta =$

$\displaystyle\frac{3}{2}\left[\theta + \frac{\sin 2\theta}{2}\right]_0^{\pi/2} = \frac{3\pi}{4}$, where $x = 3\sin\theta$

45. $\displaystyle y = x^2 \Rightarrow 1 + (y')^2 = 1 + (2x)^2 \Rightarrow L = \int_0^{\sqrt{3}/2} \sqrt{1 + (2x)^2}\,dx = \frac{1}{2}\int_0^{\pi/3} \sec^3 z\,dz =$

$\displaystyle\frac{1}{4}\left[\sec z\tan z + \ln|\sec z + \tan z|\right]_0^{\pi/3} = \frac{1}{4}\left[2\sqrt{3} + \ln(2 + \sqrt{3})\right] \approx 1.195$, by parts and

where $\tan z = 2x$

47. $\displaystyle V = \pi\int_{-2}^{11} \left(\frac{20}{\sqrt{x^2 - 2x + 17}}\right)^2 dx = 400\pi\int_{-2}^{11} \frac{1}{4^2 + (x - 1)^2}\,dx =$

$\displaystyle 400\pi\left[\frac{1}{4}\tan^{-1}\left(\frac{x-1}{4}\right)\right]_{-2}^{11} \approx 576.10245$

49. $\displaystyle M_y = \int_1^5 \frac{2x}{\sqrt{x^2 - 2x + 10}}\,dx = \int_1^5 \frac{2x}{\sqrt{(x-1)^2 + 9}}\,dx = 2\int_0^{\arctan 4/3} (3\tan\theta + 1)\sec\theta\,d\theta =$

$\displaystyle\left[6\sec\theta + 2\ln\left|\sec\theta + \tan\theta\right|\right]_0^{\arctan 4/3} = 4 + \ln 9$, where $x - 1 = 3\tan\theta$; $M_x =$

$\displaystyle\frac{1}{2}\int_1^5 \frac{4}{x^2 - 2x + 10}\,dx = 2\int_1^5 \frac{dx}{(x-1)^2 + 9} = \left[\frac{2}{3}\tan^{-1}\left(\frac{x-}{3}\right)\right]_1^5 = \frac{2}{3}\tan^{-1}\left(\frac{4}{3}\right)$; $M =$

$\displaystyle\int_1^5 \frac{2}{\sqrt{x^2 - 2x + 10}}\,dx = 2\int_1^5 \frac{dx}{\sqrt{(x-1)^2 + 9}} = 2\int_0^{\arctan 4/3} \frac{3\sec^2\theta\,d\theta}{3\sqrt{\tan^2\theta + 1}} =$

$\displaystyle 2\int_0^{\arctan 4/3} \sec\theta\,d\theta = 2\left[\ln\left(\sec\theta + \tan\theta\right)\right]_0^{\arctan 4/3} = \ln 9$, where $x - 1 = 3\tan\theta$

$\displaystyle\therefore\ \overline{x} = \frac{M_y}{M} = 1 + \frac{4}{\ln 9} \approx 2.82$ and $\overline{y} = \frac{\tan^{-1}\left(\frac{4}{3}\right)}{\ln 27} \approx 0.28$

51. $\displaystyle\int_a^3 \frac{dx}{(x+1)\sqrt{x^2+2x-3}} = \int_a^3 \frac{1}{(x+1)\sqrt{(x+1)^2-4}}\,dx = \int_{\arcsin\left(\frac{a+1}{2}\right)}^{\operatorname{arcsec} 2} d\theta =$

$\displaystyle\frac{1}{2}\left[\operatorname{arcsec} 2 - \operatorname{arcsec}\left(\frac{a+1}{2}\right)\right]$, where $x+1 = 2\sec\theta$ $\therefore \operatorname*{Lim}_{a \to 1^+} \int_a^3 \frac{dx}{(x+1)\sqrt{x^2+2x-3}} =$

$\displaystyle\frac{1}{2}\operatorname{arcsec} 2 - \frac{1}{2}\operatorname*{Lim}_{a \to 1^+}\operatorname{arcsec}\left(\frac{a+1}{2}\right) = \frac{1}{2}\operatorname{arcsec} 2$

7.5 RATIONAL FUNCTIONS AND PARTIAL FRACTIONS

1. $\displaystyle\frac{5x-13}{(x-3)(x-2)} = \frac{A}{x-3} + \frac{B}{x-2} = \frac{2}{x-3} + \frac{3}{x-2}$

3. $\displaystyle\frac{x+4}{(x+1)^2} = \frac{A}{x+1} + \frac{B}{(x+1)^2} = \frac{1}{x+1} + \frac{3}{(x+1)^2}$

5. $\displaystyle\frac{x+1}{x^2(x-1)} = \frac{A}{x} + \frac{B}{x^2} + \frac{C}{x-1} = \frac{-2}{x} + \frac{-1}{x^2} + \frac{2}{x-1}$

7. $\displaystyle\frac{x^2+8}{x^2-5x+6} = 1 + \frac{5x+2}{x^2-5x+6} = 1 + \frac{A}{x-3} + \frac{B}{x-2} = 1 + \frac{17}{x-3} + \frac{-12}{x-2}$

9. $\displaystyle\int_0^{1/2}\frac{1}{1-x^2}\,dx = \frac{1}{2}\int_0^{1/2}\frac{1}{1+x} + \frac{1}{1-x}\,dx = \frac{1}{2}\left[\ln|1+x| - \ln|1-x|\right]_0^{1/2} = \ln\sqrt{3}$

11. $\displaystyle\int\frac{x+4}{x^2+5x-6}\,dx = \frac{1}{7}\int\frac{2}{x+6} + \frac{5}{x-1}\,dx = \frac{1}{7}\left[2\ln|x+6| + 5\ln|x-1|\right] + C =$

$\displaystyle\frac{1}{7}\ln\left|(x+6)^2(x-1)^5\right| + C$

13. $\displaystyle\int_4^8\frac{y}{y^2-2y-3}\,dy = \frac{1}{4}\int_4^8\frac{3}{y-3} + \frac{1}{y+1}\,dy = \frac{1}{4}\left[3\ln|y-3| + \ln|y+1|\right]_4^8 = \ln\sqrt{15}$

15. $\displaystyle\int\frac{1}{t^3+t^2-2t}\,dt = \int\frac{-1/2}{t} + \frac{1/6}{t+2} + \frac{1/3}{t-1}\,dt = -\frac{1}{2}\ln|t| + \frac{1}{6}\ln|t+2| + \frac{1}{3}\ln|t-1| + C$

17. $\displaystyle\int\frac{x^3}{x^2+2x+1}\,dx = \int x - 2 + \frac{3}{x+1} - \frac{1}{(x+1)^2}\,dx = \frac{x^2}{2} - 2x + 3\ln|x+1| + \frac{1}{x+1} + C$

19. $\displaystyle\int\frac{1}{\left(x^2-1\right)^2}\,dx = \frac{1}{4}\int\frac{1}{x+1} + (x+1)^{-2} - \frac{1}{x-1} + (x-1)^{-2}\,dx =$

$\displaystyle\frac{1}{4}\left[\ln|x+1| - (x+1)^{-1} - \ln|x-1| - (x-1)^{-1}\right] + C = \frac{1}{4}\ln\left|\frac{x+1}{x-1}\right| - \frac{x}{2\left(x^2-1\right)} + C$

21. $\int_0^{\ln 2} \frac{e^t \, dt}{e^{2t} + 3 \, e^t + 2} = \int_1^2 \frac{dz}{z^2 + 3z + 2} = \int_1^2 \frac{1}{z + 1} - \frac{1}{z + 2} \, dz = \left[\ln\left|\frac{z + 1}{z + 2}\right|\right]_1^2 = \ln\left(\frac{9}{8}\right),$ where $z + e^t$

23. $\int \frac{\cos y}{\sin^2 y + \sin y - 6} \, dy = \int \frac{dx}{x^2 + x - 6} = \frac{1}{5}\int \frac{1}{x - 2} - \frac{1}{x + 3} \, dx = \frac{1}{5} \ln\left|\frac{x - 2}{x + 3}\right| + C,$ where $x = \sin y$

25. $\int \frac{1 - \sqrt{x}}{1 + \sqrt{x}} \, dx = 2 \int \frac{(1 - z) z}{1 + z} \, dz = -2 \int \left(z - 2 + \frac{2}{z + 1}\right) dz = -2\left[\frac{z^2}{2} - 2z + 2 \ln|z + 1|\right] + C =$

    $-x + 4\sqrt{x} - 4 \ln\left(\sqrt{x} + 1\right) + C,$ where $z = \sqrt{x}$

27. Let $u = \ln(t + 5)$ and $dv = t \, dt \Rightarrow du = \frac{dt}{t + 5}$ and $v = \frac{t^2}{2}, \int t \ln(t + 5) \, dt = \frac{t^2 \ln(t + 5)}{2} - \frac{1}{2}\int \frac{t^2}{t + t + 5} \, dt =$

    $\frac{t^2 \ln(t + 5)}{2} - \frac{1}{2}\int \left(-5 + x + \frac{25}{x + 5}\right) dx = \frac{t^2 \ln(t + 5)}{2} + \frac{5x}{2} - \frac{x^2}{4} - \frac{25}{2} \ln|x + 5| + C$

29. $\int_0^{2\sqrt{2}} \frac{x^3}{x^2 + 1} \, dx = \int_0^{2\sqrt{2}} x - \frac{x}{x^2 + 1} \, dx = \left[\frac{x^2}{2} - \frac{1}{2} \ln\left(x^2 + 1\right)\right]_0^{2\sqrt{2}} = 4 - \ln 3$

31. $\int_1^2 \frac{1}{y^3 + y} \, dy = \int_1^2 \frac{1}{y} - \frac{y}{y^2 + 1} \, dy = \left[\ln|y| - \frac{1}{2} \ln\left(1 + y^2\right)\right]_1^2 = \ln\left(\frac{2\sqrt{2}}{\sqrt{5}}\right)$

33. $\int_0^{\sqrt{3}} \frac{5x^2}{x^2 + 1} \, dx = \int_0^{\sqrt{3}} 5 - \frac{5}{1 + x^2} \, dx = \left[5x - 5 \tan^{-1} x\right]_0^{\sqrt{3}} = 5\sqrt{3} - \frac{5\pi}{3}$

35. $\int \frac{4x + 4}{x^2\left(x^2 + 1\right)} \, dx = 4\int \frac{1}{x} + x^{-2} - \frac{x + 1}{x^2 + 1} \, dx = 4 \ln|x| - 4 \, x^{-1} - 2 \ln\left(x^2 + 1\right) - 4 \tan^{-1} x + C =$

    $\ln\left(\frac{x^4}{\left(x^2 + 1\right)^2}\right) - \frac{4}{x} - 4 \tan^{-1} x + C$

37. $\int_0^1 \frac{x^3 + 1}{x^2 + 1} \, dx = \int_0^1 x + \frac{1}{x^2 + 1} - \frac{x}{x^2 + 1} \, dx = \left[\frac{x^2}{2} + \tan^{-1} x - \frac{1}{2} \ln\left(x^2 + 1\right)\right]_0^1 = \frac{2 + \pi - \ln 4}{4}$

39. $\int_0^1 \frac{y^2 + 2y + 1}{\left(y^2 + 1\right)^2} \, dy = \int_0^1 \frac{1}{y^2 + 1} + \frac{2y}{\left(y^2 + 1\right)^2} \, dy = \left[\tan^{-1} y - \frac{1}{y^2 + 1}\right]_0^1 = \frac{\pi + 2}{4}$

41. $\int_0^1 \frac{2r^3 + 3r^2 + 5r + 2}{r^2 + r + 1} \, dr = \int_0^1 2r + 1 + \frac{2r + 1}{r^2 + r + 1} \, dr = \left[r^2 + r + \ln\left|r^2 + r + 1\right|\right]_0^1 = 2 + \ln 3$

43. $\int_{-1}^0 \frac{2x + 2}{(x^2 + 1)(x - 1)^3} \, dx = \int_{-1}^0 2(x - 1)^{-3} - (x - 1)^{-2} + \frac{1}{1 + x^2} \, dx =$

    $\left[-(x - 1)^{-2} + (x - 1)^{-1} + \tan^{-1} x\right]_{-1}^0 = \frac{\pi - 5}{4}$

45. $\int \dfrac{w-1}{4w^2-4w-1}\,dw = \dfrac{1}{8}\int \dfrac{8w-8}{4w^2-4w-1}\,dw = \dfrac{1}{8}\int \dfrac{8w-4}{4w^2-4w-1} - \dfrac{1}{(w-1/2)^2-2}\,dw =$

$\dfrac{1}{8}\int \dfrac{8w-4}{4w^2-4w-1} + \dfrac{1}{\left(\frac{1}{\sqrt{2}}\right)^2-\left(w-\frac{1}{2}\right)^2}\,dw = \dfrac{1}{8}\ln\left|4w^2-4w-1\right| + \dfrac{\sqrt{2}}{8}\ln\left|\dfrac{\frac{1}{\sqrt{2}}+\left(w-\frac{1}{2}\right)}{\frac{1}{\sqrt{2}}-\left(w-\frac{1}{2}\right)}\right| + C =$

$\dfrac{1}{8}\ln\left|4w^2-4w-1\right| + \dfrac{\sqrt{2}}{8}\ln\left|\dfrac{1+\sqrt{2}w-\sqrt{2}}{1-\sqrt{2}w+\sqrt{2}}\right| + C$

47. $\int \dfrac{3\theta^2-\theta+1}{\theta^3-1}\,d\theta = \int \dfrac{1}{\theta-1} + \dfrac{2\theta}{\theta^2+\theta+1}\,d\theta = \int \dfrac{1}{\theta-1} + \dfrac{2\theta+1}{\theta^2+\theta+1} - \dfrac{1}{\theta^2+\theta+1}\,d\theta =$

$\ln\left|\theta-1\right| + \ln\left|\theta^2+\theta+1\right| - \int \dfrac{d\theta}{(\sqrt{3}/2)^2+(\theta+1/2)^2}\,d\theta = \ln\left|\dfrac{\theta-1}{\theta^2+\theta+1}\right| - \dfrac{2}{\sqrt{3}}\tan^{-1}\left(\dfrac{2\theta+1}{\sqrt{3}}\right) + C$

49. $\int \dfrac{2x^3+5x^2+8x+4}{(x^2+2x+2)^2}\,dx = \int (x^2+2x+2)^{-2}(2x+2) + \dfrac{2x+2}{x^2+2x+2} - \dfrac{1}{x^2+2x+2}\,dx =$

$-(x^2+2x+2)^{-1} + \ln\left|x^2+2x+2\right| - \int \dfrac{dx}{1+(x+1)^2} = \dfrac{-1}{x^2+2x+2} + \ln\left|x^2+2x+2\right| -$

$\tan^{-1}(x+1) + C$

51. $V = \pi \int_{1/2}^{5/2} \left(\dfrac{3}{\sqrt{3x-x^2}}\right)^2 dx = 9\pi \int_{1/2}^{5/2} \dfrac{1}{x(3-x)}\,dx = 3\pi \int_{1/2}^{5/2} \dfrac{1}{x} + \dfrac{1}{(3-x)}\,dx =$

$3\pi\left[\ln|x| - \ln|3-x|\right]_{1/2}^{5/2} = 3\pi \ln 25$

53. Let $u = \tan^{-1}x$ and $dv = x\,dx \Rightarrow du = \dfrac{dx}{1+x^2}$ and $v = \dfrac{x^2}{2}$, $M_y = \int_0^{\sqrt{3}} x\tan^{-1}x\,dx = \left[\dfrac{x^2\tan^{-1}x}{2}\right]_0^{\sqrt{3}} -$

$\dfrac{1}{2}\int_0^{\sqrt{3}} \dfrac{x^2}{1+x^2}\,dx = \dfrac{\pi}{2} - \dfrac{1}{2}\int_0^{\sqrt{3}} 1 - \dfrac{1}{1+x^2}\,dx = \dfrac{\pi}{2} - \dfrac{1}{2}\left[x-\tan^{-1}x\right]_0^{\sqrt{3}} = \dfrac{4\pi-3\sqrt{3}}{6}$; let $u = \tan^{-1}x$ and

$dv = dx \Rightarrow du = \dfrac{dx}{1+x^2}$ and $v = x$, $M = \int_0^{\sqrt{3}} \tan^{-1}x\,dx = \left[x\tan^{-1}x\right]_0^{\sqrt{3}} - \int_0^{\sqrt{3}} \dfrac{x}{1+x^2}\,dx = \dfrac{\sqrt{3}\pi}{3} -$

$\left[\dfrac{1}{2}\ln(1+x^2)\right]_0^{\sqrt{3}} = \dfrac{\sqrt{3}\pi-\ln 8}{3} \quad \therefore \quad \overline{x} = \dfrac{M_y}{M} = \dfrac{4\pi-3\sqrt{3}}{2\sqrt{3}\pi-\ln 64} \approx 1.10$

55.   a) $\dfrac{dx}{dt} = kx(N - x) \Rightarrow \displaystyle\int \dfrac{1}{x(N - x)}\,dx = \int k\,dt \Rightarrow \dfrac{1}{N}\int \dfrac{1}{x} + \dfrac{1}{N - x}\,dx = \int k\,dt \Rightarrow$

$\dfrac{1}{N}\Big[\ln|x| - \ln|N - x|\Big] = kt + C \Rightarrow \ln\left(\dfrac{x}{N - x}\right) = kNt + C,\ \text{for } x < N \Rightarrow \dfrac{x}{N - x} = A\,e^{kNt} \Rightarrow$

$x = \dfrac{NAe^{kNt}}{1 + Ae^{kNt}}$, where $k = \dfrac{1}{250}$ and $N = 1000 \Rightarrow x = \dfrac{1000Ae^{4t}}{1 + Ae^{4t}}$, where $t = 0$ and $x = 2 \Rightarrow$

$x = \dfrac{1000\,e^{4t}}{499 + e^{4t}}$

       b)   Solving $500 = \dfrac{1000\,e^{4t}}{499 + e^{4t}} \Rightarrow t = \dfrac{\ln 499}{4} \approx 1.55$ days

57. $\displaystyle\int_0^1 \dfrac{x^4(x - 1)^4}{x^2 + 1}\,dx = \int_0^1 4 - 4x^2 + 5x^4 - 4x^5 + x^6 - \dfrac{4}{1 + x^2}\,dx =$

$\left[4x - \dfrac{4x^3}{3} + x^5 - \dfrac{2x^6}{3} + \dfrac{x^7}{7} - 4\tan^{-1}x\right]_0^1 = \dfrac{22}{7} - \pi$

7.6 USING INTEGRAL TABLES; REDUCTION FORMULAS

1. $\displaystyle\int x\cos^{-1}x\,dx = \dfrac{x^2}{2}\cos^{-1}(x) + \dfrac{1}{2}\int \dfrac{x^2}{\sqrt{1 - x^2}}\,dx = \dfrac{x^2}{2}\cos^{-1}(x) +$

$\dfrac{1}{2}\left(\dfrac{1}{2}\sin^{-1}(x) - \dfrac{1}{2}x\sqrt{1 - x^2}\right) + C = \dfrac{x^2}{2}\cos^{-1}(x) + \dfrac{1}{4}\sin^{-1}(x) - \dfrac{1}{4}x\sqrt{1 - x^2} + C,$ formulas 33, 100

3. $\displaystyle\int_0^{1/2} x\tan^{-1}2x\,dx = \left[\dfrac{x^2}{2}\tan^{-1}2x\right]_0^{1/2} - \int_0^{1/2}\dfrac{x^2}{1 + 4x^2}\,dx = \dfrac{\pi}{32} - \int_0^{1/2}\dfrac{1}{4} - \dfrac{1/4}{1 + 4x^2}\,dx =$

$\dfrac{\pi}{32} - \left[\dfrac{x}{4} - \dfrac{1}{8}\tan^{-1}2x\right]_0^{1/2} = \dfrac{\pi - 2}{16}$, formula 101

5. $\displaystyle\int_4^{10}\dfrac{\sqrt{4x + 9}}{x^2}\,dx = -\left[\dfrac{\sqrt{4x + 9}}{x}\right]_4^{10} + 2\int_4^{10}\dfrac{1}{x\sqrt{4x + 9}}\,dx = -\left[\dfrac{\sqrt{4x + 9}}{x}\right]_4^{10} + 2\left[\dfrac{1}{3}\ln\left|\dfrac{\sqrt{4x + 9} - 3}{\sqrt{4x + 9} + 3}\right|\right]_4^{10} =$

$\dfrac{2}{3}\ln\left(\dfrac{8}{5}\right) + \dfrac{11}{20}$, formulas 13b, 14

7. $\displaystyle\int \dfrac{1}{x^2\sqrt{7 - x^2}}\,dx = -\dfrac{\sqrt{7 - x^2}}{7x} + C$, formula 35

9. $\displaystyle\int_{-\pi/12}^{\pi/4}\dfrac{1}{5 + 4\sin 2x}\,dx = -\dfrac{1}{3}\left[\tan^{-1}\left(\dfrac{1}{3}\tan\left(\dfrac{\pi}{4} - x\right)\right)\right]_{-\pi/12}^{\pi/4} = \dfrac{\pi}{18}$, formula 70

11. $\displaystyle\int_3^6 \frac{x}{\sqrt{x-2}}\,dx = \left[\sqrt{x-2}\left(\frac{2(x-2)}{3}+4\right)\right]_3^6 = \frac{26}{3}$, formula 7

13. $\displaystyle\int \frac{\sqrt{3x-4}}{x}\,dx = 2\sqrt{3x-4} - 4\int \frac{1}{x\sqrt{3x-4}}\,dx = 2\sqrt{3x-4} - 4\tan^{-1}\sqrt{\frac{3x-4}{4}} + C$, formula 12, 13a

15. $\displaystyle\int_0^1 \sin^{-1}\sqrt{x}\,dx = 2\int_0^1 u\left(\sin^{-1}u\right)du = 2\left[\frac{u^2}{2}\sin^{-1}u\right]_0^1 - \frac{1}{2}\int_0^1 \frac{u^2}{\sqrt{1-u^2}}\,du =$

$\displaystyle\sin^{-1}1 - \left[\frac{1}{2}\sin^{-1}u - \frac{1}{2}u\sqrt{1-u^2}\right]_0^1 = \frac{\pi}{4}$, where $u = \sqrt{x}$ and by formulas 33, 99

17. $\displaystyle\int_0^{1/2} \frac{\sqrt{x}}{\sqrt{1-x}}\,dx = 2\int_0^{1/\sqrt{2}} \frac{u^2}{\sqrt{1-u^2}}\,du = \left[\sin^{-1}u - u\sqrt{1-u^2}\right]_0^{1/\sqrt{2}} = \frac{\pi-2}{4}$,

where $u = \sqrt{x}$ and by formula 33

19. $\displaystyle M = \int_0^3 \frac{1}{\sqrt{x+1}}\,dx = 2\left[(x+1)^{1/2}\right]_0^3 = 2,\ M_x = \frac{1}{2}\int_0^3 \frac{1}{x+1}\,dx = \frac{1}{2}\left[\ln|x+1|\right]_0^3 = \ln 2,$

$\displaystyle M_y = \int_0^3 \frac{x}{\sqrt{x+1}}\,dx = \sqrt{x+1}\left[\frac{2(x+1)}{3}-2\right]_0^3 = \frac{8}{3},\ \bar{x} = \frac{M_y}{M} = \frac{4}{3}$ and $\bar{y} = \frac{M_x}{M} = \frac{\ln 2}{2} = \ln\sqrt{2}$, formula 7

21. $\displaystyle y = x^2 \Rightarrow 1+(y')^2 = 1+(2x)^2 \Rightarrow A = 2\pi\int_{-1}^1 x^2\sqrt{1+(2x)^2}\,dx = 2\pi\int_{-1}^1 x^2\sqrt{1+4x^2}\,dx =$

$\displaystyle 4\pi\int_{-1}^1 x^2\sqrt{\left(\frac{1}{2}\right)^2+x^2}\,dx = 4\pi\left[\frac{x\left(1/4+2x^2\right)\sqrt{1/4+x^2}}{8} - \frac{1}{128}\sinh^{-1}2x\right]_{-1}^1 \approx 7.62$, formula 22

23. Let $\displaystyle y = -\frac{1}{a}\sqrt{\frac{2a-x}{x}} + C = -\frac{1}{a}\left(2ax^{-1}-1\right)^{1/2} + C \Rightarrow y' = -\frac{1}{2a}\left(2axz^{-1}-1\right)^{-1/2}\left(-2ax^{-2}\right) =$

$\displaystyle\frac{|x|}{x^2\sqrt{2ax-x^2}} = \frac{1}{x\sqrt{2ax-x^2}}$, when $x > 0$ and $\displaystyle\frac{-1}{x\sqrt{2ax-x^2}}$, when $x < 0$.

Therefore, $\displaystyle\int \frac{1}{x\sqrt{2ax-x^2}}\,dx = \frac{1}{a}\sqrt{\frac{2a-x}{x}} + C$, when $x > 0$

25. $\displaystyle\int \frac{x}{(ax+b)^2}\,dx = \frac{1}{a^2}\int \frac{u-b}{u^2}\,du = \frac{1}{a^2}\int \frac{1}{u} - bu^{-2}\,du = \frac{1}{a^2}\left[\ln|u| + bu^{-1}\right] + C =$

$\displaystyle\frac{1}{a^2}\left[\ln|ax+b| + \frac{b}{ax+b}\right] + C$, where $u = ax+b$

27. $\displaystyle\int_{-\pi}^{\pi} \cos^4 x\,dx = 2\int_0^{\pi} \cos^4 x\,dx = 2\left(\left[\frac{(\cos^3 x)(\sin x)}{4}\right]_0^{\pi} + \frac{3}{4}\int_0^{\pi}\cos^2 x\,dx\right) =$

$\displaystyle\frac{3}{2}\left(\left[\frac{\cos x\,\sin x}{2}\right]_0^{\pi} + \frac{1}{2}\int_0^{\pi}dx\right) = \frac{3\pi}{4}$, formula 61

29. $\displaystyle\int_0^{\pi} \sin^4 x\,dx = -\left[\frac{\sin^3 x\cos x}{4}\right]_0^{\pi} + \frac{3}{4}\int_0^{\pi}\sin^2 x\,dx = \frac{3}{4}\left(\left[-\frac{\sin x\cos x}{2}\right]_0^{\pi} + \frac{1}{2}\int_0^{\pi}dx\right) = \frac{3\pi}{8}$, formula 6

31. $\int_0^{\pi/8} \tan^3 2x \, dx = \left[\dfrac{\tan^2 2x}{4}\right]_0^{\pi/8} + \int_0^{\pi/8} \tan 2x \, dx = \dfrac{1}{4} + \dfrac{1}{2}\Big[\ln|\cos 2x|\Big]_0^{\pi/8} = \dfrac{1 - \ln 2}{4}$, formula 86

33. $\int_{\pi/4}^{3\pi/4} \cot^4 x \, dx = \left[-\dfrac{\cot^3 x}{3}\right]_{\pi/4}^{3\pi/4} - \int_{\pi/4}^{3\pi/4} \cot^2 x \, dx = \dfrac{2}{3} - [-\cot x]_{\pi/4}^{3\pi/4} - \int_{\pi/4}^{3\pi/4} dx = \dfrac{3\pi - 8}{6}$, formula 87

35. $\int_{-\pi/3}^{\pi/3} \sec^4 x \, dx = 2\int_0^{\pi/3} \sec^4 x \, dx = 2\left(\left[\dfrac{\sec^2 x \tan x}{3}\right]_0^{\pi/3} + \dfrac{2}{3}\int_0^{\pi/3} \sec^2 x \, dx\right) =$

$2\left[\dfrac{\sec^2 x \tan x}{3} + \dfrac{2}{3}\tan x\right]_0^{\pi/3} = 4\sqrt{3}$, formula 92

37. $\int_{\pi/4}^{\pi/2} \csc^4 x \, dx = \left[-\dfrac{\csc^2 x \cot x}{3}\right]_{\pi/4}^{\pi/2} + \dfrac{2}{3}\int_{\pi/4}^{\pi/2} \csc^2 x \, dx = \left[-\dfrac{\csc^2 x \cot x}{3} - \dfrac{2}{3}\cot x\right]_{\pi/4}^{\pi/2} = \dfrac{4}{3}$, formula 93

$\left[-\dfrac{\csc^3 x \cot x}{4} - \dfrac{3 \csc x \cot x}{8} - \dfrac{3}{8}\ln|\csc x + \cot x|\right]_{\pi/4}^{\pi/2} = \dfrac{7\sqrt{2} + 3\ln\left(\sqrt{2} + 1\right)}{8}$, formula 93

39. $\int 16x^3 (\ln x)^2 \, dx = 16\left(\dfrac{x^4 (\ln x)^2}{4} - \dfrac{1}{2}\int x^3 (\ln x) \, dx\right) = 16\left(\dfrac{x^4 (\ln x)^2}{4} - \dfrac{1}{2}\left(\dfrac{x^4 (\ln x)}{4} - \dfrac{1}{4}\int x^3 \, dx\right)\right)$, where

$\int x^m (\ln x)^n \, dx = \dfrac{x^{m+1}(\ln x)^n}{m+1} - \dfrac{n}{m+1}\int x^m (\ln x)^{n-1} \, dx \therefore \int_1^e 16x^3 (\ln x)^2 \, dx =$

$\left[16\left(\dfrac{x^4(\ln x)^2}{4} - \dfrac{x^4(\ln x)}{8} + \dfrac{x^4}{32}\right)\right]_1^e = \dfrac{5e^4 - 1}{2}$

41. $\int_0^1 \left(x^2 + 1\right)^{-3/2} dx = \int_0^{\pi/4} \left((\tan u)^2 + 1\right)^{-3/2} \sec^2 u \, du = \int_0^{\pi/4} \cos u \, du = [\sin u]_0^{\pi/4} =$

$\dfrac{1}{\sqrt{2}}$, where $x = \tan u$

43. $\int_0^{3/5} \dfrac{1}{\left(1 - x^2\right)^3} dx = \int_0^{\arcsin(3/5)} \dfrac{\cos u}{\cos^6 u} du = \int_0^{\arcsin(3/5)} \sec^5 du$, where $x = \sin u \Rightarrow$

$\int_0^{3/5} \dfrac{1}{\left(1 - x^2\right)^3} dx = \left[\dfrac{\sec^3 u \tan u}{4} + \dfrac{3 \sec u \tan u}{8} + \dfrac{3}{8}\ln|\sec u + \tan u|\right]_0^{\arcsin(3/5)} =$

$\dfrac{735 + 384 \ln 2}{1024}$; see exercise 42.

45. $y = -\dfrac{\sin^{n-1} x \cos x}{n} + \dfrac{n-1}{n}\int \sin^{n-2} x \, dx \Rightarrow y' = -\dfrac{(n-1)\sin^{n-2} x \cos^2 c + \sin^n x}{n} + \dfrac{n-1}{n}\sin^{n-2} x =$

$\dfrac{(n-1)\sin^n x + \sin^n x}{n} = \sin^n x \Rightarrow$ that the formula is true

47. a)   Let $u = x^n$ and $dv = \sin x \, dx \Rightarrow du = n\, x^{n-1} \, dx$ and $v = -\cos x$

$\int x^n \sin x \, dx = -x^n \cos x + n\int x^{n-1} \cos x \, dx$

   b)   Let $u = x^n$ and $dv = \sin ax \, dx \Rightarrow du = n\, x^{n-1} \, dx$ and $v = -\dfrac{\cos ax}{a}$

$\int x^n \sin ax \, dx = -\dfrac{x^n \cos ax}{a} + \dfrac{n}{a}\int x^{n-1} \cos ax \, dx$

7.7 IMPROPER INTEGRALS

1. $\displaystyle\int_0^\infty \frac{1}{x^2+1}\,dx = \lim_{t\to\infty}\left[\tan^{-1}\right]_0^t = \left(\lim_{t\to\infty}\tan^{-1}t\right) - \tan^{-1}0 = \frac{\pi}{2}$

3. $\displaystyle\int_{-1}^1 \frac{1}{x^{2/3}}\,dx = \int_{-1}^0 x^{-2/3}\,dx + \int_0^1 x^{-2/3}\,dx = \lim_{t\to 0^-}\int_{-1}^t x^{-2/3}\,dx + \lim_{t\to 0^+}\int_t^1 x^{-2/3}\,dx =$

    $\displaystyle\lim_{t\to 0^-}\left[3x^{1/3}\right]_{-1}^t + \lim_{t\to 0^+}\left[3x^{1/3}\right]_t^1 = (0)-(-3)+(3)-(0) = 6$

5. $\displaystyle\int_0^4 \frac{1}{\sqrt{4-x}}\,dx = \lim_{t\to 4^-}\int_0^t (4-x)^{-1/2}\,dx = = -2\lim_{t\to 4^-}\left[(4-t)^{1/2}-2\right] = 4$

7. $\displaystyle\int_0^1 \frac{1}{r^{0.999}}\,dr = \lim_{t\to 0^+}\int_t^1 r^{-0.999}\,dr = 1000\lim_{t\to 0^+}\left(1^{0.001}-t^{0.001}\right) = 1000$

9. $\displaystyle\frac{2}{x^2-x} = -\frac{2}{x}+\frac{2}{x-1} \Rightarrow \int_2^\infty \frac{2}{x^2-x}\,dx = \lim_{t\to\infty}\int_2^t -\frac{2}{x}+\frac{2}{x-1}\,dx = 2\lim_{t\to\infty}\left[\ln\left|\frac{x-1}{x}\right|\right]_2^t = \ln 4$

11. $\displaystyle\int_0^{\ln 2} \frac{\sinh x}{\cosh x - 1}\,dx = \lim_{b\to 0^+}\int_b^{\ln 2} \frac{\sinh x}{\cosh x - 1}\,dx = \lim_{b\to 0^+}\left[\ln|\cosh x - 1|\right]_b^{\ln 2} = \infty,\ \text{diverges}$

13. $\displaystyle\int_0^1 \frac{\theta+1}{\sqrt{\theta^2+2\theta}}\,d\theta = \frac{1}{2}\lim_{b\to 0^+}\int_b^1 (\theta^2+2\theta)(2\theta+2)\,d\theta = \frac{1}{2}\lim_{b\to 0^+}\left[2\left(\theta^2+2\theta\right)^{1/2}\right]_b^1 = \sqrt{3},$

    converges

15. $\displaystyle\int_0^1 \ln x\,dx = \lim_{b\to 0^+}\int_b^1 \ln x\,dx = \lim_{b\to 0^+}\left[x\ln x - x\right]_b^1 = -1,\ \text{converges; integration by parts}$

17. $\displaystyle\int_0^\infty x^4 e^{-x}\,dx = \lim_{b\to\infty}\int_0^b x^4 e^{-x}\,dx = \lim_{b\to\infty}\left[\frac{-x^4-4x^3-12x^2-24x-24}{e^x}\right]_0^b = 24,\ \text{converges;}$

    by tabular integration

19. $\displaystyle\int_1^\infty \frac{1}{x^3+1}\,dx\ \text{converges, for}\ \lim_{t\to\infty}\frac{\frac{1}{x^3+1}}{\frac{1}{x^3}} = 1,\ \int_1^\infty \frac{1}{x^3}\,dx\ \text{converges and Theorem 2}$

21. $\displaystyle\int_0^\infty \frac{1}{x^{3/2}+1}\,dx = \int_0^1 \frac{1}{x^{3/2}+1}\,dx + \int_1^\infty \frac{1}{x^{3/2}+1}\,dx;\ \text{the first integral is finite and the second is}$

    dominated by $\displaystyle\int_1^\infty \frac{1}{x^{3/2}}\,dx$ which converges, $\therefore \displaystyle\int_0^\infty \frac{1}{x^{3/2}+1}\,dx$ converges

23. $\displaystyle\int_0^{\pi/2} \tan\theta\,d\theta = -\lim_{t\to\pi/2^-}\int_0^t -\frac{\sin\theta}{\cos\theta}\,d\theta = -\lim_{t\to\pi/2^-}\left(\ln|\cos t| - \ln 1\right) = \infty,\ \text{diverges}$

25. $\displaystyle\int_{-1}^{1} \frac{1}{x^{2/5}}\,dx = 2 \lim_{t \to 0^{+}} \int_{t}^{1} x^{-2/5}\,dx = \frac{10}{3} \lim_{t \to 0^{+}} \left(t^{3/5} - (-1)^{3/5}\right) = \frac{10}{3}$, converges

27. For $v > 2 \Rightarrow v^2 > v - 1 \Rightarrow v > \sqrt{v-1} \Rightarrow \dfrac{1}{v} < \dfrac{1}{\sqrt{v-1}}$ and $\displaystyle\int_{2}^{\infty} \frac{1}{v}\,dv$ diverges $\Rightarrow \displaystyle\int_{2}^{\infty} \frac{1}{\sqrt{v-1}}\,dv$

    also diverges by Theorem 1

29. $\displaystyle\int_{0}^{2} \frac{1}{1-x^2}\,dx = \frac{1}{2}\int_{0}^{2} \frac{1}{1+x}\,dx + \frac{1}{2}\int_{0}^{2} \frac{1}{1-x}\,dx = \frac{1}{2}\int_{0}^{2} \frac{1}{1+x}\,dx + \frac{1}{2}\int_{0}^{1} \frac{1}{1-x}\,dx + \frac{1}{2}\int_{1}^{2} \frac{1}{1-x}\,dx$, from

    example 2 we have that $\displaystyle\int_{0}^{1} -\frac{1}{x}\,dx$, diverges and $\displaystyle\lim_{x \to \infty} \frac{-1/x}{1/(1-x)} = 1 \therefore$ by Theorem 2,

    $\displaystyle\int_{0}^{1} \frac{1}{1-x}\,dx$ diverges $\Rightarrow \displaystyle\int_{0}^{2} \frac{1}{1-x^2}\,dx$ diverges

31. $\displaystyle\int_{1}^{\infty} \frac{dr}{(r-2)^2} = \int_{1}^{2} \frac{dr}{(r-2)^2} + \int_{2}^{3} \frac{dr}{(r-2)^2} + \int_{3}^{\infty} \frac{dr}{(r-2)^2}$, $\displaystyle\int_{2}^{3} \frac{dr}{(r-2)^2} = \lim_{b \to 2^{+}} \int_{b}^{3} (r-2)^{-2}\,dr = $

    $\displaystyle\lim_{b \to 2^{+}} \left[\frac{-1}{r-2}\right]_{b}^{3} = \lim_{b \to 2^{+}} \left(-1 + \frac{1}{b-2}\right) = \infty \Rightarrow \int_{2}^{3} \frac{dr}{(r-2)^2}$ diverges and $\therefore \displaystyle\int_{1}^{\infty} \frac{dr}{(r-2)^2}$ diverges

33. $\displaystyle\int_{0}^{\infty} \frac{1}{\sqrt{x^6+1}}\,dx = \int_{0}^{1} \frac{1}{\sqrt{x^6+1}}\,dx + \int_{1}^{\infty} \frac{1}{\sqrt{x^6+1}}\,dx \le \int_{0}^{1} 1\,dx + \int_{1}^{\infty} \frac{1}{x^3}\,dx = \frac{3}{2}$,

    since $1 \ge \dfrac{1}{\sqrt{x^6+1}}$ and $\dfrac{1}{x^3} \ge \dfrac{1}{\sqrt{x^6+1}}$; $\therefore \displaystyle\int_{0}^{\infty} \frac{1}{\sqrt{x^6+1}}\,dx$ converges by Theorem 1

35. $\displaystyle\int_{0}^{\pi} \frac{\sin\theta}{\sqrt{\pi-\theta}}\,d\theta \le \int_{0}^{\pi} \frac{d\theta}{\sqrt{\pi-\theta}} = \lim_{b \to \pi^{-}} \int_{0}^{b} (\pi-\theta)^{-1/2}\,d\theta = -2 \lim_{b \to \pi^{-}} \left[(\pi-\theta)^{1/2}\right]_{0}^{b} = 2\sqrt{\pi}$,

    converges by the domination test

37. $\dfrac{1}{x} \le \dfrac{2 + \cos x}{x}$ and $\displaystyle\int_{\pi}^{\infty} \frac{1}{x}\,dx$ diverges $\Rightarrow \displaystyle\int_{\pi}^{\infty} \frac{2 + \cos x}{x}\,dx$ diverges by Theorem 1

39. $\displaystyle\int_{6}^{\infty} \frac{1}{\sqrt{\theta+5}}\,d\theta = \lim_{t \to \infty} \int_{6}^{t} (\theta+5)^{-1/2}\,d\theta = 2 \lim_{t \to \infty} \left(\sqrt{t+5} - \sqrt{11}\right) = \infty$, diverges

41. $\displaystyle\int_{2}^{\infty} \frac{2}{t^2-1}\,dt = \lim_{b \to \infty} \int_{2}^{b} \frac{1}{t-1} - \frac{1}{t+1}\,dt = \lim_{b \to \infty} \left[\ln\left|\frac{t-1}{t+1}\right|\right]_{2}^{b} = \lim_{b \to \infty} \left(\ln\left|\frac{b-1}{b+1}\right| - \ln\left(\frac{1}{3}\right)\right) = $

    $-\ln\left(\dfrac{1}{3}\right) = \ln 3$, converges

43. $\dfrac{1}{x} \le \dfrac{1}{\ln x}$ and $\displaystyle\int_{1}^{\infty} \frac{1}{x}\,dx$ diveges $\Rightarrow \displaystyle\int_{1}^{\infty} \frac{1}{\ln x}\,dx$ diverges by Theorem 1

45. $\int_1^\infty \dfrac{1}{e^x}\,dx = \underset{t \to \infty}{\text{Lim}} \int_1^t e^{-x}\,dx = \underset{t \to \infty}{\text{Lim}} \left[-e^{-x}\right]_1^t = 1 \Rightarrow \int_1^\infty \dfrac{1}{e^x}\,dx$ converges and $\underset{t \to \infty}{\text{Lim}} \dfrac{\frac{1}{e^x - 2^x}}{\frac{1}{e^x}} =$

$\underset{t \to \infty}{\text{Lim}} \dfrac{e^x}{e^x - 2^x} = \underset{t \to \infty}{\text{Lim}} \dfrac{1}{1 - (2/e)^t} = 1 \Rightarrow \int_1^\infty \dfrac{1}{e^x - 2^x}\,dx$ converges by Theorem 2

47. $\int_0^\infty \dfrac{1}{\sqrt{x + x^4}}\,dx = \int_0^1 \dfrac{1}{\sqrt{x + x^4}}\,dx + \int_1^\infty \dfrac{1}{\sqrt{x + x^4}}\,dx;\ \dfrac{1}{\sqrt{x + x^4}} < \dfrac{1}{\sqrt{x}}$ and $\int_0^1 \dfrac{1}{\sqrt{x}}\,dx$

converges $\Rightarrow \int_0^1 \dfrac{1}{\sqrt{x + x^4}}\,dx$ converges by Theorem 1 ; $\dfrac{1}{\sqrt{x + x^4}} < \dfrac{1}{x^2}$ and

$\int_1^\infty \dfrac{1}{x^2}\,dx$ converges from remarks following example 5 $\Rightarrow \int_1^\infty \dfrac{1}{\sqrt{x + x^4}}\,dx$ converges by

Theorem 1; $\therefore \int_0^\infty \dfrac{1}{\sqrt{x + x^4}}\,dx$ must also converge

49. $\int_1^\infty \dfrac{d\theta}{\theta^2\left(|\sin \theta| + |\cos \theta|\right)} \le \int_1^\infty \dfrac{d\theta}{\theta^2\left(\sin^2\theta + \cos^2\theta\right)} = \underset{b \to \infty}{\text{Lim}} \int_1^b \theta^{-2}\,d\theta = -\underset{b \to \infty}{\text{Lim}} \left[\dfrac{1}{\theta}\right]_1^b =$

1, converges

51. $\int_{-1}^\infty \dfrac{dx}{x^2 + 5x + 6} = \underset{b \to \infty}{\text{Lim}} \int_{-1}^\infty \dfrac{1}{x + 2} - \dfrac{1}{x + 3}\,dx = \underset{b \to \infty}{\text{Lim}} \left[\ln\left|\dfrac{x + 2}{x + 3}\right|\right]_{-1}^b = \ln 2$

53. $\int_0^\infty \dfrac{dy}{\left(y^2 + 1\right)^2} = \underset{b \to \infty}{\text{Lim}} \int_0^b \dfrac{dy}{\left(y^2 + 1\right)^2} = \underset{b \to \infty}{\text{Lim}} \left[\dfrac{y}{2\left(y^2 + 1\right)} + \dfrac{1}{2}\tan^{-1}y\right]_0^b = \dfrac{\pi}{4}$, integration

by formula 17

55. $\int_3^\infty \dfrac{x + 3}{2x^3 - 8x}\,dx = \underset{b \to \infty}{\text{Lim}} \int_3^b \dfrac{5}{16(x - 2)} - \dfrac{3}{8x} + \dfrac{1}{16(x + 2)}\,dx =$

$\underset{b \to \infty}{\text{Lim}} \left[\dfrac{1}{16}\ln\left|\dfrac{(x - 2)^5(x + 2)}{x^6}\right|\right]_3^b = \dfrac{-\ln\left(\frac{5}{36}\right)}{16}$

57. $\int_0^1 \dfrac{dx}{\sqrt{x^2 + 2x}} = \underset{b \to 0^+}{\text{Lim}} \int_b^1 \dfrac{dx}{\sqrt{(x + 1)^2 - 1}} = \underset{b \to 0^+}{\text{Lim}} \left[\cosh^{-1}(x + 1)\right]_b^1 = \cosh^{-1}2$

59. $\int_0^4 \dfrac{1 - x}{\sqrt{8 + 2x - x^2}}\,dx = -\underset{b \to 3^-}{\text{Lim}} \int_{-1}^b \dfrac{z}{\sqrt{9 - z^2}}\,dz = \underset{b \to 3^-}{\text{Lim}} \left[\left(9 - z^2\right)^{1/2}\right]_{-1}^b = -\sqrt{8}$, where $z = x - 1$

61. $\int_3^\infty e^{-3x}\,dx = -\dfrac{1}{3}\underset{t \to \infty}{\text{Lim}} \left[e^{-3x}\right]_3^t = -\dfrac{1}{3}\underset{t \to \infty}{\text{Lim}} \left(\dfrac{1}{e^{3t}} - \dfrac{1}{e^9}\right) = \dfrac{1}{3\,e^9} \approx 0.000041136$; from the

Calculus Tool Kit we get 0.88617257

63. $\displaystyle\int_1^\infty \frac{1}{x^p}\,dx = \lim_{t \to \infty} \int_1^t \frac{1}{x^p}\,dx = \lim_{t \to \infty}\left[\frac{x^{(1-p)}}{1-p}\right]_1^t = \left(\frac{1}{1-p}\right)\lim_{t \to \infty}\left(\frac{1}{t^{(p-1)}} - 1\right)$

$\displaystyle\therefore \int_1^\infty \frac{1}{x^p}\,dx = \frac{1}{p-1}$ when $p > 1$ and diverges when $p < 1$

65. $\displaystyle A = \lim_{t \to \infty} \int_0^t e^{-x}\,dx = -\lim_{t \to \infty}\left[\frac{1}{e^x}\right]_0^t = -\lim_{t \to \infty}\left(\frac{1}{e^t} - 1\right) = 1$

67. $\displaystyle V = 2\pi \int_0^\infty x\,e^{-x}\,dx = 2\pi$; see exercise 66.

69. $\displaystyle\int_0^{\pi/2}(\sec x - \tan x)\,dx = \lim_{t \to (\pi/2)^-} \int_0^t\left(\sec x - \frac{\sin x}{\cos x}\right)dx =$

$\displaystyle\lim_{t \to (\pi/2)^-}\Big[\ln|\sec x + \tan x| + \ln|\cos x|\Big]_0^t = \lim_{t \to (\pi/2)^-}\ln\left|\frac{\sec t + \tan t}{\sec t}\right| = \lim_{t \to (\pi/2)^-}\ln|1 + \sin t| = \ln 2$

71. a) The statement is true since $\displaystyle\int_{-\infty}^b f(x)\,dx = \int_{-\infty}^a f(x)\,dx + \int_a^b f(x)\,dx,\ \int_b^\infty f(x)\,dx =$

$\displaystyle\int_a^\infty f(x)\,dx - \int_a^b f(x)\,dx$ and $\displaystyle\int_{-\infty}^b f(x)\,dx$ is integrable.

b) $\displaystyle\int_{-\infty}^a f(x)\,dx + \int_a^\infty f(x)\,dx = \int_{-\infty}^a f(x)\,dx + \int_a^b f(x)\,dx - \int_a^b f(x)\,dx + \int_a^\infty f(x)\,dx =$

$\displaystyle\int_{-\infty}^b f(x)\,dx + \int_b^a f(x)\,dx + \int_a^\infty f(x)\,dx = \int_{-\infty}^b f(x)\,dx + \int_b^\infty f(x)\,dx$

73. $\displaystyle\int_1^\infty \frac{1}{\sqrt{1+x^2}}\,dx$ diverges by theorem 2, since $\displaystyle\lim_{x \to \infty}\frac{\dfrac{1}{x}}{\dfrac{1}{\sqrt{x^2+1}}} = \sqrt{\lim_{x \to \infty}\frac{x^2+1}{x^2}} = 1$ and

$\displaystyle\int_1^\infty \frac{1}{x}\,dx$ diverges; $\displaystyle\therefore \int_{-\infty}^\infty \frac{1}{\sqrt{1+x^2}}\,dx$ diverges because $\displaystyle\int_{-\infty}^\infty \frac{1}{\sqrt{1+x^2}}\,dx = \int_{-\infty}^0 \frac{1}{\sqrt{1+x^2}}\,dx +$

$\displaystyle\int_0^1 \frac{1}{\sqrt{1+x^2}}\,dx + \int_1^\infty \frac{1}{\sqrt{1+x^2}}\,dx$

75. $\displaystyle\int_{-\infty}^\infty \frac{1}{e^x + e^{-x}}\,dx = 2\int_0^\infty \frac{e^x}{(e^x)^2 + 1}\,dx = 2\lim_{t \to \infty}\arctan e^x\Big|_0^t = \pi$, converges

77. $\displaystyle\int_{-\infty}^\infty e^{-|x|}\,dx = 2\int_0^\infty e^{-x}\,dx = 2\lim_{b \to \infty}\int_0^b e^{-x}\,dx = -2\lim_{b \to \infty}\left[e^{-x}\right]_0^b = 2$

79. $\displaystyle\int_{-\infty}^\infty \frac{|\sin x| + |\cos x|}{|x| + 1}\,dx = 2\int_0^\infty \frac{|\sin x| + |\cos x|}{x + 1}\,dx \geq 2\int_0^\infty \frac{\sin^2 x + \cos^2 x}{x+1}\,dx = 2\lim_{b \to \infty}\int_0^b \frac{dx}{x+1}\,dx =$

$\displaystyle 2\lim_{b \to \infty}\Big[\ln|x+1|\Big]_0^b = \infty$, diverges $\displaystyle\therefore \int_{-\infty}^\infty \frac{|\sin x| + |\cos x|}{|x| + 1}\,dx$ diverges

81. $\int_1^\infty \dfrac{2x}{1+x^2}\,dx$ diverges by theorem 2, since $\underset{x\to\infty}{\text{Lim}} \dfrac{\frac{1}{x}}{\frac{2x}{x^2+1}} = \sqrt{\underset{x\to\infty}{\text{Lim}}\dfrac{x^2+1}{2x^2}} = \dfrac{1}{\sqrt{2}}$ and

$\int_1^\infty \dfrac{1}{x}\,dx$ diverges; $\therefore \int_0^\infty \dfrac{2x}{1+x^2}\,dx$ diverges because $\int_0^\infty \dfrac{2x}{1+x^2}\,dx = \int_0^1 \dfrac{2x}{1+x^2}\,dx +$

$\int_1^\infty \dfrac{2x}{1+x^2}\,dx;\ \int_{-\infty}^\infty \dfrac{2x}{1+x^2}\,dx = \underset{t\to\infty}{\text{Lim}}\ \int_{-t}^t \dfrac{2x}{1+x^2}\,dx = \underset{t\to\infty}{\text{Lim}}\ \ln\left|1+x^2\right|\Big|_{-t}^t =$

$\underset{t\to\infty}{\text{Lim}}\ \left[\ln\left|1+(-t)^2\right| - \ln\left|1+(t)^2\right|\right] = \underset{t\to\infty}{\text{Lim}}\ \ln\left[\dfrac{1+t^2}{1+t^2}\right] = \ln(1) = 0$

83. a) $\underset{b\to\infty}{\text{Lim}} \int_{-b}^b \sin x\,dx = 0$, integrand odd

 b) $\underset{b\to 0^+}{\text{Lim}} \int_b^1 \dfrac{\cos t}{t^2}\,dt \geq \underset{b\to 0^+}{\text{Lim}} \int_b^1 \dfrac{\cos 1}{t^2}\,dt = -(\cos 1)\underset{b\to 0^+}{\text{Lim}}\left[\dfrac{1}{t}\right]_b^1 =$

$-(\cos 1)\underset{b\to 0^+}{\text{Lim}}\left[1-\dfrac{1}{b}\right] = \infty \Rightarrow \underset{b\to 0^+}{\text{Lim}} \int_b^1 \dfrac{\cos t}{t^2}\,dt = \infty \therefore \underset{x\to 0^+}{\text{Lim}}\ x \int_x^1 \dfrac{\cos t}{t^2}\,dt =$

$\underset{x\to 0^+}{\text{Lim}}\ \dfrac{\int_x^1 \dfrac{\cos t}{t^2}\,dt}{\dfrac{1}{x}} = \underset{x\to 0^+}{\text{Lim}}\ \dfrac{-\dfrac{\cos x}{x^2}}{-\dfrac{1}{x^2}} = -\underset{x\to 0^+}{\text{Lim}}\ \cos x = 1$

7.M MISCELLANEOUS EXERCISES

1. $\int \dfrac{\cos x}{\sqrt{1+\sin x}}\,dx = 2\sqrt{1+\sin x} + C$

3. $\int \dfrac{\tan x}{\cos^2 x}\,dx = \int \tan x \sec^2 x\,dx = \dfrac{\tan^2 x}{2} + C$

5. $\int_0^9 e^{\ln\sqrt{x}}\,dx = \int_0^9 \sqrt{x}\,dx = \left[\dfrac{2}{3}x^{3/2}\right]_0^9 = 18$

7. $\int \dfrac{dx}{\sqrt{x^2+2x+2}} = \int \dfrac{dx}{\sqrt{(x+1)^2+1}} = \int \sec u\,du = \ln|\sec u + \tan u| + C =$

$\ln\left|\sqrt{x^2+2x+2}+x+1\right| + C$, where $\tan u = x+1$

9. $\displaystyle\int x^2 e^x \, dx = e^x \left(x^2 - 2x + 2\right) + C$

$$
\begin{array}{ccc}
x^2 & + & e^x \\
2x & - & e^x \\
2 & + & e^x \\
0 & & e^x
\end{array}
$$

11. $\displaystyle\int \frac{dr}{1 + \sqrt{r}} = \int \frac{2z}{1 + z} \, dz = \int \left(2 - \frac{2}{1 + z}\right) dz = 2z - 2\ln|1 + z| + C = 2\sqrt{r} - 2\ln\left|1 + \sqrt{r}\right| + C,$

where $z = \sqrt{r}$

13. $\displaystyle\int \frac{4x^3 - 20x}{x^4 - 10x^2 + 9} \, dx = \int \frac{1}{x - 1} + \frac{1}{x + 1} + \frac{1}{x - 3} + \frac{1}{x + 3} \, dx = \ln\left|x^4 - 10x^2 + 9\right| + C$

15. $\displaystyle\int t^{2/3}\left(t^{5/3} + 1\right)^{2/3} dt = \frac{3}{5}\int u^{2/3} \, du = \frac{9}{25} u^{5/3} + C = \frac{9}{25}\left(t^{5/3} + 1\right)^{5/3} + C,$ where $u = t^{5/3} + 1$

17. $\displaystyle\int \frac{dt}{\sqrt{e^t + 1}} = \int \frac{2u \, du}{u(u^2 - 1)} = \int \left(\frac{1}{u - 1} - \frac{1}{u + 1}\right) du = \ln\left|\frac{u - 1}{u + 1}\right| + C,$ where $u^2 = e^t + 1$

19. $\displaystyle\int_0^{\pi/2} \frac{\cos x}{1 + \sin^2 x} \, dx = \left[\tan^{-1}(\sin x)\right]_0^{\pi/2} = \frac{\pi}{4}$

21. $\displaystyle\int_0^{\pi/2} \frac{\sin x}{1 + \cos^2 x} \, dx = -\left[\tan^{-1}(\cos x)\right]_0^{\pi/2} = \frac{\pi}{4}$

23. $\displaystyle\int_{\pi/4}^{\pi/3} \frac{1}{\sin x \cos x} \, dx = 2\int_{\pi/4}^{\pi/3} \csc(2x) \, dx = -\left[\ln|\csc 2x + \cot 2x|\right]_{\pi/4}^{\pi/3} = \ln\sqrt{3}$

25. $\displaystyle\int_{-\pi/2}^0 \sqrt{1 - \sin x} \, dx = \int_{-\pi/2}^0 \sqrt{1 - \sin x}\,\frac{\sqrt{1 + \sin x}}{\sqrt{1 + \sin x}} \, dx = \int_{-\pi/2}^0 \frac{\cos x}{\sqrt{1 + \sin x}} \, dx = \left[2\sqrt{1 + \sin x}\right]_{-\pi/2}^0 = 2$

27. $\displaystyle\int \frac{e^{2x}}{\sqrt[3]{1 + e^x}} \, dx = \int \left(3u^4 - 3u\right) du = \frac{3u^5}{5} - \frac{3u^2}{2} + C = \frac{3}{5}(1 + e^x)^{5/3} - \frac{3}{2}(1 + e^x)^{2/3} + C,$

where $u = \sqrt[3]{1 + e^x}$

29. $\displaystyle\int_{5/4}^{5/3} \frac{dx}{(x^2 - 1)^{3/2}} = \int_{\text{arcsec } 5/4}^{\text{arcsec } 5/3} \frac{\sec u \tan u}{(\tan^2 u)^{3/2}} \, du = \int_{\text{arcsec } 5/4}^{\text{arcsec } 5/3} \sin^{-2} u \cos u \, du =$

$\left[-\csc u\right]_{\text{arcsec } 5/4}^{\text{arcsec } 5/3} = \dfrac{5}{12}$

31. $\displaystyle\int \frac{dx}{x(2 + \ln x)} = \int \frac{du}{u} = \ln|u| + C = \ln|2 + \ln x| + C,$ where $u = 2 + \ln x$

33. $$\int_0^{\ln 2} \frac{e^{2x}}{\sqrt[4]{e^x + 1}}\, dx = \int_{\sqrt[4]{2}}^{\sqrt[4]{3}} \left(4u^6 - 4u^2\right) du = \left[\frac{4\,u^7}{7} - \frac{4\,u^3}{3}\right]_{\sqrt[4]{2}}^{\sqrt[4]{3}} = \frac{4}{21}\left(\sqrt[4]{8} + 2\sqrt[4]{27}\right)$$

35. $$\int_0^3 (16 + x^2)^{-3/2}\, dx = \int_0^{\arctan 3/4} (16 + 16\tan^2 u)^{-3/2}\, 4\sec^2 u\, du = \frac{1}{16}\int_0^{\arctan 3/4} \cos u\, du =$$

$$\left[\frac{\sin u}{16}\right]_0^{\arctan 3/4} = \frac{3}{80}, \text{ where } x = 4 \tan u$$

37. $$\int \sin\sqrt{x+1}\, dx = \int 2u\sin u\, du = -2u\cos u + 2\sin u + C = -2\sqrt{x+1}\,\cos\sqrt{x+1} + 2\sin\sqrt{x+1} + C,$$

where $u = \sqrt{x+1}$; integration by parts

39. $$\int_0^1 \frac{1}{4 - x^2}\, dx = \frac{1}{4}\int_0^1 \left(\frac{1}{x+2} - \frac{1}{x-2}\right) dx = \frac{1}{4}\left[\ln\left|\frac{x+2}{x-2}\right|\right]_0^1 = \frac{\ln 3}{4}$$

41. $$\int \frac{dy}{y(2y^3 + 1)^2} = \int \left(\frac{1}{y} - \frac{2y^2}{2y^3 + 1} - \frac{2y^2}{(2y^3 + 1)^2}\right) dy = \ln|y| - \frac{1}{3}\ln\left|2y^3 + 1\right| + \frac{1}{3}(2y^3 + 1)^{-1} + C,$$

partial fractions

43. $$\int \frac{dx}{x(x^2 + 1)^2} = \int \frac{\sec^2 u\, du}{(\tan u)(\sec^2 u)^2} = \int \frac{(\cos u)(1 - \sin^2 u)}{\sin u}\, du = \ln|\sin u| + \frac{\cos^2 u}{2} + C =$$

$$\ln\left|\frac{x}{\sqrt{x^2 + 1}}\right| + \frac{1}{2}(x^2 + 1)^{-1} + C, \text{ where } x = \tan u$$

45. $$\int \frac{1}{e^x - 1}\, dx = \int \frac{1}{u(u-1)}\, du = \int \left(\frac{1}{u-1} - \frac{1}{u}\right) du = \ln\left|\frac{u-1}{u}\right| + C = \ln\left|1 - e^{-x}\right| + C, \text{ where } u = e^x$$

47. $$\int \frac{x\, dx}{x^2 + 4x + 3} = \int \left(\frac{3/2}{x+3} - \frac{1/2}{x+1}\right) dx = \frac{1}{2}\ln\left|\frac{(x+3)^3}{x+1}\right| + C$$

49. $$\int \frac{4}{x^3 + 4x}\, dx = \int \left(\frac{1}{x} - \frac{x}{x^2 + 4}\right) dx = \ln\left|\frac{x}{\sqrt{x^2 + 4}}\right| + C$$

51. $$\int \frac{\sqrt{x^2 - 1}}{x}\, dx = \int \frac{\sqrt{\sec^2 u - 1}\,\sec u\tan u}{\sec u}\, du = \int \left(\sec^2 u - 1\right) dx = \tan u - u + C =$$

$$\sqrt{x^2 - 1} - \sec^{-1}|x| + C, \text{ where } x = \sec u$$

53. $$\int \frac{dx}{x(3\sqrt{x} + 1)} = \int \frac{2u}{u^2(3u + 1)}\, du = \int \left(\frac{2}{u} - \frac{6}{3u + 1}\right) du = \ln\left(\frac{u^2}{(3u + 1)^2}\right) + C = \ln\left(\frac{x}{(3\sqrt{x} + 1)^2}\right) + C,$$

where $u = \sqrt{x}$

55. $$\int_{\pi/6}^{\pi/2} \frac{\cot\theta}{1 + \sin^2\theta}\, d\theta = \int_{\pi/6}^{\pi/2} \frac{\cos\theta\, d\theta}{(\sin\theta)(1 + \sin^2\theta)} = \int_{\arctan 1/2}^{\pi/4} \frac{\cos\theta}{\sin\theta}\, d\theta =$$

$$\Big[\ln|\sin\theta|\Big]_{\text{arcstan } 1/2}^{\pi/4} = \ln\sqrt{\frac{5}{2}}, \text{ where } \tan u = \sin\theta$$

57. $\displaystyle \int \frac{e^{4t}}{(1+e^{2t})^{2/3}}\,dt = \frac{3}{2}\int \frac{(u^3-1)\,u^2}{(u^3)^{2/3}}\,du = \frac{3}{2}\int \left(u^3-1\right)du = \frac{3\,u^4}{8} - \frac{3u}{2} + C =$

$\displaystyle \frac{3(e^{2t}+1)^{4/3}}{8} - \frac{3(e^{2t}+1)^{1/3}}{2} + C, \text{ where } u^3 = 1 + e^{2t}$

59. $\displaystyle \int \frac{x^3+x^2}{x^2+x-2}\,dx = \int \left(x + \frac{4/3}{x+2} + \frac{2/3}{x-1}\right)dx = \frac{x^2}{2} + \frac{1}{3}\ln\left((x+2)^4\,(x-1)^2\right) + C$

61. $\displaystyle \int \frac{x}{(x-1)^2}\,dx = \int \left(\frac{1}{x-1} + \frac{1}{(x-1)^2}\right)dx = \ln|x-1| - \frac{1}{x-1} + C$

63. $\displaystyle \int \frac{dy}{(2y+1)\sqrt{y^2+y}} = \int \frac{dy}{(2y+1)\sqrt{(y+1/2)^2 - 1/4}} = \int \frac{\frac{1}{2}\sec u \tan u}{\frac{1}{2}\sec u \tan u}\,du = u + C =$

$\displaystyle \sec^{-1}(2y+1) + C, \text{ where } y + \frac{1}{2} = \frac{\sec u}{2}$

65. $\displaystyle \int (1-x^2)^{3/2}\,dx = \int \cos^4 u\,du = \frac{1}{4}\int \left(\frac{3}{2} + 2\cos 2u + \frac{\cos 4u}{2}\right)du = \frac{3u}{8} + \frac{\sin 2u}{4} + \frac{\sin 4u}{32} + C =$

$\displaystyle \frac{3\sin^{-1}x}{8} + \frac{x\sqrt{1-x^2}}{2} + \frac{x\sqrt{1-x^2}\left(1-2x^2\right)}{8} + C, \text{ where } x = \sin u$

67. $\displaystyle \int p\tan^2 p\,dp = \int p\sec^2 p\,dp - \int p\,dp = \int p\sec^2 p\,dp - \frac{p^2}{2} = p\tan p + \ln|\cos p| - \frac{p^2}{2} + C$

$$
\begin{array}{lll}
p & + & \sec^2 p \\
1 & - & \tan p \\
0 & & -\ln|\cos p|
\end{array}
$$

69. $\displaystyle \int_0^\pi x^2 \sin x\,dx = \left[-x^2\cos x + 2x\sin x + 2\cos x\right]_0^\pi = \pi^2 - 4$

$$
\begin{array}{lll}
x^2 & + & \sin x \\
2x & - & -\cos x \\
2 & + & -\sin x \\
0 & & \cos x
\end{array}
$$

71. $\displaystyle \int_0^1 \frac{dt}{t^4+4t^2+3}\,dt = \frac{1}{2}\int_0^1 \left(\frac{1}{t^2+1} - \frac{1}{t^2+3}\right)dt = \left[\frac{1}{2}\tan^{-1}t - \frac{\sqrt{3}}{6}\tan^{-1}\frac{t}{\sqrt{3}}\right]_0^1 = \frac{(9-2\sqrt{3})\pi}{72}$

73. Let $u = \ln(x + 2)$ and $dv = x\,dx \Rightarrow du = \dfrac{dx}{x + 2}$, $v = \dfrac{x^2}{2}$ and $\displaystyle\int x \ln\sqrt{x + 2}\,dx =$

$\dfrac{1}{2}\left(\dfrac{x^2 \ln|x + 2|}{2} - \dfrac{1}{2}\displaystyle\int \dfrac{x^2}{x + 2}\,dx\right) = \dfrac{x^2 \ln|x + 2|}{4} - \dfrac{1}{4}\displaystyle\int \left(x - 2 + \dfrac{4}{x + 2}\right) dx =$

$\dfrac{x^2 \ln|x + 2|}{4} - \dfrac{x^2}{8} + \dfrac{x}{2} - \ln|x + 2| + C. \therefore \displaystyle\int_0^2 x \ln\sqrt{x + 2}\,dx = \left[\dfrac{x^2 \ln|x + 2|}{4} - \dfrac{x^2}{8} + \dfrac{x}{2} - \ln|x + 2|\right]_0^2 =$

$\dfrac{1}{2} + \ln 2$

75. Let $u = \sec^{-1} q$ and $dv = dq \Rightarrow du = \dfrac{dq}{q\sqrt{q^2 - 1}}$, $v = q$ and $\displaystyle\int \sec^{-1} q\,dq = q \sec^{-1} q - \displaystyle\int \dfrac{dq}{\sqrt{q^2 - 1}} =$

$q \sec^{-1} q - \displaystyle\int \dfrac{\sec u \tan u}{\tan u}\,du = q \sec^{-1} q - \ln|\sec u + \tan u| + C = q \sec^{-1} q - \ln\left|q + \sqrt{q^2 - 1}\right| + C,$

where $u = \sec^{-1} q$

77. $\displaystyle\int \dfrac{x}{x^4 - 16}\,dx = \dfrac{1}{16}\displaystyle\int \left(\dfrac{1}{x - 2} + \dfrac{1}{x + 2} - \dfrac{x}{x^2 + 4}\right) dx = \dfrac{1}{16} \ln\left|\dfrac{x^2 - 4}{x^2 + 4}\right| + C$, partial fractions

79. $\displaystyle\int \dfrac{\cos x}{\sin^3 x - \sin x}\,dx = \displaystyle\int \dfrac{-\cos x}{(\sin x)(\cos^2 x)}\,dx = \displaystyle\int -2 \csc 2x\,dx = \ln|\csc 2x + \cot 2x| + C$

81. $\displaystyle\int_0^1 \dfrac{u}{1 + \sqrt{u} + u}\,du = \displaystyle\int_0^1 \dfrac{2z^3}{z^2 + z + 1}\,dz = 2\displaystyle\int_0^1 \left(z - 1 + \dfrac{1}{(z + 1/2)^2 + 3/4}\right) dz =$

$\left[z^2 - 2z + \dfrac{4}{\sqrt{3}} \tan^{-1}\left(\dfrac{2z + 1}{\sqrt{3}}\right)\right]_0^1 = \dfrac{2\pi\sqrt{3} - 9}{9}$, where $z = \sqrt{u}$

83. $\displaystyle\int \dfrac{dt}{\sec^2 t + \tan^2 t} = \displaystyle\int \dfrac{dt}{2\tan^2 t + 1} = \displaystyle\int \dfrac{1}{2\tan^2 t + 1} \dfrac{\sec^2 t}{1 + \tan^2 t}\,dt = \displaystyle\int \dfrac{du}{(2u^2 + 1)(u^2 + 1)} =$

$\displaystyle\int \left(\dfrac{1}{u^2 + 1/2} - \dfrac{1}{u^2 + 1}\right) du = \sqrt{2} \tan^{-1}(\sqrt{2}u) - \tan^{-1} u + C = \sqrt{2} \tan^{-1}\left(\sqrt{2} \tan t\right) - t + C,$

where $u = \tan t$

85. Let $u = e^t$ and $dv = e^t \cos(e^t)\,dt \Rightarrow du = e^t\,dt$, $v = \sin(e^t)$ and $\displaystyle\int e^{2t} \cos(e^t)\,dt =$

$e^t \sin(e^t) - \displaystyle\int e^t \sin(e^t)\,dt = e^t \sin(e^t) + \cos(e^t) + C$

87. Let $u = \ln(x^3 + x)$ and $dv = x\,dx \Rightarrow du = \dfrac{3x^2 + 1}{x^3 + x}\,dx$, $v = \dfrac{x^2}{2}$ and $\displaystyle\int x \ln(x^3 + x)\,dx =$

$\dfrac{x^2 \ln(x^3 + x)}{2} - \dfrac{1}{2}\displaystyle\int \left(3x - \dfrac{2x}{x^2 + 1}\right) dx = \dfrac{x^2 \ln(x^3 + x)}{2} - \dfrac{3x^2}{4} + \dfrac{\ln(x^2 + 1)}{2} + C$

89. $\displaystyle\int \dfrac{\cos x}{\sqrt{4 - \cos^2 x}}\,dx = \displaystyle\int \dfrac{\cos x}{\sqrt{3 + \sin^2 x}}\,dx = \displaystyle\int \dfrac{\sqrt{3} \sec^2 u\,du}{\sqrt{3 + 3\tan^2 u}} = \displaystyle\int \sec u\,du = \ln|\sec u + \tan u| + C =$

$\ln\left|\sqrt{3 + \sin^2 x} + \sin x\right| + C$, where $\sin x = \sqrt{3} \tan u$

91. $\int x^2 \sin(1-x)\, dx = x^2 \cos(1-x) + 2x \sin(1-x) - 2\cos(1-x) + C$

$$
\begin{array}{ll}
x^2 & + \quad \sin(1-x) \\
2x & - \quad \cos(1-x) \\
2 & + \quad -\sin(1-x) \\
0 & \quad\; -\cos(1-x)
\end{array}
$$

93. $\int \dfrac{1}{\cot^3 x}\, dx = \int \tan^3 x\, dx = \int \left(\tan x \sec^2 x - \tan x\right) dx = \dfrac{\tan^2 x}{2} + \ln|\cos x| + C$

95. $\int \dfrac{x^3}{(x^2+1)^2}\, dx = \int \dfrac{\tan^3 u \sec^2 u}{(\tan^2 u +)^2}\, du = \int \dfrac{(1-\cos^2 u)\sin u}{\cos u}\, du = \int \tan u - \cos u \sin u\, du =$

$\ln|\sec u| + \dfrac{\cos^2 u}{2} + C = \ln\left(\sqrt{x^2+1}\right) + \dfrac{1}{2(x^2+1)} + C$, where $x = \tan u$

97. $\displaystyle\int_0^4 x\sqrt{2x+1}\, dx = \int_1^3 \dfrac{(u^2-1)\,u^2}{2}\, du = \dfrac{1}{2}\int_1^3 \left(u^4 - u^2\right) du = \left[\dfrac{u^5}{10} - \dfrac{u^3}{6}\right]_1^3 = \dfrac{298}{15}$, where $2x+1 = u^2$

99. Let $u = \ln\left(x - \sqrt{x^2-1}\right)$ and $dv = dx \Rightarrow du = -\dfrac{dx}{\sqrt{x^2-1}}$, $v = x$ and $\int \ln\left(x - \sqrt{x^2-1}\right) dx =$

$x\ln\left(x - \sqrt{x^2-1}\right) + \int \dfrac{x}{\sqrt{x^2-1}}\, dx = \ln\left(x - \sqrt{x^2-1}\right) + \sqrt{x^2-1} + C$

101. $\int \ln(x + \sqrt{x})\, dx = \int 2z \ln(z^2 + z)\, dz$, where $z = \sqrt{x}$. Let $u = \ln(z^2 + z)$ and $dv = 2z\, dz \Rightarrow$

$du = \dfrac{2z+1}{z^2+z}\, dz$, $v = z^2$ and $\int 2z \ln(z^2 + z)\, dz = z^2 \ln(z^2 + z) - \int \dfrac{2z^2 + z}{z+1}\, dz = z^2 \ln(z^2 + z) -$

$\int 2z - 1 + \dfrac{1}{z+1}\, dz = z^2 \ln(z^2 + z) - z^2 + z - \ln|z+1| + C = x\ln(x + \sqrt{x}) - x + \sqrt{x} - \ln\left|\sqrt{x} + 1\right| + C$

103. Let $u = 4\sec^{-1}\sqrt{x}$ and $dv = dx \Rightarrow du = \dfrac{2x^{-1/2}}{\sqrt{x}\sqrt{x-1}}\, dx$, $v = x$ and $\displaystyle\int_2^4 4\sec^{-1}\sqrt{x}\, dx =$

$4\left[x \sec\sqrt{x}\right]_2^4 - 2\int_2^4 (x-1)^{-1/2}\, dx = \left[4x\sec^{-1}\sqrt{x} - 4\sqrt{x-1}\right]_2^4 = \dfrac{10\pi}{3} - 4\sqrt{3} + 4$

105. $\displaystyle\int_0^{\pi^2/4} \sin\sqrt{x}\, dx = \int_0^{\pi/2} 2u \sin u\, du = \left[-2u\cos u + 2\sin u\right]_0^{\pi/2} = 2$, where $u = \sqrt{x}$

$$
\begin{array}{ll}
2u & + \quad \sin u \\
2 & - \quad -\cos u \\
0 & \quad\; -\sin u
\end{array}
$$

107. $\displaystyle\int \sqrt{1-x^2}\,\sin^{-1}x\,dx = \int z\cos^2 z\,dz = \frac{z}{2}\left(z + \frac{\sin 2z}{2}\right) - \frac{1}{2}\left(\frac{z^2}{2} - \frac{\cos 2z}{4}\right) + C =$

$$
\begin{array}{ll}
z & + \quad (1 + \cos 2z)/2 \\
1 & - \quad (z + (\sin 2z)/2)/2 \\
0 & \quad\ \ (z^2/2 - (\cos 2z)/4)/2
\end{array}
$$

$\displaystyle\frac{z^2}{4} + \frac{z\sin 2z}{4} + \frac{\cos 2z}{8} + C = \frac{(\sin^{-1}x)^2}{4} + \frac{(\sin^{-1}x)(\sin(2\sin^{-1}x))}{4} + \frac{\cos(2\sin^{-1}x)}{8} + C = \frac{(\sin^{-1}x)^2}{4} +$

$\displaystyle\frac{x\sin^{-1}(x)\sqrt{1-x^2}}{2} - \frac{x^2}{4} + C$, where $x = \sin z$

109. $\displaystyle\int \frac{\tan x}{\tan x + \sec x}\,dx = \int \frac{\tan x}{\tan x + \sec x}\,\frac{\tan x - \sec x}{\tan x - \sec x}\,dx = \int \left(-\tan^2 x + \tan x \sec x\right)dx =$

$\displaystyle\int \left(1 - \sec^2 x + \tan x \sec x\right)dx = x - \tan x + \sec x + C$

111. If $u = \sin x$, then $du = \cos x\,dx$, $\displaystyle\int \frac{dx}{(\cos^2 x + 4\sin x - 5)\cos x} = \int \frac{-\cos x\,dx}{(\sin^2 x - 4\sin x + 4)\cos^2 x} =$

$\displaystyle\int \frac{-\cos\,dx}{(\sin x - 2)^2(1 - \sin^2 x)} = -\int \frac{du}{(u-2)^2(1-u^2)} = \int \left(\frac{1/3}{(u-2)^2} - \frac{4/9}{u-2} + \frac{1/2}{u-1} - \frac{1/18}{u+1}\right)dx =$

$\displaystyle\frac{-1}{3(u-2)} - \frac{4}{9}\ln|u-2| + \frac{1}{2}\ln|u-1| - \frac{1}{18}\ln|u+1| + C = \frac{-1}{3(\sin x - 2)} - \frac{4}{9}\ln|\sin x - 2| + \frac{1}{2}\ln|\sin x - 1| -$

$\displaystyle\frac{1}{18}\ln|\sin x + 1| + C$, where $u = \sin x$

113. $\displaystyle\int \sqrt{\frac{1-\cos x}{\cos \alpha - \cos x}}\,dx = 2\int \frac{\sqrt{1 - \cos 2z}}{\sqrt{(1 + \cos \alpha) - (1 + \cos 2z)}}\,dz = 2\int \frac{\sqrt{2\sin^2 z}}{\sqrt{2\cos^2\left(\frac{\alpha}{2}\right) - 2\cos^2 z}}\,dz =$

$\displaystyle 2\int \frac{\dfrac{\sin z}{\cos\left(\frac{\alpha}{2}\right)}}{\sqrt{1 - \left(\dfrac{\cos z}{\cos\left(\frac{\alpha}{2}\right)}\right)^2}}\,dz = -2\sin^{-1}\left(\frac{\cos z}{\cos\left(\frac{\alpha}{2}\right)}\right) + C = -2\sin^{-1}\left(\frac{\cos\left(\frac{x}{2}\right)}{\cos\left(\frac{\alpha}{2}\right)}\right) + C$, where $x = 2z$

115. If $u = \ln(2x^2 + 4)$ and $dv = dx$, then $du = \dfrac{2x\,dx}{x^2 + 4}$, $v = x$ and $\displaystyle\int_0^1 \ln(2x^2 + 4)\,dx = \Big[x\ln(2x^2 + 4)\Big]_0^1 -$

$\displaystyle\int_0^1 \left(2 - \frac{4}{x^2 + 2}\right)dx = \left[x\ln(2x^2 + 4) - 2x + 2\sqrt{2}\tan^{-1}\left(\frac{x}{\sqrt 2}\right)\right]_0^1 = \ln 6 - 2 + 2\sqrt{2}\tan^{-1}\left(\frac{1}{\sqrt 2}\right)$

117. $\displaystyle\int \frac{dt}{\sqrt{1 - e^{-t}}} = \int \left(\frac{1}{1-u} + \frac{1}{1+u}\right)du = \ln\left|\frac{1+u}{1-u}\right| + C = \ln\left|\frac{1 + \sqrt{1 - e^{-t}}}{1 - \sqrt{1 - e^{-t}}}\right| + C$

119. If $u = \int_0^x \sqrt{1 + (t-1)^4}\, dt$ and $dv = 3(x-1)^2 dx$, then $du = \sqrt{1 + (x-1)^4}\, dx$, $v = (x-1)^3$ and

$$\int_0^1 3(x-1)^2 \left[\int_0^x \sqrt{1 + (t-1)^4}\,(x-1)^3\right] dx = \left[(x-1)^3 \int_0^x \sqrt{1 + (t-1)^4}\, dt\right]_0^1 -$$

$$\int_0^1 \sqrt{1 + (x-1)^4}\,(x-1)^3\, dx = \left[-\frac{1}{6}\left(1 + (x-1)^4\right)^{3/2}\right]_0^1 = \frac{\sqrt{8} - 1}{6}$$

121. $\displaystyle\int_3^\infty \frac{8x + 24}{x(x^2 - 4)}\, dx = \lim_{b \to \infty} \int_3^b \left(\frac{5}{x-2} - \frac{6}{x} + \frac{1}{x+2}\right) dx = \lim_{b \to \infty} \left[\ln\left|\frac{(x-2)^5(x+2)}{x^6}\right|\right]_3^b =$

$$\lim_{b \to \infty} \left(\ln\left(\frac{(b-2)^5(b+2)}{b^6}\right) - \ln\left(\frac{(1)(5)}{729}\right)\right) = \ln\left(\frac{729}{5}\right)$$

123. If $u = y^2$ and $dv = y\, e^{-y^2}\, dy$, then $du = 2y\, dy$, $v = -\dfrac{e^{-y^2}}{2}$ and $\displaystyle\int_0^\infty y^3 e^{-y^2}\, dy =$

$$\lim_{b \to \infty} \left(\left[-\frac{y^2 e^{-y^2}}{2}\right]_0^b + \int_0^b y\, e^{-y^2}\, dy\right) = \lim_{b \to \infty}\left[-\frac{y^2 e^{-y^2}}{2} - \frac{e^{-y^2}}{2}\right]_0^b = -\frac{1}{2}\lim_{b \to \infty}\left[\frac{b+1}{e^{b^2}} - 1\right] = \frac{1}{2}$$

125. If $u = x\left(\sin^{-1}x\right)^2$ and $dv = dx$, then $du = \dfrac{2\sin^{-1}x\, dx}{\sqrt{1 - x^2}}$, $v = x$ and $\displaystyle\int \left(\sin^{-1}x\right)^2 dx = x\left(\sin^{-1}x\right)^2 -$

$$\int \frac{2x \sin^{-1}x}{\sqrt{1-x^2}}\, dx. \text{ Let } u = \sin^{-1}x \text{ and } dv = -\frac{2x\, dx}{\sqrt{1-x^2}}, \text{ then } du = \frac{1}{\sqrt{1-x^2}}, v = 2\sqrt{1-x^2} \text{ and}$$

$$\int \frac{2x \sin^{-1}x}{\sqrt{1-x^2}}\, dx = 2\left(\sin^{-1}x\right)\sqrt{1-x^2} - 2x + C. \quad \therefore \int \left(\sin^{-1}x\right)^2 dx = x\left(\sin^{-1}x\right)^2 +$$

$$2\left(\sin^{-1}x\right)\sqrt{1-x^2} - 2x + C.$$

127. If $u = \sin^{-1}x$ and $dv = x\, dx$, then $du = \dfrac{dx}{\sqrt{1-x^2}}$, $v = \dfrac{x^2}{2}$ and $\displaystyle\int x \sin^{-1}x\, dx = \frac{x^2 \sin^{-1}x}{2} -$

$$\frac{1}{2}\int \frac{x^2}{\sqrt{1-x^2}}\, dx. \text{ Let } x = \sin v, \text{ then } dx = \cos v\, dv \text{ and } -\frac{1}{2}\int \frac{x^2}{\sqrt{1-x^2}}\, dx = -\frac{1}{2}\int \sin^2 v\, dv =$$

$$-\frac{1}{2}\int \left(\frac{1 - \cos 2v}{2}\right) dv = \frac{-v + \sin v \cos v}{4}. \quad \therefore \int x \sin^{-1}x\, dx = \frac{x^2 \sin^{-1}x}{2} + \frac{x\sqrt{1-x^2} - \sin^{-1}x}{4} + C$$

129. $\displaystyle\int \frac{d\theta}{1 - \tan^2\theta} = \int \frac{\cos^2\theta}{\cos^2\theta - \sin^2\theta}\, d\theta = \int \frac{1 + \cos 2\theta}{2\cos 2\theta}\, d\theta = \frac{1}{2}\int \left(\sec 2\theta + 1\right) d\theta =$

$$\frac{\ln\left|\sec 2\theta + \tan 2\theta\right| + 2\theta}{4} + C$$

131. $\displaystyle\int \frac{dt}{t - \sqrt{1 - t^2}} = \int \frac{\cos x}{\sin x - \cos x}\, dx = \int \frac{dx}{\tan x - 1}$, where $t = \sin x$. $\displaystyle\int \frac{dx}{\tan x - 1} =$

$\displaystyle\int \frac{dh}{(h - 1)(h^2 + 1)} = \frac{1}{2}\int \left(\frac{1}{h - 1} - \frac{1}{1 + h^2} - \frac{h}{1 + h^2} \right) dh = \frac{1}{2}\left(\ln\left| \frac{h - 1}{\sqrt{1 + h^2}} \right| - \tan^{-1}h \right) + C =$

$\dfrac{1}{2}\left(\ln\left| \dfrac{\tan x - 1}{\sec x} \right| - x \right) + C = \dfrac{1}{2}\left(\ln\left(t - \sqrt{1 - t^2} \right) - \sin^{-1}t \right) + C$, where $h = \tan x$.

133. $\displaystyle\int \frac{1}{x^4 + 4}\, dx = \int \frac{1}{(x^2 + 2)^2 - 4x^2}\, dx = \int \frac{1}{(x^2 + 2x + 2)(x^2 - 2x + 2)}\, dx =$

$\dfrac{1}{16}\displaystyle\int \left(\frac{2x + 2}{x^2 + 2x + 2} + \frac{2}{(x + 1)^2 + 1} - \frac{2x - 2}{x^2 - 2x + 2} + \frac{2}{(x - 1)^2 + 1} \right) dx =$

$\dfrac{1}{16} \ln\left| \dfrac{x^2 + 2x + 2}{x^2 - 2x + 2} \right| + \dfrac{1}{8}\left(\tan^{-1}(x + 1) + \tan^{-1}(x - 1) \right) + C$

135. $\displaystyle\lim_{n \to \infty} \sum_{k = 1}^{n} \ln \sqrt[n]{1 + \frac{k}{n}} = \lim_{n \to \infty} \sum_{k = 1}^{n} \ln\left(1 + k\left(\frac{1}{n} \right) \right)\frac{1}{n} = \int_0^1 \ln(1 + x)\, dx =$

$\left[(1 + x) \ln(1 + x) - x \right]_0^1 = \ln(4) - 1$

137. a) $V = \pi \displaystyle\int_1^e 1 - (\ln x)^2\, dx = \pi e - \int_0^1 z^2 e^z\, dz = \; = \pi e - \left[\left(z^2 - 2z + 2 \right) e^z \right]_0^1 = \pi(2e - 5)$,

where $z = \ln x$

$$\begin{array}{cc} z^2 & + \quad e^z \\ 2z & - \quad e^z \\ 2 & + \quad e^z \\ 0 & \quad\quad e^z \end{array}$$

b) $V = \pi \displaystyle\int_0^1 \left(e^y \right)^2 - 1\, dy = \pi \int_0^1 e^{2y} - 1\, dy = \pi \left[\frac{e^{2y}}{2} - y \right]_0^1 = \frac{\pi\left(e^2 - 3 \right)}{2}$

c) $V = \pi \displaystyle\int_0^1 \left(e^y - 1 \right)^2 dy = \pi \int_0^1 e^{2y} - 2e^y + 1\,)dy = \pi \left[\frac{e^{2y}}{2} - 2e^y + y \right]_0^1 = \frac{\pi(5 + e^2 - 4e)}{2}$

d) $V = \pi \displaystyle\int_1^e (1 - \ln x)^2 dx = \pi \int_1^e \left(1 - 2 \ln x + (\ln x)^2 \right) dx =$

$\pi \left[x - 2x \ln x + 2x + x(\ln x)^2 - 2x(\ln x) + 2x \right]_1^e = \pi(2e - 5)$, see part a) for integration

139. $M = 2\displaystyle\int_0^1 \frac{1}{\sqrt{1-x^2}}\,dx = 2\lim_{t\to 1^-}\int_0^t \frac{1}{\sqrt{1-x^2}}\,dx = 2\lim_{t\to 1^-}\left[\sin^{-1}x\right]_0^t =$

$2\lim_{t\to 1^-}\left(\sin^{-1}t - \sin^{-1}0\right) = 2\left(\frac{\pi}{2}\right) = \pi,\ M_y = \displaystyle\int_0^1 \frac{2x}{\sqrt{1-x^2}}\,dx = -\lim_{t\to 1^-}\int_0^t \left(1-x^2\right)^{-1/2}(-2x)\,dx =$

$-2\lim_{t\to 1^-}\left[\left(1-x^2\right)^{1/2}\right]_0^t = 2\ \therefore\ \overline{x} = \frac{2}{\pi}$ and $\overline{y} = 0$, by symmetry

141. $\displaystyle\int (x\ln x)^2\,dx = \int x^2(\ln x)^2\,dx = \frac{x^3(\ln x)^2}{3} - \frac{2x^3\ln x}{9} + \frac{2x^3}{27} + C$ (see exercise 39 in section 7.6)

$\therefore\ V = \pi\displaystyle\int_0^2 (x\ln x)^2\,dx = \pi\lim_{t\to 0^+}\int_t^2 x^2(\ln x)^2\,dx = \pi\lim_{t\to 0^+}\left[= \frac{x^3(\ln x)^2}{3} - \frac{2x^3\ln x}{9} + \frac{2x^3}{27}\right]_t^2 =$

$\pi\left(\frac{8(\ln 2)^2}{3} - \frac{16(\ln 2)}{9} + \frac{16}{27}\right) - \pi\lim_{t\to 0^+}\left(\frac{t^3(\ln t)^2}{3} - \frac{2t^3\ln t}{9} + \frac{2t^3}{27}\right) = \pi\left(\frac{8(\ln 2)^2}{3} - \frac{16(\ln 2)}{9} + \frac{16}{27}\right)$

143. If $y = \ln x$. then $1 + \left(\frac{dx}{dy}\right)^2 = 1 + x^2$ and $S = 2\pi\displaystyle\int_0^1 x\sqrt{1+\left(\frac{dx}{dy}\right)^2}\,dy = 2\pi\int_0^1 e^y\sqrt{1+e^{2y}}\,dy =$

$2\pi\displaystyle\int_{\pi/4}^{\arctan e} \sec^3\theta\,d\theta = 2\pi\left[\sec\theta\tan\theta + \ln|\sec\theta + \tan\theta|\right]_{\pi/4}^{\arctan e} =$

$2\pi\left[e\sqrt{1+e^2} + \ln\left(\frac{e+\sqrt{1+e^2}}{1+\sqrt{2}}\right) - \sqrt{2}\right]$, integration by parts, where $e^y = \tan\theta$

145. $\displaystyle\int f(x)e^x\,dx = e^x\left[f(x) - f'(x) + f''(x) - \cdots + (-1)^n f^{(n)}(x)\right] + C$, tabular integration

$f(x) \quad + \quad e^x$

$f'(x) \quad - \quad e^x$

$f''(x) \quad + \quad e^x$

$\qquad\qquad\quad e^x$

$\vdots$

$f^{(n)}(x) \quad - \quad e^x$

$0 \qquad\qquad e^x$

147. $\displaystyle\int_0^\infty \frac{4x}{1+x^2}\,dx = 2\lim_{t\to\infty}\int_0^t \frac{2x}{1+x^2}\,dx = 2\lim_{t\to\infty}\ln\left|1+x^2\right|\Big|_0^t = \lim_{t\to\infty}\left(\ln\left(x^2+1\right) - \ln(1)\right) = \infty \Rightarrow$

$\displaystyle\int_{-\infty}^\infty \frac{4x}{1+x^2}\,dx$ diverges, since $\displaystyle\int_{-\infty}^\infty \frac{4x}{1+x^2}\,dx = \int_{-\infty}^0 \frac{4x}{1+x^2}\,dx + \int_0^\infty \frac{4x}{1+x^2}\,dx$

149. $\displaystyle\int_{-\infty}^{-1} \frac{e^{-x}}{e^{-x} + e^x}\,dx = \int_{\infty}^{1} \frac{-e^u}{e^{-u} + e^u}\,du = \int_{1}^{\infty} \frac{1}{1 + e^{-2u}}\,du$ where $u = -x$, $\displaystyle\int_{1}^{\infty} \frac{1}{1 + e^{-2u}}\,du$ diverges by

theorem 1, since $\dfrac{1}{1 + e^{-2u}} > \dfrac{1}{u}$ and $\displaystyle\int_{1}^{\infty} \frac{1}{u}\,du$ diverges; $\therefore \displaystyle\int_{-\infty}^{\infty} \frac{e^{-x}}{e^{-x} + e^x}\,dx$ diverges because

$$\int_{-\infty}^{\infty} \frac{e^{-x}}{e^{-x} + e^x}\,dx = \int_{-\infty}^{-1} \frac{e^{-x}}{e^{-x} + e^x}\,dx + \int_{-1}^{0} \frac{e^{-x}}{e^{-x} + e^x}\,dx + \int_{0}^{\infty} \frac{e^{-x}}{e^{-x} + e^x}\,dx$$

151.

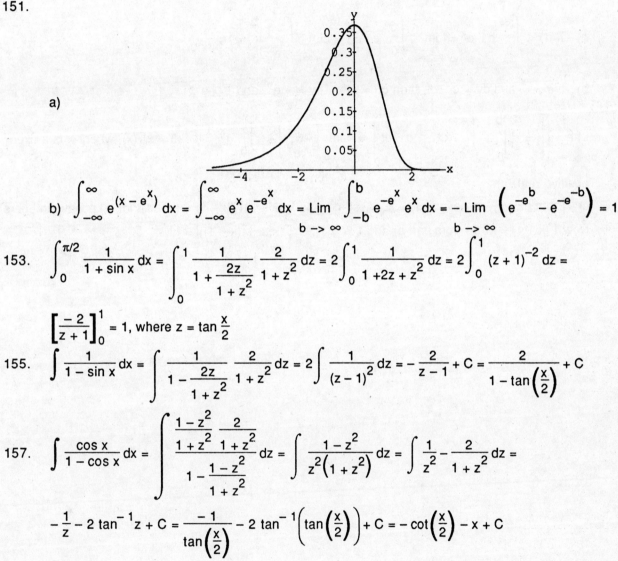

a)

b) $\displaystyle\int_{-\infty}^{\infty} e^{(x - e^x)}\,dx = \int_{-\infty}^{\infty} e^x\, e^{-e^x}\,dx = \lim_{b \to \infty} \int_{-b}^{b} e^{-e^x}\, e^x\,dx = -\lim_{b \to \infty}\left(e^{-e^b} - e^{-e^{-b}}\right) = 1$

153. $\displaystyle\int_{0}^{\pi/2} \frac{1}{1 + \sin x}\,dx = \int_{0}^{1} \frac{1}{1 + \dfrac{2z}{1 + z^2}}\, \frac{2}{1 + z^2}\,dz = 2\int_{0}^{1} \frac{1}{1 + 2z + z^2}\,dz = 2\int_{0}^{1} (z + 1)^{-2}\,dz =$

$\left[\dfrac{-2}{z + 1}\right]_{0}^{1} = 1$, where $z = \tan \dfrac{x}{2}$

155. $\displaystyle\int \frac{1}{1 - \sin x}\,dx = \int \frac{1}{1 - \dfrac{2z}{1 + z^2}}\, \frac{2}{1 + z^2}\,dz = 2\int \frac{1}{(z - 1)^2}\,dz = -\frac{2}{z - 1} + C = \frac{2}{1 - \tan\left(\dfrac{x}{2}\right)} + C$

157. $\displaystyle\int \frac{\cos x}{1 - \cos x}\,dx = \int \frac{\dfrac{1 - z^2}{1 + z^2}\, \dfrac{2}{1 + z^2}}{1 - \dfrac{1 - z^2}{1 + z^2}}\,dz = \int \frac{1 - z^2}{z^2\left(1 + z^2\right)}\,dz = \int \frac{1}{z^2} - \frac{2}{1 + z^2}\,dz =$

$-\dfrac{1}{z} - 2\tan^{-1} z + C = \dfrac{-1}{\tan\left(\dfrac{x}{2}\right)} - 2\tan^{-1}\left(\tan\left(\dfrac{x}{2}\right)\right) + C = -\cot\left(\dfrac{x}{2}\right) - x + C$

159. $\displaystyle\int \frac{1}{\sin x - \cos x}\, dx = \int \frac{\dfrac{2}{1 + z^2}}{\dfrac{2z}{1 + z^2} - \dfrac{1 - z^2}{1 + z^2}}\, dz = \int \frac{2}{(z + 1)^2 - 2}\, dz = \int \frac{2}{u^2 - 2}\, du =$

$\displaystyle\frac{1}{\sqrt{2}}\int \frac{-1}{u + \sqrt{2}} + \frac{1}{u - \sqrt{2}}\, du = \frac{1}{\sqrt{2}} \ln\left|\frac{u - \sqrt{2}}{u + \sqrt{2}}\right| + C = \frac{1}{\sqrt{2}} \ln\left|\frac{z + 1 - \sqrt{2}}{z + 1 + \sqrt{2}}\right| + C =$

$\displaystyle\frac{1}{\sqrt{2}} \ln\left|\frac{\tan\left(\frac{x}{2}\right) + 1 - \sqrt{2}}{\tan\left(\frac{x}{2}\right) + 1 + \sqrt{2}}\right| + C$, where $u = z + 1$ and $z = \tan\left(\frac{x}{2}\right)$

161. a) $\displaystyle\Gamma(1) = \int_0^\infty e^{-t}\, dt = \lim_{b \to \infty} \int_0^\infty e^{-t}\, dt = \lim_{b \to \infty} \left[\frac{-1}{e^t}\right]_0^b = 1$

b) If $u = t^x$ and $dv = e^{-t}\, dt$, then $du = xt^{x-1}$, $v = -e^{-t}$ and $\displaystyle\Gamma(x + 1) = \int_0^\infty t^x e^{-t}\, dt =$

$\displaystyle\lim_{b \to \infty} \left(\left[-\frac{t^x}{e^t}\right]_0^b + \int_0^\infty x\, t^{x-1} e^{-t}\, dt\right) = \lim_{b \to \infty} \left(-\frac{b^x}{e^b}\right) + x\int_0^\infty t^{x-1} e^{-t}\, dt = x\Gamma(x)$, where x is a fixed

positive real.

c) From part a) we have $\Gamma(1) = 1$. Assume $\Gamma(k) = (k - 1)\Gamma(k - 1) = (k - 1)!$ for all postive integers k.

Now $\Gamma(k + 1) = k\Gamma(k)$ from part b) and $\Gamma(k + 1) = k\Gamma(k) + k[(k - 1)!] = k!$

APPENDICES

APPENDIX A.4 PROOFS OF THE LIMIT THEOREMS IN CHAPTER 1

1. Let $\lim_{x \to c} f_1(x) = L_1$, $\lim_{x \to c} f_2(x) = L_2$, $\lim_{x \to c} f_3(x) = L_3$. Then $\lim_{x \to c} \left(f_1(x) + f_2(x) \right) =$
$L_1 + L_2$ by Theorem 1. Thus $\lim_{x \to c} \left(f_1(x) + f_2(x) + f_3(x) \right) = \lim_{x \to c} \left(\left(f_1(x) + f_2(x) \right) + f_3(x) \right) =$
$(L_1 + L_2) + L_3 = L_1 + L_2 + L_3$.

   Suppose functions $f_1(x), f_2(x), f_3(x), \ldots, f_n(x)$ have limits $L_1, L_2, L_3, \ldots, L_n$ as $x \to c$.
   Prove $\lim_{x \to c} \left(f_1(x) + f_2(x) + f_3(x) + \cdots + f_n(x) \right) = L_1 + L_2 + L_3 + \cdots + L_n$, n a positive integer.

   Step 1: For n = 1, we are given that $\lim_{x \to c} f_1(x) = L_1$.

   Step 2: Assume $\lim_{x \to c} \left(f_1(x) + f_2(x) + f_3(x) + \cdots + f_k(x) \right) = L_1 + L_2 + L_3 + \cdots + L_k$ for some k.
   Then $\lim_{x \to c} \left(f_1(x) + f_2(x) + f_3(x) + \cdots + f_k(x) + f_{k+1}(x) \right) =$
   $\lim_{x \to c} \left(\left(f_1(x) + f_2(x) + f_3(x) + \cdots + f_k(x) \right) + f_{k+1}(x) \right) =$
   $\left(L_1 + L_2 + L_3 + \cdots + L_k \right) + L_{k+1} = L_1 + L_2 + L_3 + \cdots + L_k + L_{k+1}$.

   $\therefore$ by Steps 1 and 2 and mathematical induction, $\lim_{x \to c} \left(f_1(x) + f_2(x) + f_3(x) + \cdots + f_n(x) \right)$
   $= L_1 + L_2 + L_3 + \cdots + L_n$

3. Given $\lim_{x \to c} x = c$. Then $\lim_{x \to c} x^n = \lim_{x \to c} \left(x \cdot x \cdot x \cdot \cdots \cdot x \right) = c \cdot c \cdot c \cdots \cdot c = c^n$ by Exercise 2.

   $\qquad\qquad\qquad\qquad\qquad\qquad\qquad\qquad$ (n factors) $\qquad\qquad$ (n factors)

5. $\lim_{x \to c} \dfrac{f(x)}{g(x)} = \dfrac{\lim_{x \to c} f(x)}{\lim_{x \to c} g(x)}$ (by Theorem 1) $= \dfrac{f(c)}{g(c)}$ if $g(c) \neq 0$ (by Exercise 4).

APPENDIX A.5 MATHEMATICAL INDUCTION

1. Step 1: For n = 1, $|x_1| = |x_1| \leq |x_1|$.

   Step 2: Assume $|x_1 + x_2 + \cdots + x_k| \leq |x_1| + |x_2| + \cdots + |x_k|$ for some positive integer k.

   Then $|x_1 + x_2 + \cdots + x_k + x_{k+1}| = |(x_1 + x_2 + \cdots + x_k) + x_{k+1}|$

   $\leq |x_1 + x_2 + \cdots + x_k| + |x_{k+1}|$ (by the triangle inequality)

   $\leq |x_1| + |x_2| + \cdots + |x_k| + |x_{k+1}|$. $\therefore |x_1 + x_2 + \cdots + x_n| \leq |x_1| + |x_2| + \cdots + |x_n|$ for all positive integers n by

   Steps 1 and 2 and mathematical induction.

3. Step 1: For $n = 1$, $\frac{d}{dx}(x) = 1 = 1 \cdot x^0$.

Step 2: Assume $\frac{d}{dx}(x^k) = kx^{k-1}$ for some positive integer k.

Then $\frac{d}{dx}(x^{k+1}) = \frac{d}{dx}(x^k x) = \frac{d}{dx}(x^k)x + x^k \frac{d}{dx}(x) = kx^{k-1}x + 1 \cdot x^k = kx^k + x^k = (k+1)x^k = (k+1)x^{(k+1)-1}$.

$\therefore \frac{d}{dx}(x^n) = nx^{n-1}$ for all positive integers n by Steps 1 and 2 and mathematical induction.

5. Step 1: For $n = 1$, $\frac{2}{3^1} = \frac{2}{3} = 1 - \frac{1}{3^1}$.

Step 2: Assume $\frac{2}{3^1} + \frac{2}{3^2} + \cdots + \frac{2}{3^k} = 1 - \frac{1}{3^k}$ for some positive integer k.

Then $\frac{2}{3^1} + \frac{2}{3^2} + \cdots + \frac{2}{3^k} + \frac{2}{3^{k+1}} = 1 - \frac{1}{3^k} + \frac{2}{3^{k+1}} = 1 - \frac{3^{k+1}}{3^k 3^{k+1}} + \frac{2(3^k)}{3^k 3^{k+1}} = 1 - \left(\frac{3^{k+1} - 2(3^k)}{3^k 3^{k+1}}\right) =$

$1 - \left(\frac{3^k(3-2)}{3^k 3^{k+1}}\right) = 1 - \frac{1}{3^{k+1}}$. $\therefore \frac{2}{3^1} + \frac{2}{3^2} + \cdots + \frac{2}{3^n} = 1 - \frac{1}{3^n}$ for all positive integers n by Steps 1 and 2 and

mathematical induction.

7.

n	1	2	3	4	5	6
2^n	2	4	8	16	32	64
n^2	1	4	9	16	25	36

Step 1: For $n = 5$, $2^5 = 32 > 25 = 5^2$.

Step 2: Assume $2^k > k^2$ for some $k \geq 5$, k a positive integer.

Then $2^{k+1} = 2^k(2) > 2k^2$. Now $k \geq 5 \Rightarrow k - 1 \geq 4 \Rightarrow (k-1)^2 \geq 16 \Rightarrow (k-1)^2 - 2 \geq 14 \Rightarrow$

$(k-1)^2 - 2 > 0$. Then $k^2 - 2k + 1 - 2 > 0 \Rightarrow k^2 - 2k - 1 > 0 \Rightarrow k^2 > 2k + 1 \Rightarrow k^2 + k^2 > k^2 + 2k + 1 \Rightarrow$

$2k^2 > (k+1)^2$.

$\therefore 2^{k+1} > (k+1)^2$. Thus $n^n > n^2$ for positive integers $n \geq 5$ by Steps 1 and 2 and mathematical

induction.

9. Step 1: For $n = 1$, $1^2 = \frac{1(1+1)(2(1)+1)}{6}$.

Step 2: Assume $1^2 + 2^2 + 3^2 + \cdots + k^2 = \frac{k(k+1)(2k+1)}{6}$ for some positive integer k.

Then $1^2 + 2^2 + 3^2 + \cdots + k^2 + (k+1)^2 = \frac{k(k+1)(2k+1)}{6} + (k+1)^2 \Rightarrow$

$1^2 + 2^2 + 3^2 + \cdots + k^2 + (k+1)^2 = \frac{k(k+1)(2k+1) + 6(k+1)^2}{6} \Rightarrow$

$1^2 + 2^2 + 3^2 + \cdots + k^2 + (k+1)^2 = \frac{(k+1)(k(2k+1) + 6(k+1))}{6} \Rightarrow$

$1^2 + 2^2 + 3^2 + \cdots + k^2 + (k+1)^2 = \frac{(k+1)(2k^2 + 7k + 6)}{6} \Rightarrow$

$1^2 + 2^2 + 3^2 + \cdots + k^2 + (k+1)^2 = \frac{(k+1)(k+2)(2k+3)}{6} \Rightarrow$

$1^2 + 2^2 + 3^2 + \cdots + k^2 + (k+1)^2 = \frac{(k+1)((k+1)+1)(2(k+1)+1)}{6}$.

$\therefore$ by Steps 1 and 2 and mathematical induction, $1^2 + 2^2 + 3^2 + \cdots + n^2 = \frac{n(n+1)(2n+1)}{6}\left(\frac{1/2}{1/2}\right) =$

9. (Continued)

$$\frac{n\left(n+\frac{1}{2}\right)(n+1)}{3} \text{ for all positive integers } n.$$

11. a) Step 1: For $n = 1$, $\displaystyle\sum_{k=1}^{1}(a_k + b_k) = a_1 + b_1 = \sum_{k=1}^{1}a_k + \sum_{k=1}^{1}b_k.$

Step 2: Assume $\displaystyle\sum_{k=1}^{j}(a_k + b_k) = \sum_{k=1}^{j}a_k + \sum_{k=1}^{j}b_k$ for some positive integer j.

Then $\displaystyle\sum_{k=1}^{j+1}(a_k + b_k) = \left(\sum_{k=1}^{j}(a_k + b_k)\right) + (a_{j+1} + b_{j+1}) = \sum_{k=1}^{j}a_k + \sum_{k=1}^{j}b_k + a_{j+1} + b_{j+1} =$

$\displaystyle\left(\sum_{k=1}^{j}a_k\right) + a_{j+1} + \left(\sum_{k=1}^{j}b_k\right) + b_{j+1} = \sum_{k=1}^{j+1}a_k + \sum_{k=1}^{j+1}b_k.$

$\therefore$ by Steps 1 and 2 and mathematical induction, $\displaystyle\sum_{k=1}^{n}(a_k + b_k) = \sum_{k=1}^{n}a_k + \sum_{k=1}^{n}b_k$ for all positive integers n.

b) Step 1: For $n = 1$, $\displaystyle\sum_{k=1}^{1}(a_k - b_k) = a_1 - b_1 = \sum_{k=1}^{1}a_k - \sum_{k=1}^{1}b_k.$

Step 2: Assume $\displaystyle\sum_{k=1}^{j}(a_k - b_k) = \sum_{k=1}^{j}a_k - \sum_{k=1}^{j}b_k$ for some positive integer j.

Then $\displaystyle\sum_{k=1}^{j+1}(a_k - b_k) = \left(\sum_{k=1}^{j}(a_k - b_k)\right) + (a_{j+1} - b_{j+1}) = \sum_{k=1}^{j}a_k - \sum_{k=1}^{j}b_k + a_{j+1} - b_{j+1} =$

$\displaystyle\left(\sum_{k=1}^{j}a_k\right) + a_{j+1} - \left(\left(\sum_{k=1}^{j}b_k\right) + b_{j+1}\right) = \sum_{k=1}^{j+1}a_k - \sum_{k=1}^{j+1}b_k.$

$\therefore$ by Steps 1 and 2 and mathematical induction, $\displaystyle\sum_{k=1}^{n}(a_k - b_k) = \sum_{k=1}^{n}a_k - \sum_{k=1}^{n}b_k$ for all positive integers n.

c) Step 1: For $n = 1$, $\displaystyle\sum_{k=1}^{1}ca_k = ca_1 = c\sum_{k=1}^{1}a_k.$

Step 2: Assume $\displaystyle\sum_{k=1}^{j}ca_k = c\sum_{k=1}^{j}a_k$ for some positive integer j.

11. (Continued)

Then $\displaystyle\sum_{k=1}^{j+1} ca_k = \sum_{k=1}^{j} ca_k + ca_{j+1} = c\left(\sum_{k=1}^{j} a_k\right) + ca_{j+1} = c\left(\left(\sum_{k=1}^{j} a_k\right) + a_{j+1}\right) = c\sum_{k=1}^{j+1} a_k.$

∴ by Steps 1 and 2 and mathematical induction, $\displaystyle\sum_{k=1}^{n} ca_k = c\sum_{k=1}^{n} a_k$ for all positive integers n.

d) Step 1: For n = 1, $\displaystyle\sum_{k=1}^{1} a_k = a_1 = c = 1\cdot c.$

Step 2: Assume $\displaystyle\sum_{k=1}^{j} a_k = jc$ for some positive integer j.

Then $\displaystyle\sum_{k=1}^{j+1} a_k = \sum_{k=1}^{j} a_k + a_{j+1} = jc + c = (j+1)c.$

∴ by Steps 1 and 2 and mathematical induction, $\displaystyle\sum_{k=1}^{n} a_k = nc$ for all positive integers n and

where $a_k = c$, a constant.